LONDON MATHEMATICAL SOCIETY LECTURE NOTE SERIES

Managing Editor: PROFESSOR I.M. JAMES, Mathematical Institute,
24-29 St Giles, Oxford

Prospective authors should contact the editor in the first instance.

Already published in this series

1. General cohomology theory and K-theory, PETER HILTON.
4. Algebraic topology: A student's guide, J.F. ADAMS.
5. Commutative algebra, J.T. KNIGHT.
7. Introduction to combinatory logic, J.R. HINDLEY, B. LERCHER and J.P. SELDIN.
8. Integration and harmonic analysis on compact groups, R.E. EDWARDS.
9. Elliptic functions and elliptic curves, PATRICK DU VAL.
10. Numerical ranges II, F.F. BONSALL and J. DUNCAN.
11. New developments in topology, G. SEGAL (ed.).
12. Symposium on complex analysis, Canterbury, 1973, J. CLUNIE and W.K. HAYMAN (eds.).
13. Combinatorics, Proceedings of the British combinatorial conference 1973, T.P. McDONOUGH and V.C. MAVRON (eds.).
14. Analytic theory of abelian varieties, H.P.F. SWINNERTON-DYER.
15. An introduction to topological groups, P.J. HIGGINS.
16. Topics in finite groups, TERENCE M.GAGEN.
17. Differentiable germs and catastrophes, THEODOR BROCKER and L. LANDER.
18. A geometric approach to homology theory, S. BUONCRISTIANO, C.P. ROURKE and B.J. SANDERSON.
19. Graph theory, coding theory and block designs, P.J. CAMERON and J.H. VAN LINT.
20. Sheaf theory, B.R. TENNISON.
21. Automatic continuity of linear operators, ALLAN M. SINCLAIR.
22. Presentations of groups, D.L. JOHNSON.
23. Parallelisms of complete designs, PETER J. CAMERON.
24. The topology of Stiefel manifolds, I.M. JAMES.
25. Lie groups and compact groups, J.F. PRICE.
26. Transformation groups: Proceedings of the conference in the University of Newcastle upon Tyne, August 1976, CZES KOSNIOWSKI.
27. Skew field constructions, P.M. COHN.
28. Brownian motion, Hardy spaces and bounded means oscillation, K.E. PETERSEN.
29. Pontryagin duality and the structure of locally compact abelian groups, SIDNEY A. MORRIS.
30. Interaction models, N.L. BIGGS.
31. Continuous crossed products and type III von Neumann algebras, A. VAN DAELE.
32. Uniform algebras and Jensen measures, T.W. GAMELIN.
33. Permutation groups and combinatorial structures, N.L. BIGGS and A.T. WHITE.
34. Representation theory of Lie groups, M.F. ATIYAH.
35. Trace ideals and their applications, BARRY SIMON.
36. Homological group theory, C.T.C. WALL (ed.).
37. Partially ordered rings and semi-algebraic geometry, GREGORY W. BRUMFIEL.
38. Surveys in combinatorics, edited by B. BOLLOBAS.
39. Affine sets and affine groups, D.G. NORTHCOTT.

London Mathematical Society Lecture Note Series. 40

Introduction to H_p Spaces

With an Appendix on Wolff's Proof of the Corona Theorem

Paul Koosis

Professor of Mathematics
University of California, Los Angeles

Typed by Charlotte Johnson

CAMBRIDGE UNIVERSITY PRESS
CAMBRIDGE
LONDON NEW YORK NEW ROCHELLE
MELBOURNE SYDNEY

Published by the Press Syndicate of the University of Cambridge
The Pitt Building, Trumpington Street, Cambridge CB2 1RP
32 East 57th Street, New York, NY 10022, USA
296 Beaconsfield Parade, Middle Park, Melbourne 3206, Australia

First published 1980

Printed in the United States of America
by BookCrafters, Inc., Chelsea, Michigan

Library of Congress Catalogue Card Number: 80-65175

ISBN 0 521 23159 0

TABLE OF CONTENTS

CONTENTS

CONTENTS

CONTENTS

CONTENTS

CONTENTS

CONTENTS

CONTENTS

CONTENTS

PREFACE

These are the lecture notes for a course I gave on the elementary theory of H_p spaces at the Stockholm Institute of Technology (tekniska högskolan) during the academic year 1977-78. The course concentrated almost exclusively on concrete aspects of the theory in its simplest cases; little time was spent on the more abstract general approach followed, for instance, in Gamelin's book. The idea was to give students knowing basic real and complex variable theory and a little functional analysis enough background to read current research papers about H_p spaces or on other work making use of their theory. For this reason, more attention was given to techniques and to what I believed were the ideas behind them than to the accumulation of a great number of results.

The lectures, about H_p spaces for the unit circle and the upper half plane, went far enough to include interpolation theory and BMO, but not as far as the corona theorem. That omission has, however, been put to rights in an appendix, thanks to T. Wolff's recent work. His proof of the corona theorem given there is a beautiful application of some of the methods developed for the study of BMO.

For Carleson's original proof of the corona theorem the reader may consult Duren's book. I have not included the more recent applications of the geometric construction Carleson devised for that proof, such as Ziskind's. Work of Douglas, Sarason, S-Y. Chang and Marshall on the algebras lying between H_∞ and L_∞ is not treated either.

Time did not allow me to cover the work of Hunt, Muckenhoupt and Wheeden on weighted mean value inequalities for harmonic conjugation. I did, however, give the proof of the Helson-Szegö theorem. Marshall's theorem (on the uniformly closed convex hull of the set of Blaschke products) is included although I did not lecture on it, and my lecture treatment of Lindelöf's theorem (on behaviour of the conformal mapping function near a point of tangency of the boundary) has been expanded.

In general, the notes stay quite close to the lectures as they were given. The style is loose and informal. Precise bibliographical references are not given in the text, nor the historical outlines at the end of each chapter that one has come to expect. A very partial bibliography is included; its purpose is to suggest further reading rather than to cover the subject thoroughly or to give due credit to all the workers in the field.

Topics not covered here, as well as the further ramifications of those I do cover, are treated in Garnett's extensive monograph now in the process of final revision. That book is recommended to the reader who wishes to go further.

I want to thank Harold Shapiro, Mats Essén and Magnus Giertz of the Stockholm Institute of Technology mathematics department for having helped me get an appointment to give this course. I want to thank the students and auditors for having successfully supported an extension of the course's length from the one semester originally planned to a full academic year. These were the students: Jockum Aniansson, Mats Lindberg, Lars Svensson and Anders Östrand. Björn Gustafsson audited most of the lectures and Dr. Stormark attended many. I was honored by Dr. G. O. Thorin's presence at all of them. To all these people, my best wishes and warmest regards.

Los Angeles
May 26, 1979

I. Functions Harmonic in $|z| < 1$

Rudiments

A. Let $U(z)$ be real and harmonic in $|z| < R$. (This means $U(z)$ is infinitely differentiable there and satisfies $\frac{\partial^2 U}{\partial x^2} + \frac{\partial^2 U}{\partial y^2} = 0$. We write throughout $z = x + iy$.) Then we can construct another real function $V(z)$, harmonic in $|z| < R$, such that

$$F(z) = U(z) + iV(z)$$

is analytic there. This function V is frequently called a harmonic conjugate of U. The construction of V is completely elementary; one way of doing it is as follows:

We want a function V, infinitely differentiable in $|z| < R$ which, with U, will satisfy the Cauchy-Riemann equations

$$\frac{\partial V}{\partial x} = \frac{\partial U}{\partial y}$$

$$\frac{\partial V}{\partial y} = -\frac{\partial U}{\partial x} .$$

(Then we will automatically have $\frac{\partial^2 V}{\partial x^2} + \frac{\partial^2 V}{\partial y^2} = 0$.)

Such a function V can be found by second year calculus if the differential $\frac{\partial U}{\partial y}\,dx - \frac{\partial U}{\partial x}\,dy$ is exact in $|z| < R$. But it is since $\frac{\partial^2 U}{\partial x^2} + \frac{\partial^2 U}{\partial y^2} = 0$! Again by second year calculus, any two functions V which we can find will differ by a constant. Frequently the constant is chosen so as to make $V(0) = 0$.

Once V is found, we have, for $|z| < R$,

$$U(z) = \Re F(z)$$

with $F(z) = \sum_0^\infty a_n z^n$, the power series expansion being uniformly convergent on compact subsets of $|z| < R$. That's because any function analytic in $|z| < R$ has such a power series development.

Writing $z = re^{i\theta}$, we easily find

$$U(re^{i\theta}) = \sum_{-\infty}^{\infty} A_n r^{|n|} e^{in\theta},$$

with

$$\begin{cases} A_n = \frac{1}{2} a_n, & n > 0 \\ A_0 = \Re a_0 & \\ A_n = \frac{1}{2} \bar{a}_{-n}, & n < 0. \end{cases}$$

<u>Thus, any function</u> $U(z)$ <u>harmonic in</u> $|z| < R$ <u>has a series representation</u>

$$U(re^{i\theta}) = \sum_{-\infty}^{\infty} A_n r^{|n|} e^{in\theta}$$

<u>uniformly convergent on compact subsets of</u> $|z| < R$.

B. The formula derived in the last section can be put in closed form. If $R > 1$ we easily find

$$U(re^{i\theta}) = \frac{1}{2\pi} \int_{-\pi}^{\pi} U(e^{it}) \sum_{-\infty}^{\infty} r^{|n|} e^{in(\theta - t)} dt.$$

By summing two geometric series we easily find

$$\sum_{-\infty}^{\infty} r^{|n|} e^{in\varphi} = \frac{1 - r^2}{1 + r^2 - 2r\cos\varphi} \quad \text{if} \quad 0 \leq r < 1.$$

<u>Thus we have derived Poisson's representation:</u>

<u>If</u> $U(z)$ <u>is harmonic for</u> $|z| < R$, <u>if</u> $R > 1$, <u>and if</u> $0 \leq r < 1$,

$$U(re^{i\theta}) = \frac{1}{2\pi} \int_{-\pi}^{\pi} \frac{(1 - r^2)U(e^{it})}{1 + r^2 - 2r\cos(\theta - t)}\, dt.$$

This formula is basic for the whole course - we shall soon see that it holds under much more general conditions than the one stated above. We call

$$P_r(\theta) = \frac{1 - r^2}{1 + r^2 - 2r\cos\theta}$$

the <u>Poisson kernel</u> for $|z| < 1$.

C. Suppose we merely know that $U(z)$ is harmonic in $|z| < 1$. It is remarkable that some version of the Poisson representation will frequently hold for U in that circle.

<u>Theorem</u>. Let $p > 1$, let $U(z)$ be harmonic in $|z| < 1$, and suppose the means

$$\int_{-\pi}^{\pi} |U(re^{i\theta})|^p \, d\theta$$

are <u>bounded for</u> $r < 1$. Then there is an $F \in L_p(-\pi, \pi)$ with

$$U(re^{i\theta}) = \frac{1}{2\pi} \int_{-\pi}^{\pi} \frac{1 - r^2}{1 + r^2 - 2r\cos(\theta - t)}\, F(t)dt$$

for $r < 1$.

<u>Proof</u>. For $p > 1$, L_p is the dual of L_q, where $\frac{1}{p} + \frac{1}{q} = 1$. The functions $U_n(\theta) = U((1 - \frac{1}{n})e^{i\theta})$ (instead of $1 - \frac{1}{n}$, <u>any</u> sequence of r_n tending to 1

from below will do!) have $\|U_n\|_p \le C$ ($\|\ \|_p$ is here taken over $[-\pi,\pi]$, of course!), so, by the Cantor diagonal process, we can extract a subsequence U_{n_j} of them such that

$$LG = \lim_{j\to\infty} \int_{-\pi}^{\pi} G(\theta)U_{n_j}(\theta)d\theta$$

exists for G ranging over a <u>countable dense subset of</u> L_q. Since $\|U_{n_j}\|_p \le C$, this limit, LG, will actually exist for <u>all</u> $G \in L_q$ (easy exercise), and LG is then a bounded linear functional on L_q. So, since L_p is the dual of L_q, there <u>is</u> an $F \in L_p$ with

$$LG = \int_{-\pi}^{\pi} F(\theta)G(\theta)d\theta$$

for all $G \in L_q$.

Now for each n, $u_n(z) = U((1 - \frac{1}{n})z)$ is harmonic for $|z| < \frac{1}{1-(1/n)}$, so, if $r < 1$,

$$u_{n_j}(re^{i\theta}) = \frac{1}{2\pi} \int_{-\pi}^{\pi} P_r(\theta - t)u_{n_j}(e^{it})dt$$

$$= \frac{1}{2\pi} \int_{-\pi}^{\pi} P_r(\theta - t)U_{n_j}(t)dt.$$

<u>Fix</u> any $r < 1$ and any θ, and use $G(t) = P_r(\theta - t)$; $G \in L_q$. Then

$$\lim_{j\to\infty} \int_{-\pi}^{\pi} P_r(\theta - t)U_{n_j}(t)dt = LG = \int_{-\pi}^{\pi} G(t)F(t)dt = \int_{-\pi}^{\pi} P_r(\pi - t)F(t)dt.$$

The leftmost member is

$$\lim_{j\to\infty} 2\pi u_{n_j}(re^{i\theta}) = 2\pi U(re^{i\theta}).$$

Thus,

$$U(re^{i\theta}) = \frac{1}{2\pi}\int_{-\pi}^{\pi} P_r(\theta - t)F(t)dt,$$

where $F \in L_p$.

Q.E.D.

Remark. The same result holds, with the same proof, if $p = \infty$, if we change the statement slightly:

Theorem. If $U(z)$ is harmonic and bounded in $|z| < 1$, there is an $F \in L_\infty(-\pi,\pi)$ with

$$U(re^{i\theta}) = \frac{1}{2\pi}\int_{-\pi}^{\pi} \frac{1 - r^2}{1 + r^2 - 2r\cos(\theta - t)} F(t)dt.$$

If I am not mistaken, this result was proved by Fatou, in his famous thesis, Séries trigonométriques et séries de Taylor, published just before the First World War. The result is indeed the starting point of the whole subject treated here. Many of the ideas in the first half of this course have their origin in Fatou's thesis.

What if $p = 1$? $L_1(-\pi,\pi)$ is, unfortunately not the dual of anything. But M - the space of finite signed measures μ on $[-\pi,\pi]$ - with $\|\mu\|$ = total variation of μ - is the dual of $C[-\pi,\pi]$ - the space of continuous functions on $[-\pi,\pi]$. If $p \in L_1[-\pi,\pi]$, we can associate to p a signed measure μ_p by putting

$$\int_{-\pi}^{\pi} G(t)d\mu_p(t) = \int_{-\pi}^{\pi} G(t)p(t)dt;$$

then $\|\mu_p\| = \|p\|_1$.

Here, then, the argument used in proving the first theorem of this section gives:

Theorem. If $U(z)$ is harmonic in $|z| < 1$ and the means

$$\int_{-\pi}^{\pi} |U(re^{i\theta})|\,d\theta$$

are bounded for $r < 1$, there is a finite signed measure μ on $[-\pi,\pi]$ with

$$U(re^{i\theta}) = \frac{1}{2\pi}\int_{-\pi}^{\pi} \frac{1 - r^2}{1 + r^2 - 2r\cos(\theta - t)}\,d\mu(t)$$

for $0 \leq r < 1$.

Corollary (Evans). Let $U(z)$ be harmonic in $|z| < 1$ and positive (here and henceforth "positive" just means "non-negative") there. Then there is a finite positive measure μ on $[-\pi,\pi]$ with

$$U(re^{i\theta}) = \frac{1}{2\pi}\int_{-\pi}^{\pi} P_r(\theta - t)\,d\mu(t), \qquad 0 \leq r < 1.$$

Proof. For $r < 1$ (using, e.g., the expansion $U(re^{i\theta}) = \sum_{-\infty}^{\infty} a_n r^{|n|} e^{in\theta}$ valid in $|z| < 1$), we have

$$2\pi U(0) = \int_{-\pi}^{\pi} U(re^{i\theta})\,d\theta = \int_{-\pi}^{\pi} |U(re^{i\theta})|\,d\theta,$$

since $U \geq 0$. Now just apply the theorem. The measure μ is positive because here (look again at the proof of the first theorem in this section) $\int_{-\pi}^{\pi} G(t)\,d\mu(t)$ comes out positive for each positive $G \in \mathbf{C}$ - it's the limit of positive things!

D. If we have one of the representations

$$U(re^{i\theta}) = \frac{1}{2\pi}\int_{-\pi}^{\pi} \frac{1 - r^2}{1 + r^2 - 2r\cos(\theta - t)} F(t)dt$$

$$U(re^{i\theta}) = \frac{1}{2\pi}\int_{-\pi}^{\pi} \frac{1 - r^2}{1 + r^2 - 2r\cos(\theta - t)} d\mu(t)$$

derived in the previous section, we should examine the connection between $U(z)$ and the function $F(t)$ or the measure $d\mu(t)$.

1. We first obtain some crude results which are sufficient for many investigations.

The Poisson kernel

$$P_r(\varphi) = \frac{1 - r^2}{1 + r^2 - 2r\cos\varphi} = \sum_{-\infty}^{\infty} r^{|n|}e^{in\varphi}$$

has the following properties:

a) $P_r(\varphi) > 0, \quad r < 1$

b) $P_r(\varphi + 2\pi) = P_r(\varphi)$

c) For each $r < 1$, $\int_{-\pi}^{\pi} P_r(t)dt = 2\pi$.

Of these, a) and b) are evident, and c) follows from the series development for $P_r(\varphi)$.

<u>If</u> $F \in L_p[-\pi,\pi]$ <u>it is convenient to assume</u> F <u>defined on all of</u> $\mathbb{R}$ <u>by periodicity</u> $F(t + 2\pi) = F(t)$. <u>We henceforth assume this</u>. <u>First we have converses to the representation theorems given in Section</u> C.

<u>Theorem</u>. If $p \geq 1$ and $F \in L_p[-\pi,\pi]$ and $U(re^{i\theta}) = \frac{1}{2\pi}\int_{-\pi}^{\pi} P_r(\theta-t)F(t)dt$, then $U(z)$ <u>is</u> harmonic in $|z| < 1$ and $\int_{-\pi}^{\pi} |U(re^{i\theta})|^p d\theta \leq \text{const}, \ r < 1$.

Proof. Let $\frac{1}{2\pi}\int_{-\pi}^{\pi} e^{-int}F(t)dt = A_n$. Then, for $0 \leq r < 1$,

$$U(re^{i\theta}) = \sum_{-\infty}^{\infty} A_n e^{|n|} e^{in\theta},$$

which is harmonic in $|z| < 1$ by inspection, because the series converges uniformly in the interior (meaning uniformly on compact subsets - complex variable language!) of that region. (If F is real, the series is clearly the real part of an analytic function which can be easily written down.)

Given $r < 1$, by property b) and 2π-periodicity of F we can also write

$$U(re^{i\theta}) = \frac{1}{2\pi}\int_{-\pi}^{\pi} F(\theta - s)P_r(s)ds.$$

Now take $G \in L_q[-\pi,\pi]$, $\|G\|_q = 1$, so that (with any given fixed r - G of course will depend on r)

$$\sqrt[p]{\int_{-\pi}^{\pi} |U(re^{i\theta})|^p d\theta} = \int_{-\pi}^{\pi} U(re^{i\theta})G(\theta)d\theta.$$

By Fubini's theorem, the integral on the right is

$$\frac{1}{2\pi}\int_{-\pi}^{\pi}\int_{-\pi}^{\pi} P_r(s)F(\theta - s)G(\theta)d\theta\, ds$$

which is in modulus $\leq$

$$\leq \frac{1}{2\pi}\int_{-\pi}^{\pi} P_r(s)\|F\|_p\|G\|_q ds = \|F\|_p$$

by choice of G and property c).

In fine,

$$\int_{-\pi}^{\pi} |U(re^{i\theta})|^p \, d\theta \le \|F\|_p^p$$

and we are done.

<u>Theorem</u>. Let μ be a finite signed measure on $[-\pi,\pi]$. Then

$$U(re^{i\theta}) = \frac{1}{2\pi} \int_{-\pi}^{\pi} P_r(\theta - t)d\mu(t)$$

is harmonic in $|z| < 1$ and

$$\int_{-\pi}^{\pi} |U(re^{i\theta})| d\theta \le \text{const.}, \qquad r < 1.$$

<u>Proof</u>. <u>Harmonicity</u> is established as above. Given $r < 1$, let $G \in L_\infty$, $\|G\|_\infty = 1$, be such that

$$\int_{-\pi}^{\pi} |U(re^{i\theta})| d\theta = \int_{-\pi}^{\pi} G(\theta)U(re^{i\theta})d\theta.$$

The right hand integral is, by Fubini's theorem, equal to

$$\frac{1}{2\pi} \int_{-\pi}^{\pi} \int_{-\pi}^{\pi} P_r(\theta - t)G(\theta)d\theta \, d\mu(t)$$

which, by properties a), b) and c) is modulus

$$\le \frac{1}{2\pi} \int_{-\pi}^{\pi} \int_{-\pi}^{\pi} P_r(\theta - t)\|G\|_\infty \, d\theta |d\mu(t)|$$

$$= \|G\|_\infty \int_{-\pi}^{\pi} |d\mu(t)| = \int_{-\pi}^{\pi} |d\mu(t)|.$$

We are done.

2. The Poisson kernel $P_r(\theta)$ has a fourth property:

d) Given any $\delta > 0$,

$$P_r(\theta) \to 0 \quad \text{uniformly for}$$

$$\delta \le |\theta| \le \pi \quad \text{as} \quad r \to 1.$$

This is obvious from the formula for $P_r(\theta)$.

Theorem. Let F be continuous on $\mathbb{R}$ and $F(t + 2\pi) = F(t)$. Let

$$U(re^{i\theta}) = \frac{1}{2\pi} \int_{-\pi}^{\pi} P_r(\theta - t)F(t)dt.$$

Then $U(z) \to F(\varphi)$ as $z \to e^{i\varphi}$, and the convergence is uniform in φ.

Proof. The result goes back to Poisson himself, who thought it showed that the Fourier series of a function converges to that function (it doesn't show that!).

Write

$$U(re^{i\theta}) = \frac{1}{2\pi} \int_{-\pi}^{\pi} F(\theta - t)P_r(t)dt.$$

Given any φ, we have, by property c),

$$F(\varphi) = \frac{1}{2\pi} \int_{-\pi}^{\pi} F(\varphi)P_r(t)dt.$$

Therefore

$$U(re^{i\theta}) - F(\varphi) = \frac{1}{2\pi} \int_{-\pi}^{\pi} [F(\theta - t) - F(\varphi)]P_r(t)dt.$$

Let $\delta < \pi/2$ be such that $|F(s) - F(\varphi)| < \varepsilon$ for $|s - \varphi| < 2\delta$; δ depends only on ε and not on φ here by (uniform!) continuity of F.

Write the last right hand integral as a sum of two:

$$|U(re^{i\theta}) - F(\varphi)| \leq \frac{1}{2\pi} \int_{|t|\leq\delta} |F(\theta - t) - F(\varphi)| P_r(t)dt$$

$$+ \frac{1}{2\pi} \int_{\delta\leq t\leq\pi} |F(\theta - t) - F(\varphi)| P_r(t)dt.$$

If $|\theta - \varphi| < \delta$, the first integral on the right is

$$\leq \frac{\varepsilon}{2\pi} \int_{|t|\leq\delta} P_r(t)dt < \varepsilon.$$

Let M be a bound on $|F(t)|$. Then the second integral is

$$\leq \frac{M}{2\pi} \int_{\delta\leq|t|\leq\pi} P_r(t)dt$$

which is $< \varepsilon$, say, if r is close enough to 1, by property d).

So $|U(re^{i\theta}) - F(\varphi)| < 2\varepsilon$ if $|\theta - \varphi| < \delta$ and r is close enough to 1.

Q.E.D.

Remark. Properties a), b), c), and d) together constitute the so-called approximate identity property of $\frac{1}{2\pi} P_r(\theta)$. The above theorem holds good because of them - all kinds of other kernels besides the Poisson kernel would work to yield similar theorems.

Theorem. Let $F \in L_1(-\pi,\pi)$, and suppose $F(t)$ is continuous at θ_0. Then $U(re^{i\theta}) = \frac{1}{2\pi} \int_{-\pi}^{\pi} P_r(\theta - t)F(t)dt$ tends to $F(\theta_0)$ as $re^{i\theta}$ tends to $e^{i\theta_0}$.

Proof. Similar to that of the above theorem.

<u>Theorem</u>. Let $F \in L_p$, $1 \le p < \infty$ (sic!) and let

$$U(re^{i\theta}) = \frac{1}{2\pi}\int_{-\pi}^{\pi} P_r(\theta - t)F(t)dt.$$

<u>Then</u> $\int_{-\pi}^{\pi} |U(re^{i\theta}) - F(\theta)|^p d\theta \to 0$ as $r \to 1$; i.e., $U(re^{i\theta})$ approaches $F(\theta)$ in L_p-norm as $r \to 1$.

<u>Proof</u>. Write $F_r(\theta) = U(re^{i\theta})$; then

$$F_r(\theta) - F(\theta) = \frac{1}{2\pi}\int_{-\pi}^{\pi} [F(\theta - t) - F(\theta)]P_r(t)dt.$$

Using properties a) and c) (we think of $F_r(\theta) - F(\theta)$ as a limit of convex combinations of the functions $F(\theta - t) - F(\theta)$, taking t as a parameter and θ as the variable), we have, by an evident generalization of the triangle inequality

$$\sqrt[p]{\int_{-\pi}^{\pi}|F_r(\theta) - F(\theta)|^p d\theta} \le \frac{1}{2\pi}\int_{-\pi}^{\pi} \sqrt[p]{\int_{-\pi}^{\pi}|F(\theta - t) - F(\theta)| d\theta} \cdot P_r(t)dt.$$

That is, if we write

$$\Phi(t) = \sqrt[p]{\int_{-\pi}^{\pi}|F(\theta - t) - F(\theta)|^p d\theta},$$

$$\|F_r - F\|_p \le \frac{1}{2\pi}\int_{-\pi}^{\pi} \Phi(t)P_r(t)dt.$$

<u>But</u> $\Phi(t) \to 0$ <u>as</u> $t \to 0$! <u>That is because translation is continuous in the L_p-norm for</u> $1 \le p < \infty$. This, in turn, follows from the rudiments of real variable theory as follows: given $F \in L_p(-\pi,\pi)$ and $\varepsilon > 0$ take a <u>continuous</u> G, <u>periodic of period</u> 2π, with $\|F - G\|_p < \varepsilon$. Then $\int_{-\pi}^{\pi} |G(\theta - t) - G(\theta)|^p \, d\theta$ <u>obviously</u> $< \varepsilon^p$ for $|t| < \delta$, say, by uniform continuity, so

$$\|F(\theta - t) - F(\theta)\|_p < 3\varepsilon \quad \text{for} \quad |t| < \delta.$$

Anyway, $\Phi(t)$ is continuous at 0 where it equals zero.

So, by a previous result, $\|F_r - F\|_p \to 0$ as $r \to 1$.

If $p = \infty$ all we have is weak* convergence:

Theorem. If $F \in L_\infty$, and

$$U(re^{i\theta}) = \frac{1}{2\pi} \int_{-\pi}^{\pi} P_r(\theta - t)F(t)dt,$$

then $U(re^{i\theta}) \to F(\theta)$ w^* as $r \to 1$.

Proof. Take any $G \in L_1(-\pi,\pi)$. We are to prove that

$$\int_{-\pi}^{\pi} U(re^{i\theta})G(\theta)d\theta \to \int_{-\pi}^{\pi} F(\theta)G(\theta)d\theta$$

for $r \to 1$. But this is true, because ($P_r(\varphi)$ being even!)

$$\int_{-\pi}^{\pi} F(\theta)P_r(\theta - t)d\theta = \int_{-\pi}^{\pi} P_r(t - \theta)F(\theta)d\theta$$

tends in L_1-norm to $F(t)$ as $r \to 1$ by the preceding theorem. We need then only apply Fubini's theorem.

Similarly

Theorem. Let $U(re^{i\theta}) = \frac{1}{2\pi}\int_{-\pi}^{\pi} P_r(\theta - t)d\mu(t)$, μ a finite signed measure on $[-\pi,\pi]$. Then $U(re^{i\theta})d\theta \to d\mu(\theta)$ w^* as $r \to 1$, i.e., for any continuous $G(\theta)$, periodic and of period 2π,

$$\int_{-\pi}^{\pi} U(re^{i\theta})G(\theta)d\theta \to \int_{-\pi}^{\pi} G(\theta)d\mu(\theta)$$

as $r \to 1$.

Proof. Use Fubini's theorem together with the first result of this number.

3. If $U(z)$, harmonic in $|z| < 1$, has one of the representations

$$U(re^{i\theta}) = \frac{1}{2\pi}\int_{-\pi}^{\pi} P_r(\theta - t)F(t)dt, \qquad F \in L_p$$

$$U(re^{i\theta}) = \frac{1}{2\pi}\int_{-\pi}^{\pi} P_r(\theta - t)d\mu(t),$$

we have still to discuss the pointwise behaviour of $U(z)$ as z tends to points $e^{i\theta}$ on the boundary of the unit circle. Study of such behaviour cannot proceed on the basis of the approximate identity properties a) - d) alone, but requires a more detailed examination of $P_r(\theta)$.

Both representations written above for $U(re^{i\theta})$ are subsumed in the second one, for if $F \in L_p(-\pi,\pi)$, and we take $d\mu(\theta) = F(\theta)d\theta$, then μ is in fact a finite signed measure on $[-\pi,\pi]$. In dealing with such a measure, it is convenient to introduce the function $\mu(\theta)$ of bounded variation on $[-\pi,\pi]$ given by

$$\mu(\theta) = \int_0^{\theta} d\mu(t)$$

(with usual interpretation of the integral if $\theta < 0$). Then we have the

Theorem (Fatou). Let $-\pi < \varphi_0 < \pi$ and suppose that the derivative $\mu'(\varphi_0)$ exists and is finite. Then $U(re^{i\theta}) = \frac{1}{2\pi}\int_{-\pi}^{\pi} P_r(\theta - t)d\mu(t)$ tends to $\mu'(\varphi_0)$ for $re^{i\theta}$ tending to $e^{i\varphi_0}$ within any sector of the form $|\theta - \varphi_0| \le c(1 - r)$.

Remark 1. Thus, $z = re^{i\theta}$ is required to tend to $e^{i\varphi_0}$ from within sectors of opening $< 180°$ having vertex at $e^{i\varphi_0}$, and symmetric about the radius from 0 out to $e^{i\varphi_0}$. We frequently say $U(re^{i\theta}) \to \mu'(\varphi_0)$ as $re^{i\theta}$ tends

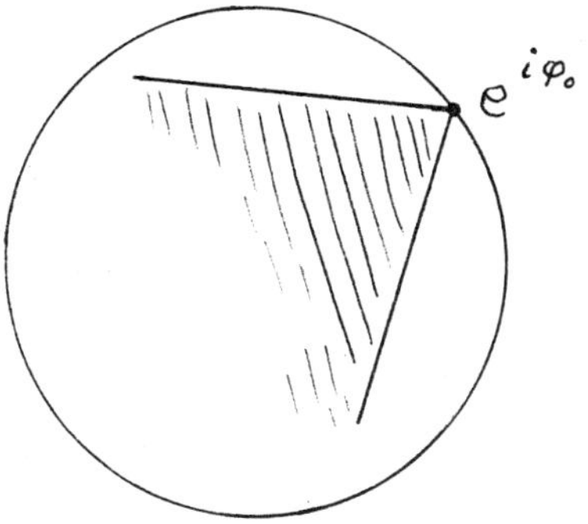

to $e^{i\varphi_0}$ non-tangentially, and write

$$U(re^{i\theta}) \to \mu'(\varphi_0) \quad \text{for} \quad re^{i\theta} \xrightarrow[\measuredangle]{} e^{i\varphi_0}.$$

Remark 2. A similar result holds for $\varphi_0 = \pm\pi$ provided $\mu'(\varphi_0)$ exists in a properly defined sense there. Let the reader figure out what this sense should be.

Proof. To simplify the notation, take $\varphi_0 = 0$. Then, if $\mu'(0)$ exists and is finite, and $|\theta_r| \le c(1 - r)$, to show that

$$\frac{1}{2\pi}\int_{-\pi}^{\pi} \frac{(1 - r^2)}{1 + r^2 - 2r\cos(t - \theta_r)}\, d\mu(t) \to \mu'(0)$$

as $r \to 1$. Without loss of generality, assume $\mu'(0) = 0$; otherwise work with $d\mu(t) - \mu'(0)dt$ instead of $d\mu(t)$; $\frac{1}{2\pi}\int_{-\pi}^{\pi} P_r(\theta_r - t)\mu'(0)dt$ is $= \mu'(0)$.

Let $\delta > 0$ be such that $|\mu(t)| \le \varepsilon|t|$ for $|t| \le \delta$. If $1 - r$ is very close to 0, so that $2|\theta_r|$ is much smaller than δ,

$$\frac{1}{2\pi}\int_{-\pi}^{\pi} \frac{1-r^2}{1+r^2-2r\cos(t-\theta_r)}\,d\mu(t)$$

$$= o(1) + \frac{1}{2\pi}\int_{-\delta}^{\delta} \frac{1-r^2}{1+r^2-2r\cos(t-\theta_r)}\,d\mu(t)$$

where $o(1) \to 0$ as $r \to 1$. Integrate $\frac{1}{2\pi}\int_{-\delta}^{\delta} \frac{1-r^2}{1+r^2-2r\cos(t-\theta_r)}\,d\mu(t)$ by parts to get an integrated term (is $o(1)$) plus $\frac{1}{2\pi}\int_{-\delta}^{\delta} \frac{2(1-r^2)r\sin(t-\theta_r)}{[1+r^2-2r\cos(t-\theta_r)]^2}\,\mu(t)dt$. Assuming, without loss of generality, that $\theta_r > 0$, we break up the last integral as

$$\frac{1}{2\pi}\left[\int_{-\delta}^{0} + \int_{0}^{2\theta_r} + \int_{2\theta_r}^{\delta}\right] \frac{2r(1-r^2)\sin(t-\theta_r)\mu(t)}{[1+r^2-2r\cos(t-\theta_r)]^2}\,dt = I + II + III, \quad \text{say.}$$

Then

$$|II| \le \frac{1}{2\pi}\int_{0}^{2\theta_r} \frac{4\theta_r}{(1-r)^3}\cdot \varepsilon t\,dt \le \frac{4\varepsilon\theta_r^3}{\pi(1-r)^3} \le \frac{4}{\pi}c^3\varepsilon, \quad \text{since } 0 \le \theta_r \le c(1-r).$$

For $2\theta_r \le t \le \delta$, $|F(t)| \le \varepsilon t \le 2\varepsilon(t-\theta_r)$, so

$$|III| \le \frac{\varepsilon}{\pi}\int_{2\theta_r}^{\delta} \frac{2(1-r^2)\,r\sin(t-\theta_r)}{[1+r^2-2r\cos(t-\theta_r)]^2}\,(t-\theta_r)dt =$$

$$= \frac{\varepsilon}{\pi}\int_{\theta_r}^{\delta-\theta_r} \frac{2r(1-r^2)\sin t}{(1+r^2-2r\cos t)^2}\,t\,dt \le \frac{\varepsilon}{\pi}\int_{0}^{\pi} \frac{2r(1-r^2)\sin t}{(1+r^2-2r\cos t)^2}\,t\,dt.$$

This last is integrated by parts (in the opposite direction to our first integration by parts!) to give

$$\frac{\varepsilon}{\pi}\left[o(1) + \int_{0}^{\pi} \frac{(1-r^2)dt}{1+r^2-2r\cos t}\right] = \varepsilon + o(1).$$

Similarly $|I| \le \frac{\varepsilon}{2} + o(1)$. So $|I + II + III| \le (4c^3/\pi + 3/2)\varepsilon + o(1)$ as $r \to 1$, and since $\varepsilon > 0$ is arbitrary, we are done.

Remark. What makes the above proof work is the monotoneity of $P_r(\theta)$ on each of the intervals $[-\pi, 0]$ and $[0, \pi]$.

Theorem (Fatou). If $-\pi < \varphi_0 < \pi$ and if $\mu'(\varphi_0)$ exists and is infinite, then, for $U(re^{i\theta}) = \frac{1}{2\pi}\int_{-\pi}^{\pi} P_r(\theta - t)d\mu(t)$, $U(re^{i\varphi_0}) \to \mu'(\varphi_0)$ as $r \to 1$.

Remark. Thus, even when $\mu'(\varphi_0)$ infinite, we still have

$$U(z) \to \mu'(\varphi_0)$$

when z goes out radially to $e^{i\varphi_0}$.

Proof. Take $\varphi_0 = 0$ and assume $\mu'(0) = +\infty$. Choose $\delta > 0$ so small that $\mu(t)\operatorname{sgn} t \ge M|t|$ for $|t| \le \delta$. Then, computing as in the proof of the preceding theorem,

$$U(r) = o(1) + \frac{1}{2\pi}\int_{-\delta}^{\delta} \frac{2(1-r)r \sin t}{(1+r^2-2r\cos t)^2}\mu(t)dt$$

$$\ge o(1) + \frac{1}{\pi}M\int_0^{\delta} \frac{2(1-r)r\sin t}{(1+r^2-2r\cos t)^2}t\, dt.$$

Doing the reverse integration by parts, this last is seen to be $o(1) + M$ for r close enough to 1.

Scholium. Can we replace "$z \to e^{i\varphi_0}$ radially" by "$z \underset{\measuredangle}{\longrightarrow} e^{i\varphi_0}$" in the theorem just proved? We can, if $U(z) \ge 0$ for $|z| < 1$, i.e. if μ is a positive measure.

<u>Lemma</u> (Harnack's Theorem). Let $U(z)$ be positive in $|z| < 1$. Then

$$U(z) \geq \frac{1 - |z|}{1 + |z|} \; U(0).$$

<u>Proof</u>.

$$U(re^{i\theta}) = \frac{1}{2\pi} \int_{-\pi}^{\pi} \frac{1 - r^2}{1 + r^2 - 2\cos(\theta - t)} \, d\mu(t)$$

$$\geq \frac{1 - r}{1 + r} \cdot \frac{1}{2\pi} \int_{-\pi}^{\pi} d\mu(t) = \frac{1 - r}{1 + r} \, U(0)$$

if $d\mu(t) \geq 0$.

Now, assume $U(z) \geq 0$ and $U(r) \to \infty$ for $r \to 1$. If S is any sector with vertex at 1, symmetric about the positive real axis, of opening $< 180^\circ$, let S' be a similar but <u>slightly larger</u> sector:

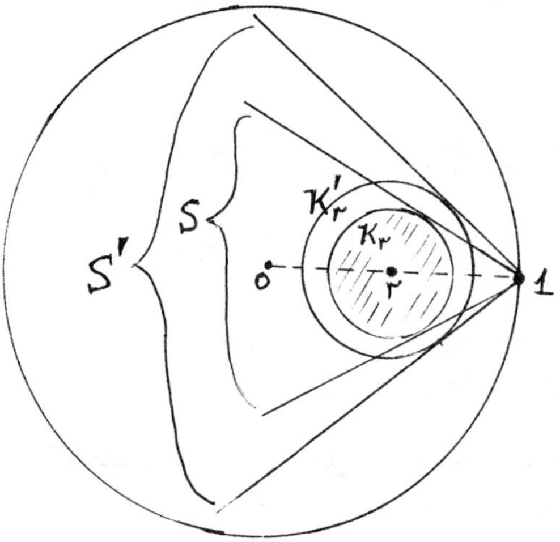

If $0 < r < 1$, let K_r be the circle about r tangent to the sides of S, and K'_r be the circle about r tangent to the sides of S'. Clearly

γ = radius of K_r/radius of K'_r is <u>independent</u> of r and < 1. Then, $U(z)$ is positive in K'_r, so, if z is in K_r, by the lemma,

$$U(z) \geq \frac{1-\gamma}{1+\gamma} U(r),$$

proving $U(z) \to \infty$ as $z \to 1$ <u>from within</u> S.

<u>However, a complete generalization of the second Fatou theorem for nontangential movement of z out to the boundary is false.</u>

That is the content of the following problem.

In order to simplify the computations, we have made a change of variable corresponding to the conformal representation

$$w = re^{i\theta} = \frac{i-z}{i+z}, \qquad z = x + iy,$$

which makes the upper half plane $y > 0$ correspond to the circle $|w| < 1$ and $z = 0$ correspond to $w = 1$.

Then, if $e^{i\tau} = \frac{i-t}{i+t}$, it is easily checked that

$$\frac{1-r^2}{1+r^2-2r\cos(\theta-\tau)} d\nu(\tau) = \frac{y}{(x-t)^2+y^2} d\mu(t),$$

where

$$\frac{2\,d\mu(t)}{1+t^2} = d\nu(\tau).$$

<u>Problem 1</u>. To construct a signed measure μ on $[0,1]$ such that $\mu'(0+) = \infty$, but if

$$U(z) = \int_0^1 \frac{y}{(x-t)^2+y^2} d\mu(t),$$

$U(x+ix)$ <u>does not</u> tend to ∞ as $x \to 0+$.

Procedure. Obtain inductively the positive numbers $t_0 = 1$, $x_0 = 1/2$, $t_1 < x_0$, $x_1 < t_1$, $t_2 < x_1$, & c, and construct μ on each of the intervals $[t_1,t_0]$, $[t_2,t_1)$, $[t_3,t_2)$,... one after the other.

μ is to be a discrete measure, and when constructed will satisfy

(1) $\mu(t) \geq \sqrt{t}$, $0 \leq t \leq 1$

(2) $\mu_+([t_k,t_{k-1})) = 9[\sqrt{t_{k-1}} - \sqrt{t_k}]$ and $\mu_-([t_k,t_{k-1})) = 8[\sqrt{t_{k-1}} - \sqrt{t_k}]$, where μ_+ and μ_- denote the positive and negative parts of μ.

Start by taking $t_1 < x_0/10$ so small that $9\sqrt{t_1} < 1 - \sqrt{t_1} = \Delta_1$, say, and, on $[t_1,t_0]$, put $d\mu(t) = 9\Delta_1 d\delta(t - t_1) - 8\Delta_1 d\delta(t - x_0)$, where δ denotes the unit point mass at 0. Show that, if, at the end of the construction, δ satisfies (2), then

$$\int_0^1 \frac{x_0}{(x_0 - t)^2 + x_0^2}\, d\mu(t) \leq -\frac{3\Delta_1}{x_0} \leq \frac{3}{2\sqrt{x_0}} .$$

Next, take $x_1 < t_1$ so small that

$$\int_{t_1}^1 \frac{x_1^2}{(t - x_1)^2 + x_1^2}\, d\mu(t) \leq \sqrt{t_1} - \sqrt{x_1}$$

(on $[t_1,1]$, μ is already constructed!). Assuming that μ will satisfy (2) at the end of the construction, how should we choose $t_2 < x_1$, and how should we define μ on $[t_2,t_1)$ so as to ensure

$$\int_0^1 \frac{x_1 d\mu(t)}{(x_1 - t)^2 + x_1^2} \leq -\frac{2\Delta_2}{x_1} \leq -\frac{1}{\sqrt{x_1}} ,$$

writing $\Delta_2 = \sqrt{t_1} - \sqrt{t_2}$?

Show how the construction can be carried out so as to get a measure μ on $[0,1]$ satisfying (1) and (2) with $\mu'(0+) = \infty$ but

$$\int_0^1 \frac{x_k \, d\mu(t)}{(t - x_k)^2 + x_k^2} \leq - \frac{1}{\sqrt{x_k}} .$$

Let $F \in L_p[-\pi,\pi]$, $p \geq 1$ and let

$$U(re^{i\theta}) = \frac{1}{2\pi} \int_{-\pi}^{\pi} P_r(\theta - t)F(t)dt.$$

A classical theorem of Lebesgue says that

$$\frac{d}{d\theta} \int_0^{\theta} F(t)dt \text{ exists a.e. and equals } F(\theta).$$

In conjunction with the first theorem of this subsection we thus see that

$$U(re^{i\theta}) \to F(\varphi) \text{ a.e. as } re^{i\theta} \xrightarrow[\measuredangle]{} e^{i\varphi}.$$

Combined with a theorem of Section C, we thus have

Theorem. Let $1 < p \leq \infty$ (sic!) and let $U(z)$ be harmonic in $|z| < 1$,

$$\sqrt[p]{\int_{-\pi}^{\pi} |U(re^{i\theta})|^p d\theta} \leq C$$

for $0 \leq r < 1$. Then, for almost all θ, $U(z)$ tends to a finite limit, say $U(e^{i\theta})$, as $z \xrightarrow[\measuredangle]{} e^{i\theta}$, $U(e^{i\theta}) \in L_p(-\pi,\pi)$, and, for $0 \leq r < 1$,

$$U(e^{i\theta}) = \frac{1}{2\pi} \int_{-\pi}^{\pi} \frac{1 - r^2}{1 + r^2 - 2r \cos(\theta - t)} U(e^{it})dt.$$

Notation. $U(e^{i\theta})$ is called the (non-tangential) boundary value function for $U(z)$; we frequently write $U(e^{i\theta}) = \lim_{z \xrightarrow[\measuredangle]{} e^{i\theta}} U(z)$ a.e.

In future, whenever we have a function U harmonic in $|z| < 1$, satisfying the hypothesis of the above theorem (for $p > 1$), we assume it to be automatically extended a.e. to $|z| = 1$ in the manner described.

In case $p = 1$ the theorem is not completely true. In that case we have a measure $d\mu(\theta)$. A decomposition theorem of Lebesgue says that then the derivative $\mu'(\theta)$ still exists and is finite a.e., that $\mu'(\theta) \in L_1(-\pi,\pi)$, but that $d\mu(\theta)$ is not in general $\mu'(\theta)d\theta$. Instead,

$$d\mu(\theta) = \mu'(\theta)d\theta + d\sigma(\theta)$$

where σ is a singular measure, i.e., one supported on a set of Lebesgue measure zero.

Thus, if we only know that the means

$$\int_{-\pi}^{\pi} |U(re^{i\theta})|\,d\theta$$

are bounded for $r < 1$, we still have a.e. existence of the finite non-tangential

$$\lim_{z \xrightarrow{\angle} e^{i\theta}} U(z) = \mu'(\theta),$$

but we cannot recover $U(z)$ from this boundary value function. Instead, we have

$$U(re^{i\theta}) = \frac{1}{2\pi}\int_{-\pi}^{\pi} P_r(\theta - t)\mu'(t)\,dt + \frac{1}{2\pi}\int_{-\pi}^{\pi} P_r(\theta - t)\,d\sigma(t)$$

with some singular measure σ.

The simplest cases show that a representation with nonzero σ can actually occur; one such is the ordinary Poisson kernel

$$U(re^{i\theta}) = \frac{1 - r^2}{1 + r^2 - 2r\cos\theta} \quad !$$

Indeed, $\lim_{z \xrightarrow[\measuredangle]{} e^{i\theta}} U(z) = 0$ save for $\theta = 0$, and

$$U(re^{i\theta}) = \frac{1}{2\pi}\int_{-\pi}^{\pi} P_r(\theta - t) \cdot 2\pi \, d\delta_0(t)$$

where δ_0 is the unit point mass at 0.

This distinction between the cases $p = 1$ and $p > 1$ is one of the fundamental complications of the theory, and will be seen to have deep and far-reaching implications in its further development.

E. Given a function $U(z)$ harmonic in $|z| < 1$, having one of the representations studied in this chapter, we proceed to investigate the pointwise boundary behaviour of its harmonic conjugate. At the beginning of this chapter, we said that a harmonic function $V(z)$ is a harmonic conjugate of $U(z)$ if $U(z) + iV(z)$ is analytic in $|z| < 1$. Harmonic conjugates are only defined to within an additive constant; working in the unit circle, it is customary to require $V(0) = 0$; the resulting harmonic conjugate $V(z)$ of U is denoted by $\tilde{U}(z)$. The tilde notation is customary in the designation of harmonic conjugates.

1. Suppose that

$$U(re^{i\theta}) = \sum_{-\infty}^{\infty} A_n r^{|n|} e^{in\theta}, \qquad 0 \le r < 1,$$

then

$$\tilde{U}(re^{i\theta}) = -\sum_{-\infty}^{\infty} i \operatorname{sgn} n \, A_n r^{|n|} e^{in\theta},$$

where $\operatorname{sgn} 0$ means 0.

Indeed, $\tilde{U}(re^{i\theta})$ is harmonic in $|z| < 1$ by inspection (the series converges absolutely there), and $\tilde{U}(0) = 0$. And also

$$U(re^{i\theta}) + i\tilde{U}(re^{i\theta}) = A_0 + \sum_{1}^{\infty} 2A_n r^n e^{in\theta}$$

is analytic (by inspection!) in $|z| < 1$.

Now if $U(re^{i\theta}) = \frac{1}{2\pi} \int_{-\pi}^{\pi} P_r(\theta - t)\,d\mu(t)$ with a measure μ on $[-\pi,\pi]$, then the above series development for U is valid with coefficients $A_n = \frac{1}{2\pi} \int_{-\pi}^{\pi} e^{-int}\,d\mu(t)$. Looking at the series development for $\tilde{U}$, we see that

$$\tilde{U}(re^{i\theta}) = -\frac{1}{2\pi}\int_{-\pi}^{\pi} \sum_{-\infty}^{\infty} i \text{ sgn } n \, r^{|n|} e^{in(\theta-t)} d\mu(t).$$

Call

$$-\sum_{-\infty}^{\infty} i \text{ sgn } n \, r^{|n|} e^{in\theta} = Q_r(\theta),$$

the <u>conjugate Poisson kernel</u>. <u>By direct summation of two geometric series, we find</u>

$$Q_r(\theta) = \frac{2r \sin \theta}{1 + r^2 - 2r \cos \theta}.$$

Thus:

<u>Theorem</u>. If $U(re^{i\theta}) = \frac{1}{2\pi}\int_{-\pi}^{\pi} [(1-r^2)/(1 + r^2 - 2r\cos(\theta - t))]d\mu(t)$ with a measure μ, then the harmonic conjugate $\tilde{U}$ of U is given by

$$\tilde{U}(re^{i\theta}) = \frac{1}{2\pi}\int_{-\pi}^{\pi} \frac{2r \sin(\theta - t)}{1 + r^2 - 2r \cos(\theta - t)} d\mu(t).$$

2. We are especially interested in the boundary behaviour of $\tilde{U}(z)$ when $U(re^{i\theta}) = \frac{1}{2\pi}\int_{-\pi}^{\pi} P_r(\theta - t)F(t)dt$ <u>with a function</u> F (belonging, say, to $L_p(-\pi,\pi)$, $p \geq 1$).

Assuming, as usual the definition of F to be extended to $\mathbb{R}$ so as to make F 2π-periodic, we have

$$\tilde{U}(re^{i\theta}) = \frac{1}{2\pi}\int_{-\pi}^{\pi} \frac{2\, r \sin(\theta - t)}{1 + r^2 - 2r \cos(\theta - t)} F(t)dt$$

$$= \frac{1}{\pi}\int_{0}^{\pi} \frac{r \sin s}{1 + r^2 - 2r \cos s} [F(\theta - s) - F(\theta + s)]ds.$$

We have $(r \sin s)/(1 + r^2 - 2r \cos s) = (2r \sin \frac{s}{2} \cos \frac{s}{2})/(1 - r^2 + 4r \sin^2 \frac{s}{2})$, so if $\int_{\theta}^{\pi} (|F(\theta - s) - F(\theta + s)|/s)ds < \infty$, we clearly have, for $r \to 1$,

$$\tilde{U}(re^{i\theta}) \to \frac{1}{2\pi}\int_0^{\pi} \frac{F(\theta - s) - F(\theta + s)}{\tan \frac{s}{2}}\, ds,$$

the integral on the right being absolutely convergent.

<u>This certainly happens if</u> $F'(\theta)$ <u>exists and is finite</u>.

Indeed, if $F'(\theta)$ <u>is continuous for</u> $\alpha < \theta < \beta$, say, <u>then</u> $\tilde{U}(z)$ <u>has a continuous extension up to any closed subarc of the open arc</u> $\{e^{i\theta};\ \alpha < \theta < \beta\}$ on $|z| = 1$, <u>and for such</u> θ,

$$\tilde{U}(e^{i\theta}) = \lim_{r \to 1} \tilde{U}(re^{i\theta}) = \frac{1}{\pi}\int_0^{\pi} \frac{F(\theta - t) - F(\theta + t)}{2 \tan \frac{t}{2}}\, dt,$$

<u>the integral being absolutely convergent</u>.

<u>Problem</u> 2. <u>Prove the statement just made</u>. (Hint: The mean value theorem for derivatives comes in here.) <u>Also prove the following</u>:

<u>If</u> $F \geq 0$ <u>is</u> 2π-<u>periodic and continuous as a mapping into</u> $[0,\infty]$ (that is, F is allowed to take the value ∞, but if $F(\theta_0) = \infty$, then $F(\theta) \to \infty$ for $\theta \to \theta_0$), <u>and if</u> $F \in L_1(-\pi, \pi)$, <u>then</u> $U(re^{i\theta}) = \frac{1}{2\pi}\int_{-\pi}^{\pi} P_r(\theta - t)F(t)dt$ <u>extends continuously up</u> to $|z| = 1$ <u>as a mapping to</u> $[0,\infty]$.

3. It turns out that the radial boundary value

$$\tilde{U}(e^{i\theta}) = \frac{1}{\pi}\int_0^{\pi} \frac{F(\theta - t) - F(\theta + t)}{2 \tan \frac{t}{2}}\, dt$$

of $\tilde{U}(re^{i\theta}) = \frac{1}{2\pi}\int_{-\pi}^{\pi} Q_r(\theta - t)F(t)dt$ exists a.e. in θ <u>for very general functions</u> F. No differentiability is really required for F. Of course, the integral $\int_0^{\pi} \frac{F(\theta - t) - F(\theta + t)}{2 \tan (t/2)}\, dt$ must be interpreted properly. Just taking it as the limit of

$$\int_{\varepsilon}^{\pi} \frac{F(\theta - t) - F(\theta + t)}{2 \tan(t/2)}\, dt$$

for $\varepsilon \to 0+$ will, as we shall see, be enough.

Definition. Let F be 2π-periodic, and in $L_1(-\pi,\pi)$. We say θ is in the Lebesgue set for F if

$$\lim_{h\to 0} \frac{1}{h}\int_{-h}^{h} |F(\theta + t) - F(\theta)|\, dt \to 0$$

as $h \to 0$.

Theorem (Lebesgue!). Almost every θ is in the Lebesgue set of F.

Proof. Given any rational number r, the function $|F(t) - r|$ is in L_1, hence equal to the derivative of its indefinite integral a.e. I.e., for almost all θ,

$$\lim_{h\to 0} \frac{1}{h}\int_{\theta}^{\theta+h} |F(t) - r|\, dt = |F(\theta) - r|$$

holds simultaneously for all rational r as long as $\theta \notin E$, say, E being a set of measure zero.

Let $\theta \notin E$ and let $\varepsilon > 0$ be given; if r is a rational number with $|F(\theta) - r| < \varepsilon$, then

$$\frac{1}{h}\int_{0}^{h} |F(\theta + t) - F(\theta)|\, dt$$

$$\leq \frac{1}{h}\int_{\theta}^{\theta+h} |F(t) - r|\, dt + |F(\theta) - r| \leq \varepsilon + \frac{1}{h}\int_{\theta}^{\theta+h} |F(t) - r|\, dt.$$

The thing on the right tends to $\varepsilon + |F(\theta) - r| \leq 2\varepsilon$ as $h \to 0$, so

$$\limsup_{h\to 0} \frac{1}{h}\int_{0}^{h} |F(\theta + t) - F(\theta)|\, dt \leq 2\varepsilon$$

if $\theta \notin E$. Since ε is arbitrary > 0,

$$\frac{1}{h}\int_0^h |F(\theta + t) - F(\theta)|dt \to 0$$

as $h \to 0$, and this holds for all $\theta \notin E$, i.e., for almost all θ.

Now we have the basic

Theorem. Let $F(t + 2\pi) = F(t)$, $F \in L_1(-\pi,\pi)$. Then

$$\frac{1}{\pi}\int_{-\pi}^{\pi} \frac{r\sin t}{1+r^2-2r\cos t} F(\theta - t)dt - \frac{1}{\pi}\int_{1-r}^{\pi} \frac{F(\theta - t) - F(\theta + t)}{2\tan t/2} dt$$

tends to zero as $r \to 1$ for all θ in the Lebesgue set of F, i.e., almost everywhere.

Remark. The idea is that $\frac{1}{\pi}\int_{\varepsilon}^{\pi} \frac{F(\theta - t) - F(\theta + t)}{2\tan t/2} dt$ can, for almost all θ, be compared with $\tilde{U}((1-\varepsilon)e^{i\theta})$ - a value of the harmonic conjugate inside the unit circle - when ε is small.

Proof. $(r\sin t)/(1+r^2-2r\cos t)$ is an odd function of t so the difference in question is unchanged if F is everywhere replaced by $F - F(\theta)$.

Do this replacement. Then the difference breaks up into two parts, of which the first is

$$I = \frac{1}{\pi}\int_{-(1-r)}^{(1-r)} \frac{r\sin t}{1+r^2-2r\cos t}[F(\theta - t) - F(\theta)]dt.$$

We have

$$\left|\frac{r\sin t}{1+r^2-2r\cos t}\right| \le \frac{|\sin t|}{(1-r)^2} \le \frac{1}{1-r}$$

for $|t| \leq 1 - r$, therefore

$$|I| \leq \frac{1}{\pi(1 - r)} \int_{-(1-r)}^{(1-r)} |F(\theta - t) - F(\theta)|dt$$

which $\to 0$ as $r \to 1$, if θ is in the Lebesgue set of F.

It is convenient to write $\Delta = 1 - r$. Then the rest of our difference is

$$II = \frac{1}{\pi} \int_{\Delta \leq |t| \leq \pi} \left\{ \frac{r \sin t}{\Delta^2 + 4r \sin^2 t/2} - \frac{\sin t}{4 \sin^2 t/2} \right\} [F(\theta - t) - F(\theta)]dt.$$

The expression in { } works out to

$$\frac{-\Delta^2 \sin t}{4(\Delta^2 + 4r \sin^2 t/2)\sin^2 t/2}$$

which, for $r \geq 1/2$, say, is in absolute value $\leq C((1-r)^2/|t|^3)$, C being a numerical constant.

Thus,

$$|II| \leq \frac{C\Delta^2}{\pi} \int_{\Delta \leq |t| \leq \pi} \frac{|F(\theta - t) - F(\theta)|}{|t|^3} dt.$$

To evaluate, for instance,

$$\Delta^2 \int_{\Delta}^{\pi} \frac{|F(\theta - t) - F(\theta)|}{t^3} dt,$$

integrate by parts, getting

$$\frac{\Delta^2}{\pi^3} \int_0^{\pi} |F(\theta - t) - F(\theta)|dt - \frac{1}{\Delta} \int_0^{\Delta} |F(\theta - t) - F(\theta)|dt$$
$$+ 4\Delta^2 \int_{\Delta}^{\pi} \frac{\int_0^s |F(\theta - t) - F(\theta)|dt}{s^4} ds.$$

The first two terms obviously tend to 0 as $\Delta \to 0$, θ being in the Lebesgue set of F.

Given $\varepsilon > 0$, pick a <u>fixed</u> $\eta > 0$ such that $\frac{1}{s}\int_0^s |F(\theta - t) - F(\theta)|dt < \varepsilon$ for $0 < s < \eta$; <u>the last term is then</u>

$$\leq 4\Delta^2 \int_\Delta^\eta \frac{\varepsilon}{s^3}\, ds + 4\Delta^2 \int_\eta^\pi \frac{1}{\eta^4} \int_0^s |F(\theta - t) - F(\theta)|dt\, ds$$

$$\leq \frac{2\Delta^2}{\Delta^2}\,\varepsilon + \frac{4\pi\Delta^2}{\eta^4} \int_0^\pi |F(\theta - t) - F(\theta)|dt = 2\varepsilon + \Delta^2 \frac{M}{\eta^4}$$

when $0 < \Delta < \eta$. This is $< 4\varepsilon$ if Δ is small enough, so, since ε is arbitrary, we see that $\Delta^2 \int_\Delta^\pi (|F(\theta - t) - F(\theta)|/t^3)dt$, and hence II, tends to 0 as $\Delta \to 0$.

The theorem is completely proved.

4. Now we shall use material from Sections C and D together with the comparison theorem of Subsection 3 above in order to show that

$$\lim_{\varepsilon \to 0+} \int_\varepsilon^\pi \frac{f(\theta - t) - f(\theta + t)}{2 \tan t/2}\, dt$$

exists a.e. whenever $f \in L_2(-\pi,\pi)$! This is the first really <u>deep</u> theorem in the whole chapter.

First, a quick remark. The relations

$$\frac{1}{2\pi} \int_{-\pi}^{\pi} e^{in\theta} e^{-im\theta} d\theta = \begin{cases} 0, & n \neq m \\ 1, & n = m \end{cases}$$

show, together with absolute convergence, that if $U(re^{i\theta}) = \sum_{-\infty}^{\infty} A_n r^{|n|} e^{in\theta}$ is harmonic in $|z| < 1$,

$$\int_{-\pi}^{\pi} |U(re^{i\theta})|^2 d\theta = 2\pi \sum_{-\infty}^{\infty} |A_n|^2 r^{2|n|}$$

for $0 \leq r < 1$. From this relation together with material of Sections C, D we immediately get the

<u>Lemma</u>. If $U(z)$ is harmonic in $|z| < 1$,

$$U(re^{i\theta}) = \frac{1}{2\pi}\int_{-\pi}^{\pi} \frac{1 - r^2}{1 + r^2 - 2r\cos(\theta - t)} F(t)dt$$

with $F \in L_2(-\pi,\pi)$ iff

$$U(re^{i\theta}) = \sum_{-\infty}^{\infty} A_n r^{|n|} e^{in\theta}$$

with

$$\sum_{-\infty}^{\infty} |A_n|^2 < \infty.$$

<u>Theorem</u>. Let $F \in L_2(-\pi,\pi)$ be 2π-periodic. Then

$$\tilde{F}(\theta) = \lim_{\varepsilon \to 0} \frac{1}{\pi}\int_{\varepsilon}^{\pi} \frac{F(\theta - t) - F(\theta + t)}{2 \tan t/2} dt$$

exists for almost all θ, $\tilde{F} \in L_2(-\pi,\pi)$, and $\|\tilde{F}\|_2 \leq \|F\|_2$.

If

$$U(re^{i\theta}) = \frac{1}{2\pi}\int_{-\pi}^{\pi} P_r(\theta - t)F(t)dt,$$

then

$$\tilde{U}(re^{i\theta}) = \frac{1}{2\pi}\int_{-\pi}^{\pi} Q_r(\theta - t)F(t)dt$$

is <u>also equal to</u>

$$\frac{1}{2\pi}\int_{-\pi}^{\pi} P_r(\theta - t)\tilde{F}(t)dt.$$

Proof. By the Lemma,

$$U(re^{i\theta}) = \sum_{-\infty}^{\infty} A_n r^{|n|} e^{in\theta}$$

with

$$\sum_{-\infty}^{\infty} |A_n|^2 < \infty ,$$

so therefore, again by the Lemma,

$$\tilde{U}(re^{i\theta}) = -i \sum_{-\infty}^{\infty} \operatorname{sgn} n \; A_n r^{|n|} e^{in\theta}$$

must in fact equal

$$\frac{1}{2\pi}\int_{-\pi}^{\pi} P_r(\theta - t)G(t)dt$$

for some $G \in L_2(-\pi,\pi)$.

By Section D,

$$\tilde{U}(z) \longrightarrow G(\theta) \quad \text{as} \quad z \xrightarrow[\measuredangle]{} e^{i\theta}$$

for almost all θ, in particular, $\lim_{r\to 1} \tilde{U}(re^{i\theta}) = G(\theta)$ a.e. By the last theorem of Subsection 3 we now see that

$$\frac{1}{\pi}\int_{1-r}^{\pi} \frac{F(\theta - t) - F(\theta + t)}{2 \tan t/2}\, dt$$

must also tend to $G(\theta)$ for almost all θ as $r \to 1$. So most of the theorem holds already if we just write $\tilde{F}(\theta) = G(\theta)$!

It remains to check the norm inequalities. But that's easy. By work in Section D,

$$\|F\|_2^2 = \lim_{r\to 1} \int_{-\pi}^{\pi} |U(re^{i\theta})|^2 d\theta$$

which equals $2\pi \sum_{-\infty}^{\infty} |A_n|^2$ according to the computation done above; at the same time,

$$\|\tilde{F}\|_2^2 = \lim_{r\to 1} \int_{-\pi}^{\pi} |\tilde{U}(re^{i\theta})|^2 d\theta = 2\pi \sum_{n\neq 0} |A_n|^2 \quad \text{(remember sgn } 0 = 0!\text{)}$$

That does it.

Discussion. Where it exists, the limit

$$\tilde{F}(\theta) = \lim_{\varepsilon\to 0} \frac{1}{\pi} \int_{\varepsilon}^{\pi} \frac{F(\theta - t) - F(\theta + t)}{2 \tan t/2} dt$$

is the same as

$$\lim_{\varepsilon\to 0} \frac{1}{\pi} \left\{ \int_{-\pi}^{\theta-\varepsilon} + \int_{\theta+\varepsilon}^{\pi} \right\} \frac{F(t)dt}{2 \tan\frac{\theta-t}{2}} .$$

This exhibits it as a Cauchy principal value. We frequently write

$$\tilde{F}(\theta) = \frac{1}{\pi} ⨍_{-\pi}^{\pi} \frac{F(t)dt}{2 \tan \frac{\theta-t}{2}}$$

to emphasize that the expression is evaluated by omitting a small interval having θ as its midpoint, and then having the width of that interval shrink to zero. The symmetry is crucial here.

$$⨍_{-\pi}^{\pi} \frac{F(t)dt}{2 \tan\frac{(\theta-t)}{2}}$$

is usually not an integral in any ordinary sense.

We also write

$$\tilde{F}(\theta) = \frac{1}{\pi}\int_{0+}^{\pi} \frac{F(\theta - t) - F(\theta + t)}{2 \tan t/2}\, dt.$$

Although $\tilde{F}(\theta)$ is obtained from F by such a delicate **limiting** process (the symmetry of the omitted interval which shrinks down to θ!), see how strongly it is bound to F, in the metric sense! For $\tilde{F}$ depends linearly on F, and $\|\tilde{F}\|_2 \le \|F\|_2$!

Taken by themselves, the statements about $\tilde{F}(\theta)$ constitute a purely real variable result. Yet we used a lot of complex function theory (harmonic functions and their conjugates) to establish it, for we made the passage

$F \longrightarrow$	$U(z)$ $\longrightarrow$	$\tilde{U}(z)$ $\longrightarrow$	$\tilde{F}$
	harmonic in unit circle	harmonic in unit circle	boundary function

during its proof. And we used a differentiability property of something related to F in order to compare $\tilde{U}((1-\varepsilon)e^{i\theta})$ with $\frac{1}{\pi}\int_{\varepsilon}^{\pi} \frac{F(\theta-t)-F(\theta+t)}{2 \tan t/2}dt$. Altogether, a most intricate business. At the beginning of this century, Lusin wanted to find a more direct proof - one which didn't bring in complex variable theory. He thought that doing this would lay bare more of the real mechanism of the interference of translates of functions on the real line which must be operating in order to make

$$\int_{0+}^{\pi} \frac{F(\theta - t) - F(\theta + t)}{2 \tan t/2}\, dt$$

exist a.e. For it is indeed a real process of interference which is taking place here (see next subsection!).

The nth partial sum of the Fourier series of $F(\theta)$ is essentially given by $\int_{-\pi}^{\pi} F(\theta - t)\frac{\sin nt}{t}\,dt$ - an expression very like the one for $\tilde{F}(\theta)$, except for the factor $\sin nt$ in the integrand. Lusin thought that if one could really see why the limit defining $\tilde{F}(0)$ exists a.e., one might stand a chance of proving that the Fourier series of an L_2 function converges a.e. In a way, he was right. Carleson's celebrated proof of this convergence (published in 1966) depends greatly on delicate properties of the operation taking F to $\tilde{F}$.

Long before Carleson did this, real variable proofs of the existence of $\tilde{F}(\theta)$ were found. Lusin himself obtained one. They are harder than the classical one given above.

5. It was said in the last subsection that the existence a.e. of the limit

$$\tilde{f}(\theta) = \frac{1}{\pi}\int_{0+}^{\pi} \frac{f(\theta - t) - f(\theta + t)}{2 \tan t/2}\, dt$$

is <u>deep</u>, and comes from some complicated <u>interference</u> phenomenon. The existence of the limit really comes from <u>cancellation</u> of <u>positive</u> and <u>negative</u> contributions, and not from the <u>smallness</u> of $|f(\theta - t) - f(\theta + t)|$ <u>even when</u> f is <u>continuous</u>. This is shown by the following

<u>Theorem</u>. There exists a continuous function G, periodic of period 2π, such that

$$\int_0^{\pi} \frac{|G(\theta + t) - G(\theta - t)|}{t}\, dt = \infty$$

for <u>every</u> θ.

<u>Proof</u>. By contradiction, using the Baire category theorem.

Denote by $\mathbf{C}$ the space of continuous functions of period 2π. Using the usual norm $\|f\| = \sup_{\theta}|f(\theta)|$, $\mathbf{C}$ becomes a <u>complete</u> normed space.

Assume that for every $f \in \mathbf{C}$ there is <u>at least one</u> θ with

$$\int_0^{\pi} \frac{|f(\theta + t) - f(\theta - t)|}{t}\, dt < \infty;$$

we shall arrive at a contradiction. For each $n = 1,2,3,\ldots$ let

$$E_n = \{f \in \mathbf{C};\ \text{for at least one } \theta,\ \int_0^{\pi} \frac{|f(\theta+t) - f(\theta-t)|}{t}\, dt \leq n\}.$$

By hypothesis, $\bigcup_1^{\infty} E_n = \mathbf{C}$. Now each E_n is <u>closed</u> in the norm topology of $\mathbf{C}$. Indeed, suppose $f_k \in E_n$ and $\|f_k - f\| \underset{k}{\to} 0$ with $f \in \mathbf{C}$. To prove $f \in E_n$. For each f_k there is a θ_k with

$$\int_0^{\pi} \frac{|f_k(\theta_k + t) - f_k(\theta_k - t)|}{t}\, dt \leq n.$$

By 2π-periodicity, we may take $0 \leq \theta_k \leq 2\pi$, and then there is no loss of generality in assuming $\theta_k \underset{k}{\to} \theta$, $0 \leq \theta \leq 2\pi$ (otherwise just go to a subsequence). We have

$$|(f(\theta + t) - f(\theta - t)) - (f_k(\theta_k + t) - f_k(\theta_k - t)|$$

$$\leq |(f(\theta_k + t) - f(\theta_k - t)) - (f_k(\theta_k + t) - f_k(\theta_k - t)|$$

$$+ |f(\theta + t) - f(\theta_k + t) - f(\theta - t) + f(\theta_k + t)|$$

which $\underset{k}{\to} 0$ uniformly in t, since $f \in \mathbf{C}$. In particular,

$$|f(\theta + t) - f(\theta - t)| = \lim_{k\to\infty} |f_k(\theta_k + t) - f_k(\theta_k - t)|$$

for all t, so, by Fatou's Lemma,

$$\int_0^{\pi} \frac{|f(\theta - t) - f(\theta - t)|}{t}\, dt \leq \liminf_{k\to\infty} \int_0^{\pi} \frac{|f_k(\theta_k + t) - f_k(\theta_k - t)|}{t}\, dt$$

which is $\leq n$. So $f \in E_n$.

Because $\bigcup_1^{\infty} E_n = \mathbf{C}$, closure of the E_n implies, by the Baire category theorem, that at least one E_n contains an entire sphere. That is, there is an $F \in \mathbf{C}$ and a $\rho > 0$ such that $f \in \mathbf{C}$ and $\|f - F\| < 2\rho$ imply $f \in E_n$. Pick an F_0 (e.g., a trigonometric polynomial) with $\|F_0 - F\| < \rho$ such that $F_0'(\theta)$ exists everywhere and is finite. If, say, $\|F_0'\| \leq K$ (K may of course be enormous!) we have

$$\int_0^{\pi} \frac{|F_0(\theta + t) - F_0(\theta - t)|}{t}\, dt \leq 2\pi K$$

for all θ by the mean value theorem.

Now let $g \in \mathbf{C}$, and let $\|g\| < \rho$. Then $\|F_0 + g - F\| < 2\rho$, so $F_0 + g \in E_n$. In view of the previous inequality, this means that there must be at least one θ with

$$\int_0^{\pi} \frac{|g(\theta + t) - g(\theta - t)|}{t}\, dt \leq n + 2\pi K,$$

i.e., $\|g\| < \rho$ implies that for at least one θ, $\int_0^{\pi}(|g(\theta + t) - g(\theta - t)|/t)dt \leq$ some fixed number, say M.

But this cannot be. Take any continuous function h of period π such that $h(\theta + t) - h(\theta - t)$ is not identically zero in t for any θ. Here is an example of such a function:

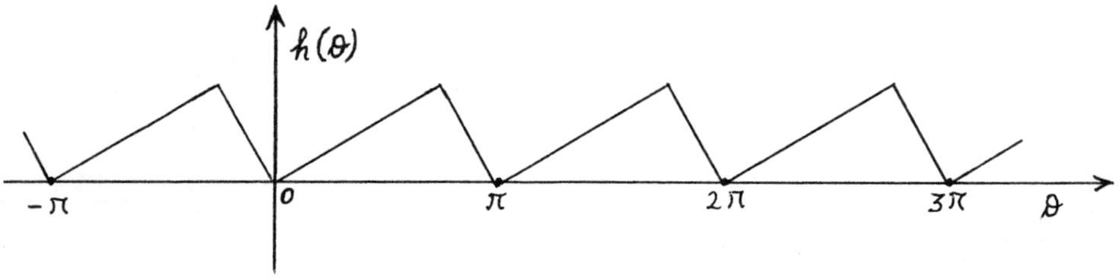

Then for all θ, $\int_0^{\pi} |h(\theta + t) - h(\theta - t)|dt$ is $\geq$ say $\alpha > 0$. Given $m = 2,3,4,\ldots,$ let $\theta_m \in [0,\pi]$ be such that $m\theta - \theta_m$ is a multiple of π. Then

$$\int_0^{\pi} \frac{|h(m(\theta + t)) - h(m(\theta - t))|}{t}\, dt = \int_0^{m\pi} \frac{|h(\theta_m + s) - h(\theta_m - s)|}{s}\, ds$$

$$\geq \int_0^{\pi} |h(\theta_m + s) - h(\theta_m - s)|ds + \frac{1}{2}\int_{\pi}^{2\pi} |h(\theta_m + s) - h(\theta_m - s)|ds + \ldots$$

$$\ldots + \frac{1}{m}\int_{(m-1)\pi}^{m\pi} |h(\theta_m + s) - h(\theta_m - s)|ds$$

$$\geq \alpha(1 + \tfrac{1}{2} + \ldots + \tfrac{1}{m}) \sim \alpha \log m.$$

Taking such an h with $\|h\| < \rho$, we see that for $g(t) = h(mt)$ there is _no_ θ with

$$\int_0^{\pi} \frac{|g(\theta + t) - g(\theta - t)|}{t} \, dt \le M,$$

provided m _is large enough_. _This contradicts our previous statement, and proves the theorem._

II. Theorem of the Brothers Riesz.

Introduction to the Space H_1

A. Let μ be a (signed or in general complex-valued) measure on $[-\pi,\pi]$. In 1917, in the proceedings of the fourth Scandanavian mathematical congress (this is a paper every analyst should read!), F. and M. Riesz published the celebrated

Theorem. If $\int_{-\pi}^{\pi} e^{in\theta}d\mu(\theta) = 0$ for $n = 1,2,3,\ldots,$ then μ is absolutely continuous with respect to Lebesgue measure.

The theorem is not only deep, but is also the basis for much of the following work. We shall go through three proofs of this theorem, two here, and a third in Chapter IV.

1. The original proof of F. and M. Riesz is perhaps the simplest, and goes as follows.

Suppose μ is not absolutely continuous. Then there is some set $E \subset [-\pi,\pi]$ of Lebesgue measure zero but such that

$$\int_E e^{i\theta}d\mu(\theta) \neq 0.$$

Notation. As is customary, we henceforth denote the Lebesgue measure of a set E by $|E|$.

From elementary measure theory, we know that we can take E to also be closed.

This being granted, we proceed thus. Denote the (disjoint) open intervals complementary to E - the so-called contiguous intervals - by (α_n,β_n) - there are at most countably many of them. We mean by this that the arcs $\{e^{i\theta};\ \alpha_n < \theta < \beta_n\}$ are disjoint, and that, together with $\{e^{i\theta};\ \theta \in E\}$, they precisely fill up the unit circumference $|z| = 1$.

Because $|E| = 0$, $\sum_n (\beta_n - \alpha_n) = 2\pi$. Let $p_n \underset{n}{\to} \infty$, $p_n > 0$, but in such fashion that $\sum_n p_n(\beta_n - \alpha_n) < \infty$. Such a sequence can always be found. Then define a function $F(\theta)$ as follows:

i) $F(\theta + 2\pi) = F(\theta)$

ii) $F(\theta) = \infty$, $\theta \in E$

iii) If $\alpha_n < \theta < \beta_n$,

$$F(\theta) = p_n \frac{\ell_n}{\sqrt{\ell_n^2 - (\theta - \gamma_n)^2}},$$

where $\ell_n = \frac{1}{2}(\beta_n - \alpha_n)$ and $\gamma_n = \dfrac{\beta_n + \alpha_n}{2}$.

Then $\int_{\alpha_n}^{\beta_n} F(\theta)d\theta = \pi\, p_n \ell_n$, so $F \in L_1(-\pi,\pi)$ by choice of the p_n, since $|E| = 0$. $F(\theta)$ is also continuous from $\mathbb{R}$ to $[0,\infty]$ (with ∞ included). Indeed, if $\theta_0 \in (\alpha_n,\beta_n)$ for some n, continuity of F at θ_0 is obvious. If $\theta_0 \in E$, and a sequence of θ's tends to θ_0, then we can split the sequence into two parts:

One part lying in finitely many of the complementary intervals (α_n,β_n) - in fact, in at most two of them - and tending to some common endpoint of them. For this part of the sequence it is clear by the formula in iii) that $F(\theta) \to \infty$.

Another part running through infinitely many of the (α_n,β_n). Since, by construction, $F(\theta) \geq p_n$ on (α_n,β_n) and $p_n \to \infty$, we again have $F(\theta) \to \infty$. (That's the reason for the introduction of the p_n!)

We now put

$$U(re^{i\theta}) = \frac{1}{2\pi} \int_{-\pi}^{\pi} \frac{1 - r^2}{1 + r^2 - 2r\cos(\theta - t)} F(t)dt,$$

which we can do since $F \in L_1$. By part of problem 2 above, (Chapter I, Section D) $U(z) \to F(e^{i\theta})$ for $z \to e^{i\theta}$, i.e., U is continuous from $|z| \leq 1$ into $[0,\infty]$. We also take

$$\tilde{U}(re^{i\theta}) = \frac{1}{2\pi} \int_{-\pi}^{\pi} \frac{2r\sin(\theta - t)}{1 + r^2 - 2r\cos(\theta - t)} F(t)dt.$$

By the same problem 2, $\tilde{U}(z)$ extends continuously up to each open arc $\{e^{i\theta};\ \alpha_n < \theta < \beta_n\}$, for $F(\theta)$ is $\mathcal{C}_1$ (and even $\mathcal{C}_\infty$) on each such open arc.

Now put

$$\varphi(z) = \frac{U(z) + i\tilde{U}(z)}{U(z) + i\tilde{U}(z) + 1}, \qquad |z| < 1.$$

Since $F(\theta) > 0$, $U(z) > 0$, so $\varphi(z)$ is analytic in $|z| < 1$, and $|\varphi(z)| < 1$ there. It even extends continuously up to $|z| = 1$. For if $\theta_0 \notin E$, we have just seen that $U(z)$ and $\tilde{U}(z)$ tend to continuous limits as z tends to points in some small arc about $e^{i\theta_0}$. And if $\theta_0 \in E$, $U(z) \to \infty$ as $z \to e^{i\theta_0}$, so, independently of how $\tilde{U}(z)$ behaves,

$$\varphi(z) = \frac{U(z) + i\tilde{U}(z)}{U(z) + i\tilde{U}(z) + 1} \longrightarrow 1$$

for $z \to e^{i\theta_0}$. For $\theta \notin E$ but tending to a point θ_0 of E we also clearly have for the boundary limit $\varphi(e^{i\theta})$,

$$\varphi(e^{i\theta}) = \frac{F(\theta) + i\tilde{F}(\theta)}{F(\theta) + i\tilde{F}(\theta) + 1} \longrightarrow 1,$$

since $F(\theta) \to \infty$. Taking $\varphi(z)$ to be thus extended by continuity up to $|z| = 1$, we now see that $|\varphi(z)| < 1$ everywhere for $|z| \leq 1$ (sic !), save when $z = e^{i\theta}$ with $\theta \in E$, where $\varphi(e^{i\theta}) = 1$. And this function φ is continuous on $|z| \leq 1$ and analytic in $|z| < 1$. The first construction of such a function was essentially made by Fatou in his thesis - the Riesz brothers just used it. (But Fatou never thought of the theorem they proved by its help!)

Let $k = 1,2,\ldots$. Then $[\varphi(z)]^k$ is also analytic in $|z| < 1$, continuous in $|z| \leq 1$, so, by Section C of Chapter I, $[\varphi(re^{i\theta})]^k \to$ $\to [\varphi(e^{i\theta})]^k$ uniformly for $r \to 1$. ($[\varphi(z)]^k$ is a harmonic function plus i times another harmonic function - therefore Poisson's representation holds for it - with complex valued, but continuous boundary values!) So for each k,

$$\int_{-\pi}^{\pi} [\varphi(re^{i\theta})]^k e^{i\theta}\, d\mu(\theta) \to \int_{-\pi}^{\pi} [\varphi(e^{i\theta})]^k e^{i\theta}\, d\mu(\theta)$$

as $r \to 1$.

Now for $r < 1$, by analyticity,

$$[\varphi(re^{i\theta})]^k = \sum_0^\infty a_n r^n e^{in\theta},$$

the series being uniformly convergent.

So by the hypothesis of the theorem to be proved

$$\int_{-\pi}^{\pi} [\varphi(re^{i\theta})]^k e^{i\theta}\, d\mu(\theta) = 0.$$

Making $r \to 1$, we have

$$\int_{-\pi}^{\pi} [\varphi(e^{i\theta})]^k e^{i\theta}\, d\mu(\theta) = 0, \quad k = 1,2,\ldots \quad .$$

But now, by construction, $\varphi(e^{i\theta}) = 1$ for $\theta \in E$ whereas elsewhere, $|\varphi(e^{i\theta})| < 1$. Therefore by bounded convergence,

$$\int_{-\pi}^{\pi} [\varphi(e^{i\theta})]^k e^{i\theta}\, d\mu(\theta) \to \int_E e^{i\theta} d\mu(\theta)$$

as $k \to \infty$, i.e., $\int_E e^{i\theta} d\mu(\theta) = 0$ for any closed E with $|E| = 0$. Since, as stated at the beginning, if μ were not absolutely continuous we could get such an E with $\int_E e^{i\theta} d\mu(\theta) \neq 0$, we have reached a contradiction, proving the theorem.

2. Here is a modern proof of the F. and M. Riesz theorem, due to Helson and Lowdenslager, ca. 1958.

Given a finite positive measure ν on $[-\pi, \pi]$, we let $L_2(d\nu)$ be the Hilbert space obtained by defining

$$\langle f, g \rangle_\nu = \int_{-\pi}^{\pi} f(\theta)\overline{g(\theta)} d\nu(\theta)$$

for continuous 2π-periodic functions f and g, and then forming the completion in the usual way, using the norm $\|f\|_\nu = \sqrt{\langle f, f \rangle_\nu}$.

i) Write $d\mu(\theta) = h(\theta)d\theta + ds(\theta)$ where $h \in L_1$ and s is singular, and suppose $\int_{-\pi}^{\pi} e^{in\theta} d\mu(\theta) = 0$, $n = 1, 2, \ldots$. Our main object is to prove that $\int_{-\pi}^{\pi} e^{in\theta} ds(\theta) = 0$, $n = 1, 2, \ldots$.

Take $d\nu(\theta) = (1 + |h(\theta)|)d\theta + |ds(\theta)|$.

We consider $\mathfrak{A}$ = subspace of $L_2(d\nu)$ (sic!) spanned by $\{e^{in\theta};\ n = 1, 2, \ldots\ .\}$.

Let $F \in \mathfrak{A}$ be the element of $\mathfrak{A}$ making $\int_{-\pi}^{\pi} |1 - F(\theta)|^2 d\nu(\theta)$ as small as possible. Since $d\nu(\theta) \geq d\theta$,

$$\int_{-\pi}^{\pi} |1 - F(\theta)|^2 d\nu(\theta) \geq \inf\{ \int_{-\pi}^{\pi} |1 - P(\theta)|^2 d\theta;\ P(\theta) =$$
$$= a_1 e^{i\theta} + a_2 e^{2i\theta} + \ldots + a_p e^{pi\theta}\} = 2\pi.$$

By elementary Hilbert space geometry, <u>in</u> $L_2(d\nu)$, $1-F \perp \mathfrak{A}$, in particular

$$1 - F \perp e^{in\theta}(1 - F(\theta)) \quad \text{in} \quad L_2(d\nu)$$

for $n = 1,2,\ldots$.

$$\therefore \int_{-\pi}^{\pi} |1 - F(\theta)|^2 e^{-in\theta} d\nu(\theta) = 0, \qquad n = 1,2,\ldots,$$

so, since $d\nu(\theta) \geq 0$, by <u>conjugation</u>, $\boxed{|1-F(\theta)|^2 d\nu(\theta) = c\,d\theta.}$. By the previous, $c > 0$.

ii) From $|1-F(\theta)|^2 d\nu(\theta) = c\,d\theta$, $|1-F(\theta)|^2 d\nu(\theta)$ is absolutely continuous, so

$$|1 - F(\theta)|^2 |ds(\theta)| = 0, \quad \text{or} \quad F(\theta) \equiv 1 \quad \text{a.e.} \quad |ds|.$$

$$\therefore |1-F(\theta)|^2 d\nu(\theta) = |1-F(\theta)|^2 (1+|h(\theta)|)d\theta.$$

$$\therefore |1-F|^2(1 + |h|) = c \quad \text{or}$$

$$\frac{1}{|1-F|^2} = \frac{1 + |h|}{c} \in L_1(d\theta), \quad \text{so}$$

$$\boxed{\frac{1}{1-F} \in L_2},$$

and

$$\boxed{|1-F|(1 + |h|) = \frac{c}{|1-F|} \in L_2}$$

iii) Since $F(\theta) = 1$ a.e. $|ds|$,

$$(1 - \overline{F(\theta)})d\nu(\theta) = (1 - \overline{F(\theta)})(1 + |h(\theta)|)d\theta.$$

Now $1 - F \perp \mathfrak{A}$ in $L_2(d\nu)$ makes, <u>with this</u>,

$$\int_{-\pi}^{\pi} e^{in\theta}[1 - \overline{F(\theta)}](1 + |h(\theta)|)d\theta = 0, \qquad n = 1,2,\ldots,$$

i.e., since $\frac{1}{1-F} = (1 - \overline{F})(1 + |h(\theta)|)$,

$$\int_{-\pi}^{\pi} \frac{1}{1-F(\theta)} e^{in\theta} d\theta = 0, \qquad n = 1,2,3,\ldots \quad .$$

With $\frac{1}{1-F} \in L_2$, this last implies there is a sequence of polynomials $G_n(z)$ with $\|G_n(e^{i\theta}) - \frac{1}{1-F(\theta)}\|_2 \xrightarrow[n]{} 0$.

iv) From $\int_{-\pi}^{\pi} e^{in\theta} d\mu(\theta) = 0$; $n = 1,2,\ldots$ and $F(\theta)$ being in the closure of the set of linear combinations of the $e^{in\theta}$, $n \geq 1$, in $L_2(d\nu)$ with $d\nu(\theta) \geq d\mu(\theta)$, we surely have

$$\int_{-\pi}^{\pi} e^{ik\theta}(1 - F(\theta))d\mu(\theta) = 0, \qquad k = 1,2,\ldots \quad .$$

But $d\mu(\theta) = h(\theta)d\theta + ds(\theta)$, and $1 - F(\theta) \equiv 0$ a.e. $|ds(\theta)|$.

So

$$\boxed{\int_{-\pi}^{\pi} e^{ik\theta}[1 - F(\theta)]h(\theta)d\theta = 0, \qquad k = 1,2,\ldots \quad .}$$

By ii), $(1-F)h \in L_2$, so, if G_n is the sequence of polynomials from iii), by Schwarz,

$$\int_{-\pi}^{\pi} e^{ik\theta}h(\theta)d\theta = \int_{-\pi}^{\pi} e^{ik\theta} \frac{1}{1-F(\theta)} [1 - F(\theta)]h(\theta)d\theta$$

$$= \lim_{n\to\infty} \int_{-\pi}^{\pi} e^{ik\theta}G_n(e^{i\theta})(1 - F(\theta))h(\theta)d\theta.$$

But each of these last integrals is zero by previous boxed remark, the $G_n(z)$ being polynomials in z.

Conclusion:

$$\int_{-\pi}^{\pi} e^{ik\theta}h(\theta)d\theta = 0, \qquad k = 1,2,\ldots \quad .$$

v) The result just obtained shows, with the hypothesis, that

$$\int_{-\pi}^{\pi} e^{ik\theta} ds(\theta) = 0, \qquad k = 1,2,3,\ldots \;.$$

Since $F \in \mathfrak{A}$ is in the closure of the set of linear combinations of $e^{i\theta}, e^{2i\theta}$, &c in $L_2(d\nu)$ where $d\nu \geq |ds|$, we have

$$\int_{-\pi}^{\pi} F(\theta) ds(\theta) = 0,$$

i.e., $\int_{-\pi}^{\pi} ds(\theta) = 0$, since $F(\theta) = 1$ a.e. $|ds|$ by step ii). We thus see that we even have

$$\int_{-\pi}^{\pi} e^{ik\theta} \cdot e^{-i\theta} ds(\theta) = 0 \quad \text{for} \quad k = 1,2,3,\ldots \;.$$

Now apply the argument in steps i) - iv) using a new measure μ given by $d\mu(\theta) = e^{-i\theta} ds(\theta)$. Since this μ is already entirely singular, we will obtain for it

$$\int_{-\pi}^{\pi} e^{ik\theta} e^{-i\theta} d\mu(\theta) = 0, \qquad k = 1,2,3,\ldots,$$

i.e.,

$$\int_{-\pi}^{\pi} e^{ik\theta} e^{-2i\theta} ds(\theta) = 0, \qquad k = 1,2,3,\ldots \;.$$

Taking now a new (singular) μ given by $d\mu(\theta) = e^{-2i\theta} ds(\theta)$ and repeating the process, we see that

$$\int_{-\pi}^{\pi} e^{ik\theta} e^{-3i\theta} ds(\theta) = 0, \qquad k = 1,2,\ldots \;.$$

By repetition, we thus have $\int_{-\pi}^{\pi} e^{i\ell\theta} ds(\theta) = 0$ for all integers ℓ, whence $ds(\theta) = 0$.

We have thus proven that, for our given original μ, $d\mu(\theta) = h(\theta)d\theta$. Thus μ is absolutely continuous.

Q.E.D.

B. Definition and Basic Properties of H_1.

Admission of complex valued harmonic functions into the discussion is rather straightforward - we simply denote by that name any complex linear combination of ordinary (real valued) harmonic functions. (This notion was avoided up to now because we wanted to be able to say that a harmonic function was the real part of an analytic one!) It becomes convenient now to consider complex valued harmonic functions because we can look upon an analytic function as harmonic. The representation theorems of Chapter I, Section C and the boundary behavior theorems of Chapter I, Section D do, of course hold for complex valued harmonic functions - their extension is trivial.

Definition. $F(z)$, analytic for $|z| < 1$ is said to be in H_1 if

$$\int_{-\pi}^{\pi} |F(re^{i\theta})|\,d\theta \quad \text{is bounded for} \quad r < 1.$$

1. Let $F \in H_1$. Since $F(z)$ is, in particular, harmonic for $|z| < 1$ we have, by a theorem in Chapter I, Section C,

$$F(re^{i\theta}) = \frac{1}{2\pi}\int_{-\pi}^{\pi} \frac{1 - r^2}{1 + r^2 - 2r\cos(\theta - t)}\,d\mu(t)$$

for some measure (here complex valued!) μ on $[-\pi,\pi]$. By Chapter I, Section D,

$$F(re^{i\theta})d\theta \to d\mu(\theta) \quad w^*$$

as $r \to 1$. Since F is analytic in $|z| < 1$, Cauchy's theorem (or straightforward manipulation with a power series!) here yields, for $r < 1$

$$\int_{-\pi}^{\pi} e^{in\theta} F(re^{i\theta})\, d\theta = 0, \qquad n = 1,2,\ldots .$$

Therefore

$$\int_{-\pi}^{\pi} e^{in\theta} d\mu(\theta) = 0, \qquad n = 1,2,3,\ldots .$$

Now the theorem of the Brothers Riesz, given above in Section A, guarantees that μ is absolutely continuous, i.e., $d\mu(\theta) = h(\theta)d\theta$ for some $h \in L_1(-\pi,\pi)$.

So in this case we really have

$$F(re^{i\theta}) = \frac{1}{2\pi} \int_{-\pi}^{\pi} P_r(\theta - t)h(t)\,dt$$

with a function h. The distinction between this case (for analytic F) and the more general one treated in Chapter 1, Section C (F merely harmonic) is very important for the whole development of the theory.

By Chapter I, Section D we now have $F(z) \to h(e^{i\theta})$ a.e. for $z \underset{\measuredangle}{\longrightarrow} e^{i\theta}$, so if, as mentioned at the end of Chapter I, Section D, we call

$$F(e^{i\theta}) = \lim_{z \underset{\measuredangle}{\longrightarrow} e^{i\theta}} F(z),$$

we have

$$\boxed{F(re^{i\theta}) = \frac{1}{2\pi} \int_{-\pi}^{\pi} \frac{1 - r^2}{1 + r^2 - 2r\cos(\theta - t)} F(e^{it})\,dt}$$

for $F \in H_1$.

2. From the boxed formula in subsection 1, and the elementary approximate identity property of $P_r(\theta)$ (Chapter I, Section D, 2):

$$\boxed{\int_{-\pi}^{\pi} |F(re^{i\theta}) - F(e^{i\theta})|\,d\theta \to 0 \quad \text{as} \quad r \to 1.}$$

3. By Cauchy's formula, for $|z| < R < 1$,

$$F(z) = \frac{1}{2\pi i} \int_{|\zeta|=R} \frac{F(\zeta)}{\zeta - z}\,d\zeta.$$

Now if $|z| < 1$ is <u>fixed</u> and we make $R \to 1$, we conclude from subsection 2 that

$$\boxed{F(z) = \frac{1}{2\pi i} \int_0^{2\pi} \frac{F(e^{it})\,d(e^{it})}{e^{it} - z}},$$

which is Cauchy's formula for functions in H_1.

C. <u>Digression on conformal mapping theory</u>.

A simply connected region whose boundary contains more than two points can be mapped conformally on the interior of the unit circle. This is the Riemann mapping theorem, whose proof belongs to the basic course on complex variable theory.

It is also true that if the simply connected region's boundary is a <u>Jordan curve</u>, then the function mapping the region onto $\{|z| < 1\}$ has a continuous one-one extension up to the boundary, which it takes onto $\{|z| = 1\}$.

This fact is often given without proof in complex variable courses. Because we <u>use</u> it sometimes in the <u>present course</u>, we give a complete proof in Subsection 1 below. The proof uses the Jordan curve theorem, here assumed known, but is otherwise self-contained.

1. Definition. A Jordan curve is a continuous one-one image of $\{|\zeta| = 1\}$ in $\mathbb{C}$. The continuous one-one function mapping $\{|\zeta| = 1\}$ onto the Jordan curve is called a parameterization of the curve.

By means of its parameterization, any Jordan curve has a natural order defined on it, in an obvious fashion. We can thus speak of arcs on a Jordan curve, and so forth.

Definition. A Jordan arc is a continuous one-one image of an interval of $\mathbb{R}$ (with or without endpoints) in $\mathbb{C}$.

A Jordan curve can be a very complicated object:

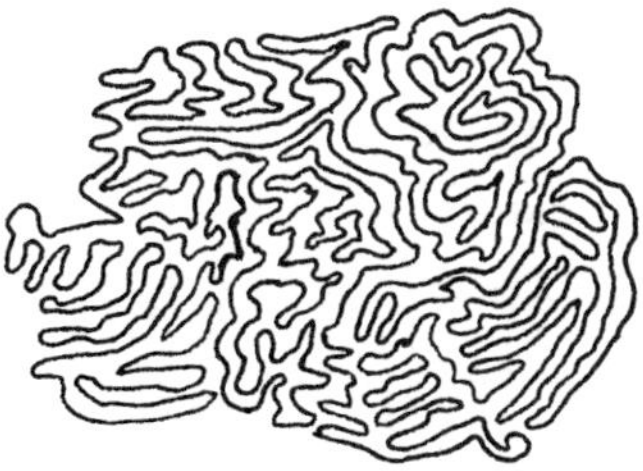

We admit without proof the following Jordan Curve Theorem. Let Γ be a Jordan curve. Then $\mathbb{C} \sim \Gamma$ consists of two connected components, $\mathcal{O}$ and Ω, one of which contains all z with sufficiently large modulus. If $\omega \in \Gamma$, any neighborhood of ω has in it points of $\mathcal{O}$ and points of Ω.

Definition. Let Ω be the component of $\mathbb{C} \sim \Gamma$ which contains all points of sufficiently large modulus. Ω is called the outside of Γ. The other component, $\mathcal{O}$, is called the inside of Γ.

In complex variable theory, one uses the following definition:

A *connected open set* in $\mathbb{C}$ is called *simply* connected if, whenever any Jordan polygon Π lies in the *open set*, the *inside* of Π also lies in it.

(A *Jordan polygon* is a Jordan curve made up of straight segments, i.e., a *simple closed broken-line path*. The Jordan curve theorem is *elementary* for Jordan polygons (although not trivial), and can be proved by induction.)

Lemma. Let Γ be a Jordan curve, and $\mathfrak{G}$ its *inside*. $\mathfrak{G}$ is simply connected.

Proof. Assume not. Then there is a Jordan polygon $\Pi \subseteq \mathfrak{G}$ whose *inside*, $\mathfrak{G}'$, is *not* in $\mathfrak{G}$. Therefore there is a $z \in \mathfrak{G}'$ with $z \in \Gamma$ or $z \in \Omega$, the *outside* of Γ. If $z \in \Gamma$, let $\mathcal{N}$ be a neighborhood of z with $\mathcal{N} \subseteq \mathfrak{G}'$. By Jordan curve theorem, $\mathcal{N}$ contains a point z' of Ω. So in any case, if result is *false*, $\mathfrak{G}'$ has in it a point $z' \in \Omega$. Since an open set is arcwise connected, we can join z' to ∞ by a path Λ lying in Ω, in particular, *not touching* Γ. Since $z' \in \mathfrak{G}'$, *inside* of Π, Λ must touch Π, say in z''. Then $z'' \in \mathfrak{G}$. But Λ joins z'' to ∞ without touching Γ, which is impossible. So the lemma is true.

Remark. The above argument will be *used again* in what follows. We shall refer to it as the *Jordan curve argument*.

Let now Γ be a Jordan curve, and let $\mathcal{D}$ be its inside. The Riemann mapping theorem says there is a conformal mapping Φ of $\{|z| < 1\}$ onto $\mathcal{D}$ since, by the lemma, $\mathcal{D}$ is simply connected. Our problem is to show that Φ can be continuously extended up to $\{|z| \leq 1\}$ so as to take $\{|z| = 1\}$ in continuous one-one fashion onto Γ.

Lemma. There is a function $\eta(\delta)$ defined for all sufficiently small $\delta > 0$, with $\eta(\delta) \to 0$ as $\delta \to 0$, such that, given a & $b \in \Gamma$ with $|a - b| \leq \delta$, there is one and only one arc of Γ having endpoints a & b whose diameter is $\leq \eta(\delta)$.

Proof. Let $\psi(e^{it})$ be a parameterization of Γ, and let $\delta_0 > 0$ be so small that whenever $|\psi(\zeta) - \psi(\zeta')| \leq \delta_0$, $|\zeta - \zeta'| < 2$. For ζ & ζ' with $|\psi(\zeta) - \psi(\zeta')| \leq \delta_0$, let σ be the (unique!) shorter arc of $\{|z| = 1\}$ having endpoints ζ & ζ', and call $\gamma = \psi(\sigma)$. By continuity of ψ and its inverse, $\operatorname{diam} \gamma \to 0$ uniformly for $|\psi(\zeta) - \psi(\zeta')| \to 0$, whatever ζ, ζ' we take on unit circle with $|\psi(\zeta) - \psi(\zeta')| \leq \delta_0$. So for $\delta < \delta_0$, call

$$\eta(\delta) = \sup\{\operatorname{diam} \gamma; |\psi(\zeta) - \psi(\zeta')| \leq \delta\}$$

Then $\eta(\delta) \to 0$ as $\delta \to 0$. Let $\delta_1 < \delta_0$ be so small that $\eta(\delta_1) < \frac{1}{2} \operatorname{diam} \Gamma$. Then the lemma holds for $\delta \leq \delta_1$.

Definition. Let $a, b \in \Gamma$ with $|a - b|$ sufficiently small. The unique arc of Γ with endpoints a & b having diameter $\leq \eta(|a - b|)$ is called the smaller arc of Γ joining a to b.

Lemma. Let Φ map $\{|z| < 1\}$ conformally onto $\mathcal{D}$ bounded by Jordan curve Γ. Let $|\zeta| = 1$. Then $\lim_{z \to \zeta} \Phi(z)$ exists and is on Γ.

This is the <u>main part</u> of the solution of our problem, and will be done in a series of steps.

i) Wlog, $\zeta = 1$, and, instead of looking at Φ, we look at $F(z)$ which maps $\Re z > 0$ onto $\mathfrak{D}$. <u>We are to prove</u> that $\lim_{z\to 0} F(z)$ <u>exists</u> and is on Γ; this will be enough.

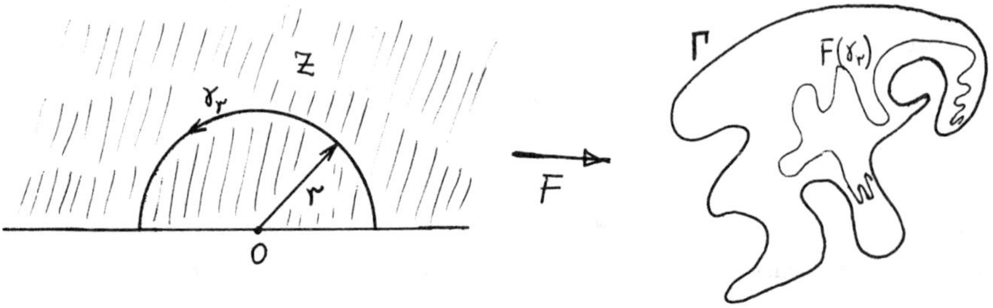

With γ_r the half-circle $re^{i\theta}$, $0 < \theta < \pi$, consider the (un-ended) Jordan arc $F(\gamma_r) \subset \mathfrak{D}$.

$$\text{length } F(\gamma_r) = \int_0^{\pi} |F'(re^{i\theta})|\, r\, d\theta.$$

But $\infty > \text{area } \mathfrak{D} = \iint_{\Im z>0} |F'(z)|^2 dx\, dy = \int_0^{\infty}\int_0^{\pi} |F'(re^{i\theta})|^2\, r\, d\theta\, dr$ and this by Schwarz' inequality is $\geq \int_0^{\infty} \{[\text{length } F(\gamma_r)]^2/\pi r\} dr$. So

$$\int_0^{\infty} \frac{1}{r}\, [\text{length } F(\gamma_r)]^2 dr < \infty.$$

<u>Since for any</u> $p > 0$, $\int_0^{p} \frac{dr}{r} = \infty$, there must be a sequence $r_n \downarrow 0$ with length $F(\gamma_{r_n}) \to 0$. (N.B. Examples show that it is <u>not true</u> that length $F(\gamma_r) \to 0$ for $r \to 0$!) <u>Choose such a sequence</u> $\{r_n\}$ <u>once and for all, and call</u> $\gamma_{r_n} = \gamma_n$.

ii) As soon as length $F(\gamma_n) < \infty$, the limits $a = \lim_{\theta \to 0} F(r_n e^{i\theta})$ and $b = \lim_{\theta \to \pi} F(r_n e^{i\theta})$ exist. They are on Γ. Assume that, e.g., $a \in \mathfrak{D}$, and let η be a neighborhood of a with $\eta \subset \mathfrak{D}$. Let $z_0 = F^{-1}(a)$; $\Im z_0 > 0$, and choose η so small that $F^{-1}(\eta)$ is a proper part of $\Im z > 0$. This is possible because F^{-1} is continuous! Then, for $\theta > 0$ sufficiently close to 0, $F(r_n e^{i\theta})$ is in η, so $r_n e^{i\theta}$ in $F^{-1}(\eta)$, which is false for θ close enough to 0. So a and b are on Γ. We adjoin a and b to $F(\gamma_n)$ in an obvious fashion so as to get either:

a) A closed Jordan arc $\overline{F(\gamma_n)}$ if $a \neq b$

b) A Jordan curve if $a = b$.

In case a), $|a - b| \leq$ length $F(\gamma_n)$ is small if n is large; therefore by above lemma there is a unique smaller arc Γ_{ab} of Γ going from a to b. In this case $F(\gamma_n)$ and Γ_{ab} form, together a Jordan curve. The proof of this consists in writing down a parameterization for $F(\gamma_n) \cup \Gamma_{ab}$ in terms of a readily available one for $F(\gamma_n)$ and the one for Γ, and is so straightforward and routine that we omit it.

iii) Let $\mathfrak{G}_n$ be the inside of the Jordan curve $F(\gamma_n) \cup \Gamma_{ab}$ in case a) above, or of $\overline{F(\gamma_n)}$ in case b) above. Then $\mathfrak{G}_n \subseteq \mathfrak{D}$.

Proof. Note that $F(\gamma_n) \subseteq \mathfrak{D}$, and repeat the Jordan curve argument already used in proving the first lemma.

iv) Let n be sufficiently large, and let S_n and T_n be the regions shown in $\Im z > 0$:

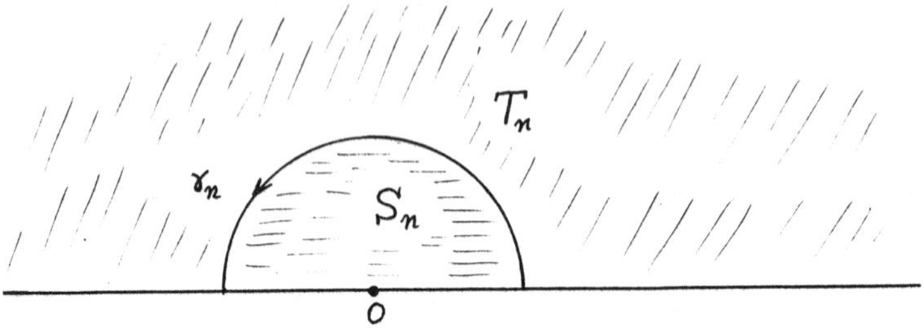

Then $F(S_n) \subseteq \mathfrak{G}_n$, the inside of $F(\gamma_n) \cup \Gamma_{ab}$ (or of $\overline{F(\gamma_n)}$ if $a = b$).

Proof. By iii), $\mathfrak{G}_n \subseteq \mathfrak{D}$. Take any $w_0 \in \mathfrak{G}_n$, then $w_0 = F(z_0)$ where $\Im z_0 > 0$. Since $w_0 \notin F(\gamma_n)$, $z_0 \in S_n$ or $z_0 \in T_n$.

a) If $z_0 \in S_n$, $F(S_n) \subseteq \mathfrak{G}_n$, and we're done. Indeed, S_n is open and connected, so $F(S_n)$ is also. $F(S_n) \subseteq \mathfrak{D}$ $(\subseteq \mathbb{C} \sim \Gamma)$ and $S_n \cap \gamma_n = \emptyset$, so $F(S_n)$ lies entirely in the complement of the Jordan curve bounding $\mathfrak{G}_n$ ($\overline{F(\gamma_n)}$ or $F(\gamma_n) \cup \Gamma_{ab}$ as the case may be). $F(S_n)$ has a point in common, namely w_0, with the connected component $\mathfrak{G}_n$ of that complement. Therefore $F(S_n)$ must lie entirely in that component, i.e., $F(S_n) \subseteq \mathfrak{G}_n$.

b) If $z_0 \in T_n$ then we prove as in a) that $F(T_n) \subseteq \mathfrak{G}_n$. However, this we now show to be impossible for sufficiently large n. Indeed, area $F(T_n)$ = area $\mathfrak{D}$ - area $F(S_n)$ does $\rightarrow$ area $\mathfrak{D}$ as $n \rightarrow \infty$ since $\bigcap_n S_n = \emptyset$. On the other hand, if length $F(\gamma_n) = \delta_n$, we have $\delta_n \rightarrow 0$ by i), and $|a - b| \leq \delta_n$ so surely Γ_{ab} has diameter $\leq \eta(\delta_n)$ which $\rightarrow 0$ as $n \rightarrow \infty$, by second lemma. (We count Γ_{ab} as $\{a\}$ if $a = b$. a and b and Γ_{ab} depend, of course, on n, but we do not show this dependence in the notation, in order to have easier symbols to read!) For any given large n, let Δ_n be a disk of radius $\delta_n + \eta(\delta_n)$ about the end a of $\overline{F(\gamma_n)}$, then the whole Jordan curve $F(\gamma_n) \cup \Gamma_{ab}$ lies in Δ_n:

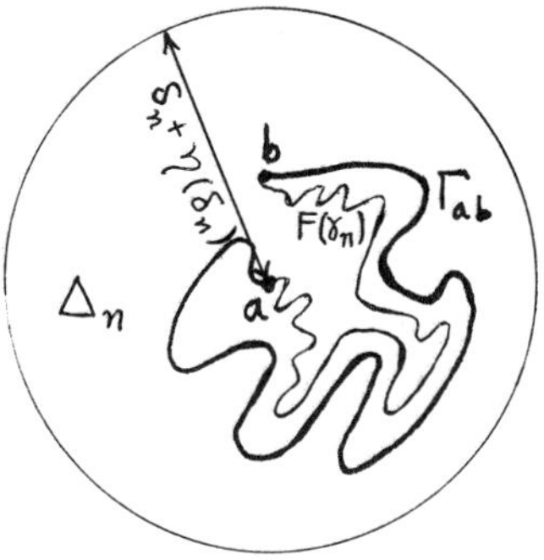

From this it follows that the inside $\mathfrak{G}_n$ of $F(\gamma_n) \cup \Gamma_{ab}$ also lies in Δ_n - this is proven by a repetition of the Jordan curve argument.

So area $\mathfrak{G}_n \leq \pi(\delta_n + \eta(\delta_n))^2 \xrightarrow[n]{} 0$ as $n \to \infty$, whilst, as we have seen, area $F(T_n) \xrightarrow[n]{}$ area $\mathfrak{D} > 0$. Therefore, for sufficiently large n, $F(T_n) \subseteq \mathfrak{G}_n$ is impossible, and we must have $F(S_n) \subseteq \mathfrak{G}_n$.

Q.E.D.

v) We have $S_n \supseteq S_{n+1} \supseteq \cdots$. By iv), for sufficiently large n, $F(S_n) \subseteq \mathfrak{G}_n$, and as we saw in proving iv), diam $\mathfrak{G}_n \leq \delta_n + \eta(\delta_n) \xrightarrow[n]{} 0$.

Now suppose $\Im z_k > 0$ and $z_k \xrightarrow[k]{} 0$. The sequence $\{F(z_k)\}$ is eventually in every $F(S_n)$ and diam $F(S_n) \xrightarrow[n]{} 0$. Therefore $F(z_k)$ converges to a definite limit, w, say, by Cauchy's convergence criterion, and w is independent of the particular sequence $\{z_k\}$ chosen of points tending to 0. By the proof in iv), the distance of any point in $F(S_n) \subseteq \mathfrak{G}_n$ to Γ is $\leq \delta_n + \eta(\delta_n)$ which $\xrightarrow[n]{} 0$, so $w \in \Gamma$. (This could also be seen by the argument at the beginning of ii).)

We have completed the proof of our third lemma.

Go back to $\Phi(z)$ which maps $\{|z| < 1\}$ *conformally* on $\mathfrak{D}$. The lemma just *proven* shows that for *each* ζ, $|\zeta| = 1$, $\lim_{z \to \zeta} \Phi(z)$ *exists* and *is on* Γ.

Notation. Call $\lim_{z \to \zeta} \Phi(z) = \Phi(\zeta)$ for $|\zeta| = 1$.

Lemma. $\Phi(\zeta)$ as thus defined in continuous on $\{|\zeta| = 1\}$.

The proof is easy and routine, so we omit it.

Lemma. $\Phi(\zeta)$ is one-one on $|\zeta| = 1$.

Proof. Assume not. Then, for, say, α and β on unit circle, $\alpha \neq \beta$, we have $\Phi(\alpha) = \Phi(\beta)$. Let σ be the path shown, and consider $\Phi(\sigma)$.

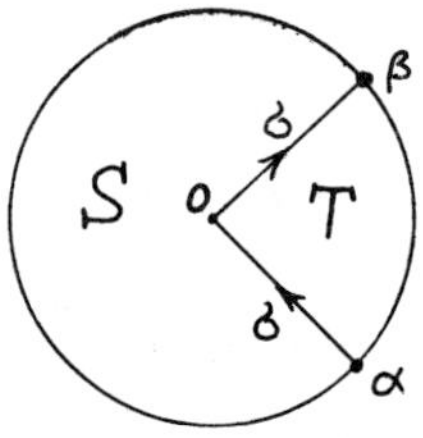

Since Φ is one-one in $\{|z| < 1\}$, it is easy to see that $\Phi(\sigma)$ is a Jordan curve lying entirely in $\mathfrak{D}$ except for the one point $\Phi(\alpha) = \Phi(\beta)$ which lies on Γ. A repetition of the *Jordan curve argument shows* that the *inside*, $\mathfrak{G}$, of $\Phi(\sigma)$ lies in $\mathfrak{D}$. Let S and T be the two sectors shown of $\{|z| < 1\}$. The argument used in step iv) of the proof of the *third* lemma shows that $\Phi(S) \subseteq \mathfrak{G}$ or $\Phi(T) \subseteq \mathfrak{G}$. Wlog, let $\Phi(S) \subseteq \mathfrak{G}$. *Then*, if $z \in S$ and $|z| \to 1$, we *must have* $\Phi(z) \to \Phi(\alpha)$. Indeed, all the limit points of $\Phi(z)$ *must be* on Γ if $|z| \to 1$, by the argument in ii) above. Also, $\Phi(z)$ stays in $\mathfrak{G}$ whose *only limit point* on Γ is $\Phi(\alpha)$.

(Proof: $\mathfrak{G} \subseteq \mathfrak{D}$ so if w is a limit point of $\mathfrak{G}$ which is on Γ (hence not in $\mathfrak{D}$), w is surely $\notin \mathfrak{G}$, so w is on the boundary of $\mathfrak{G}$, $\Phi(\sigma)$. But the only point of $\Phi(\sigma)$ which lies on Γ is $\Phi(\alpha)$!) So $\Phi(z)$ necessarily $\to \Phi(\alpha)$.

Since the boundary of S includes a whole arc of $|\zeta| = 1$ (from α to β), and $\Phi(z)$ is analytic in $|z| < 1$, it must be in fact constant in $\{|z| < 1\}$ and equal to $\Phi(\alpha)$. This is absurd, and the lemma is proven.

With $\Phi(z)$ extended as indicated above to $\{|z| < 1\}$, it becomes continuous on that closed disk and takes $\{|z| = 1\}$ in one-one fashion into Γ.

Lemma. Φ takes $\{|z| = 1\}$ onto Γ.

Proof. Let $w_0 \in \Gamma$. By the Jordan curve theorem there is a sequence of points $w_n \in \mathfrak{D}$ with $w_n \xrightarrow[n]{} w_0$. Let z_n, $|z_n| < 1$, be such that $\Phi(z_n) = w_n$, then $\Phi(z_n) \xrightarrow[n]{} w_0$. Wlog, $z_n \xrightarrow[n]{} z_0$, $|z_0| \le 1$. $|z_0| < 1$ is impossible, for then $w_0 = \lim_{n\to\infty} \Phi(z_n) = \Phi(z_0)$ would be in $\mathfrak{D}$. So $|z_0| = 1$, and, by the above-described extension of Φ, $\Phi(z_0) = w_0$.

Theorem. The conformal mapping Φ of $\{|z| < 1\}$ onto the inside $\mathfrak{D}$ of Γ has a continuous one-one extension up to $\{|z| = 1\}$ and when thus extended takes $\{|z| = 1\}$ onto Γ.

Proof. Combine the last 4 lemmas.

Remark. The result is due to Caretheodory, ca. 1912 or 1914. The above proof I learned, in part, from my teacher, Rafael Robinson - he gave the analytic argument of i) in his course in Berkeley, in the autumn of 1951. I suspect that the whole argument consists of very well-known ideas.

2. Suppose $\mathcal{D}$ is a simply connected region bounded by a simple Jordan curve Γ, and suppose, without loss of generality, that $0 \in \Gamma$. Suppose $f(z)$ maps $\{|z| < 1\}$ conformally onto $\mathcal{D}$; as we have seen, f has a continuous one-one extension up to $|z| = 1$, and, when thus extended, takes that circumference onto Γ, yielding a parameterization of Γ. We suppose, without loss of generality, that $f(1) = 0$:

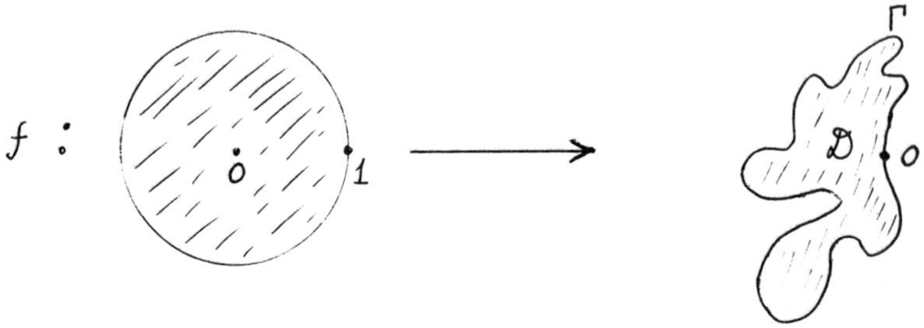

In 1917 (in the same Scandinavian Congress proceedings where the famous paper of F. and M. Riesz appeared), Lindelöf published a theorem about the case where Γ has a tangent at 0. This theorem states that for $|z| < 1$, $z \to 1$, we have

$$\arg f(z) - \arg(1 - z) \to \text{constant}.$$

It means that the conformal images of sectors in $|z| < 1$ with their vertices at 1 are asymptotically like sectors in $\mathcal{D}$ of the same opening with their vertices at 0:

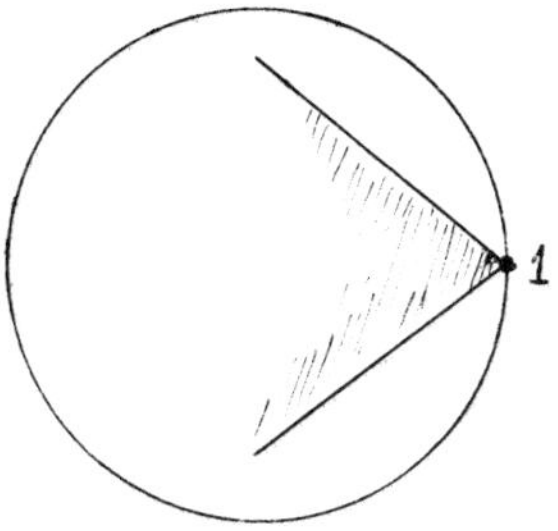

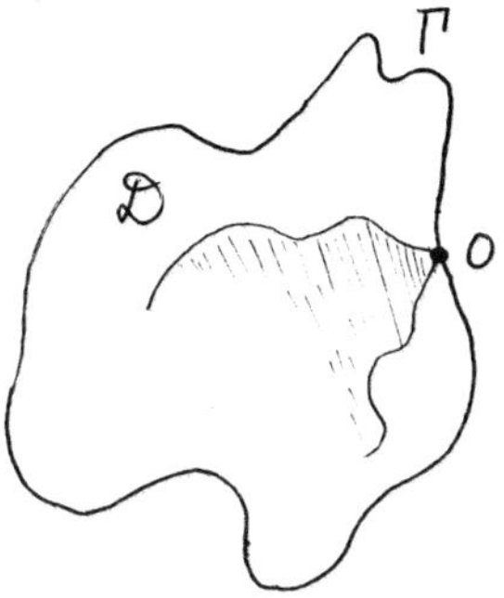

Let us, for the moment, grant the following:

Lemma. $\arg f(z)$ is bounded in $|z| < 1$.

Then the proof of Lindelöf's theorem runs as follows:

Since $f(e^{i\theta})$ is continuous and $f(e^{i\theta}) \neq 0$ for $e^{i\theta} \neq 1$ (Γ being a simple Jordan curve!), $\arg f(e^{i\theta})$ (defined as $\lim_{z \to e^{i\theta}} \arg f(z)$) is continuous for $e^{i\theta}$ away from 1; the lemma says it's bounded.

For Γ to have a tangent at 1 means that $\arg f(e^{i\theta})$ tends to a certain limit, say α, as $\theta \to 0+$, and tends to α + an odd multiple of π as $\theta \to 0-$. It is reasonable from considerations of orientation that this multiple be simply π; granted this (which will be proved after the lemma), we see that $\arg f(e^{i\theta}) - \arg(1 - e^{i\theta})$ is continuous at 0, where it equals $\alpha - \frac{\pi}{2}$. Since $\arg f(z) - \arg(1 - z)$ is harmonic and bounded in $|z| < 1$, we have, by Chapter I, Section C,

$$\arg f(re^{i\theta}) - \arg(1 - re^{i\theta}) = \frac{1}{2\pi}\int_{-\pi}^{\pi} \frac{1 - r^2}{1 + r^2 - 2r\cos(\theta - t)} \arg\left[\frac{f(e^{it})}{1 - e^{it}}\right] dt$$

for $|z| < 1$. The abovementioned continuity now shows that

$$\arg f(z) - \arg(1 - z) \longrightarrow \alpha - \frac{\pi}{2}$$

as $z \to 1$, by a theorem in Chapter 1, Section D,2.

To make this argument rigorous, the lemma and the statement about $\arg f(e^{\pm i0})$ must be proved. A rigorous proof uses the Jordan curve theorem.

Assume without loss of generality that the tangent to Γ at 0 is vertical. Then, given $\eta > 0$ there is a $\delta > 0$ such that all points $w = f(e^{i\theta})$ on Γ with $|w| < \delta$ lie in one of the two sectors $|\arg w \pm \frac{\pi}{2}| < \eta$:

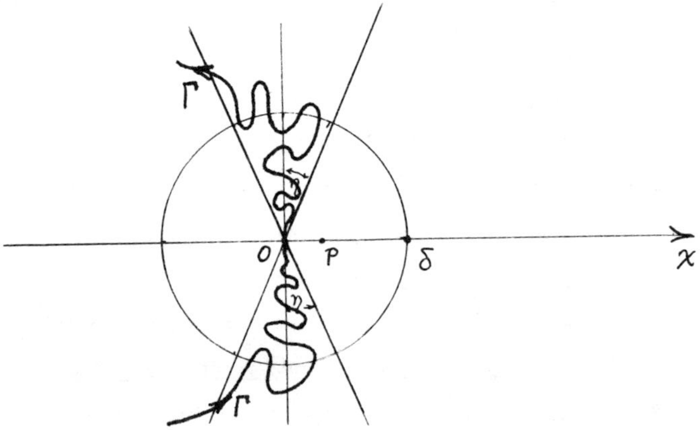

There are points P on the x-axis arbitrarily close to 0 with $P \notin \mathcal{D}$. Indeed, let Φ be a conformal mapping of $|z| > 1$ (sic!) onto the outside of Γ, and consider the curves $\gamma_n = \Phi(\lambda_n)$, where λ_n is a sequence of small arcs about 1 lying in $|z| > 1$, whose radius tends to 0:

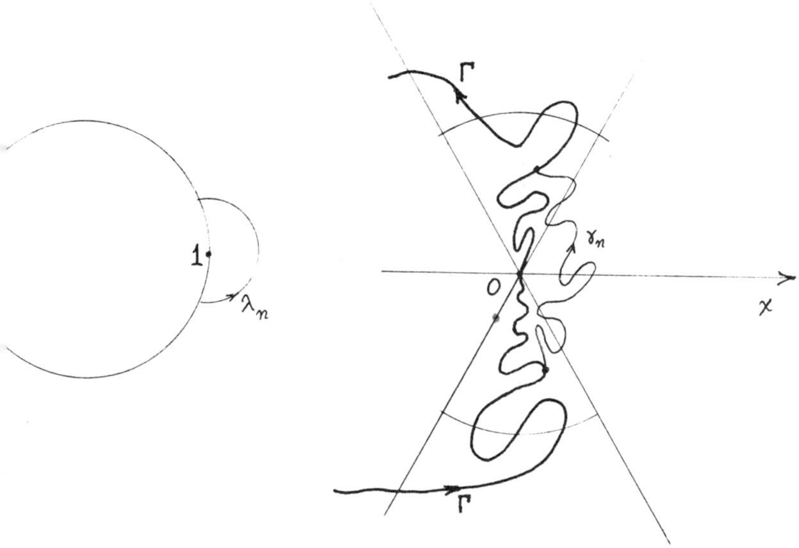

Since Φ is continuous and one-one up to $|z| = 1$, each γ_n is a Jordan arc lying in the outside of Γ, except for its two endpoints, which are on Γ. These endpoints are distinct, and equal to points $\Phi(e^{\pm i\alpha})$ say, where $\alpha > 0$ is small if λ_n has sufficiently small radius. $\Phi(e^{i\theta})$ is a parameterization of Γ, so, since Γ has a vertical tangent at 0, $\Phi(e^{i\alpha})$ and $\Phi(e^{-i\alpha})$ must lie on opposite sides of 0 in the union of the two vertical sectors $|\arg w \pm \frac{\pi}{2}| < \eta$, if $\alpha > 0$ is small. That means that γ_n starts on one side of the x-axis and ends on the other side. Therefore it crosses it at, say, P, so P lies in the outside of Γ.

Now if we do this construction with radius λ_n sufficiently small, we get $0 < |P| < \delta$.

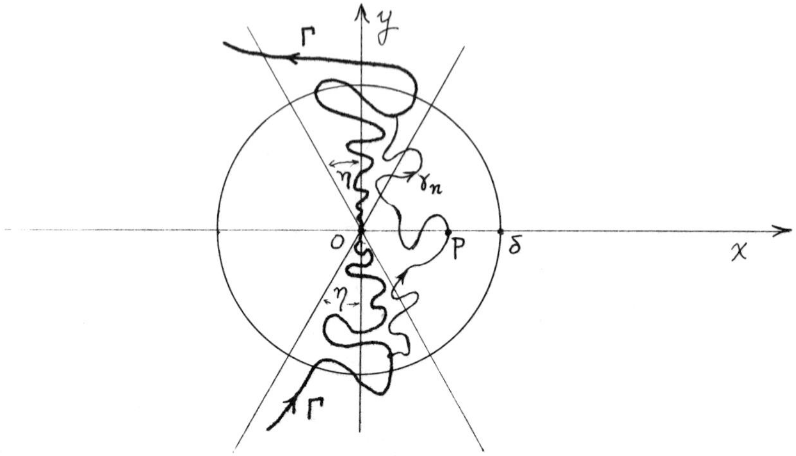

If, then, $P > 0$, the whole sector $|\arg w| < \frac{\pi}{2} - \eta$ in $|w| < \delta$ lies in the outside of Γ, for any w there can be joined to P by a line not touching Γ. A similar statement of course holds if $P < 0$.

Now $\{|z| < 1\}$ is simply connected and $f(z)$ is analytic, bounded and $\neq 0$ there, so we have a single valued function $\log f(z)$, analytic there, and bounded on any set where $|f(z)| \geq \delta$, say. But we have just seen that all the w, $|w| < \delta$, with $|\arg w| \leq \frac{\pi}{2} - \eta$ are in the outside of Γ, i.e., not in $\mathfrak{D}$ - that means that if $|f(z)| < \delta$, $f(z)$ must lie in the complementary sector

$$\frac{\pi}{2} - \eta \leq \arg f(z) \leq \frac{3\pi}{2} + \eta.$$

So at least $\arg f(z) = \Im \log f(z)$ continues to remain bounded for $|f(z)| < \delta$, and the lemma is proved.

Supposing always a vertical tangent, let us see why $\arg f(e^{-i\theta}) \cong \arg f(e^{i\theta}) + \pi$ if $\theta > 0$ is small. We have just seen that for $|z| < 1$, $|z-1| < \varepsilon$, say, the variation of $\arg f(z)$ is at most $\pi + 2\eta$, because then $|f(z)| < \delta$ and $f(z)$ is confined to a sector of opening $\pi + 2\eta$. So, since $|\arg f(e^{\pm i\theta}) \pm \frac{\pi}{2}| < \eta$ if θ is small, and Γ has a tangent at 0, the only two possibilities are

$$\arg f(e^{-i\theta}) \cong \arg f(e^{i\theta}) + \pi$$

and

$$\arg f(e^{-i\theta}) \cong \arg f(e^{i\theta}) - \pi.$$

We must confirm the first possibility and invalidate the second.

$f(z)$ is a conformal mapping of $|z| < 1$ onto $\mathcal{D}$, therefore, if $P \notin \mathcal{D}$ and $r < 1$, the variation of $\arg(f(z) - P)$ around the circle $z = re^{i\theta}$ is zero. Since $f(z)$ is continuous up to $|z| = 1$, we can make $r \to 1$, and we see that

$$\Delta_\Gamma \arg(w - P) = 0, \qquad P \notin \mathcal{D}.$$

Since, on the other hand, $f(z)$ is conformal, by the principle of argument, if $P' \in \mathcal{D}$, for $r < 1$ sufficiently close to 1, the variation of $\arg(f(z) - P')$ around $z = re^{i\theta}$ is 2π. Making $r \to 1$, we get

$$\Delta_\Gamma \arg(w - P') = 2\pi$$

if $P' \in \mathcal{D}$.

We just saw in the proof of the lemma that there are $P \notin \mathcal{D}$ with $0 < |P|$ arbitrarily small and P on the x-axis; without loss of generality let these points be on the positive x-axis. Then, the whole circular sector $|\arg w| < \frac{\pi}{2} - \eta$, $|w| < \delta$ lies outside $\mathfrak{D}$. An entirely similar argument shows that there are points P' in $\mathfrak{D}$ on the x-axis with $|P'|$ arbitrarily small - these must then lie on the negative x-axis. We see that a circular sector of the form $\frac{\pi}{2} + \eta < \arg w < \frac{3\pi}{2} - \eta$, $|w| < \delta$ lies in $\mathfrak{D}$.

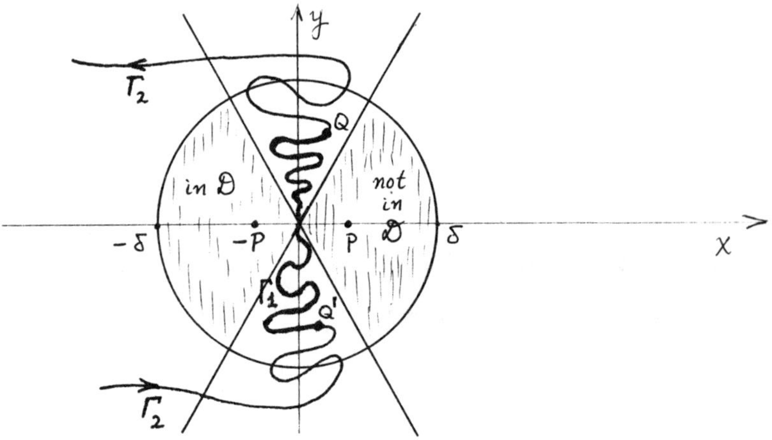

Now take a small fixed $\alpha > 0$, and put $Q = f(e^{i\alpha})$, $Q' = f(e^{-i\alpha})$. Let Γ_1 be the arc of Γ from Q' to Q (corresponding to $-\alpha \leq \theta \leq \alpha$) and Γ_2 the complementary arc corresponding to $\alpha \leq \theta \leq 2\pi - \alpha$. Let $P > 0$, P small, then

$$\Delta_\Gamma \arg(w - P) = 0,$$

$$\Delta_\Gamma \arg(w + P) = 2\pi,$$

as we have just seen.

So

$$\Delta_{\Gamma_1} \arg(w - P) + \Delta_{\Gamma_2} \arg(w - P) = 0$$

$$\Delta_{\Gamma_1} \arg(w + P) + \Delta_{\Gamma_2} \arg(w + P) = 2\pi.$$

If P is close to zero, the two *second* terms are sensibly equal - Γ_2 stays away from 0, so we find

$$\Delta_{\Gamma_1} \arg(w + P) - \Delta_{\Gamma_1} \arg(w - P) \to 2\pi$$

as $P \to 0+$.

Now if $Q' = f(e^{-i\alpha})$ lies *below* 0 in the double sector

$$\left|w \pm \frac{\pi}{2}\right| < \eta$$

and $Q = f(e^{i\alpha})$ *lies above*, then for P close to 0,

$$\left|\Delta_{\Gamma_1} \arg(w + P) - \pi\right| < 2\eta$$

and

$$\left|\Delta_{\Gamma_1} \arg(w - P) + \pi\right| < 2\eta.$$

Whereas, if Q' lies *above* 0 and Q lies *below*

$$\left|\Delta_{\Gamma_1} \arg(w + P) + \pi\right| < 2\eta,$$

$$\left|\Delta_{\Gamma_1} \arg(w - P) - \pi\right| < 2\eta.$$

The *first possibility gives*

$$\Delta_{\Gamma_1} (\arg(w + P) - \arg(w - P)) \cong 2\pi,$$

and the *second*

$$\Delta_{\Gamma_1} (\arg(w + P) - \arg(w - P)) \cong -2\pi,$$

So the first is correct;

$$f(e^{i\alpha}) \text{ lies above } 0$$

and

$$f(e^{-i\alpha}) \text{ below.}$$

Since points on the negative x-axis close to 0 are in $\mathfrak{D}$, we see that (as determined by analytic continuation in $|z| < 1$ of $\log f(z)$),

$$|\arg f(e^{i\alpha}) - \frac{\pi}{2}| < 2\eta,$$

$$|\arg f(e^{-i\alpha}) + \frac{\pi}{2}| < 2\eta,$$

and, since η is finally arbitrary, $\arg f(e^{-i\alpha}) - \arg f(e^{i\alpha}) \to \pi$ as $\alpha \to 0+$.

Lindelöf's theorem is thus completely proved.

An important application is the establishment of boundary behaviour results for certain functions analytic in regions bounded by rectifiable simple Jordan curves. A rectifiable Jordan curve has a tangent at almost every one of its points.

Therefore, at almost every boundary point of such a region (i.e., one bounded by a simple rectifiable Jordan curve), the notion of nontangential limit transfers over from the unit circle by conformal mapping. Such a conformal mapping takes a sector of opening $< 180°$ in the unit circle, with its vertex at the pre-image of a boundary point, to a subdomain which is asymptotically like a sector of the same opening having its vertex at the boundary point.

This fact, together with the important theorem of F. and M. Riesz to be proved in Section D below, enables us to extend many of the results proven in the next chapter about boundary behaviour of functions analytic in $|z| < 1$ to corresponding results about boundary behaviour of functions analytic in a region bounded by a rectifiable Jordan curve.

D. Consider now a domain $\mathscr{D}$ bounded by a rectifiable Jordan curve.

Let Φ be a conformal mapping of $\{|z| < 1\}$ onto $\mathscr{D}$, a region bounded by the rectifiable Jordan curve Γ:

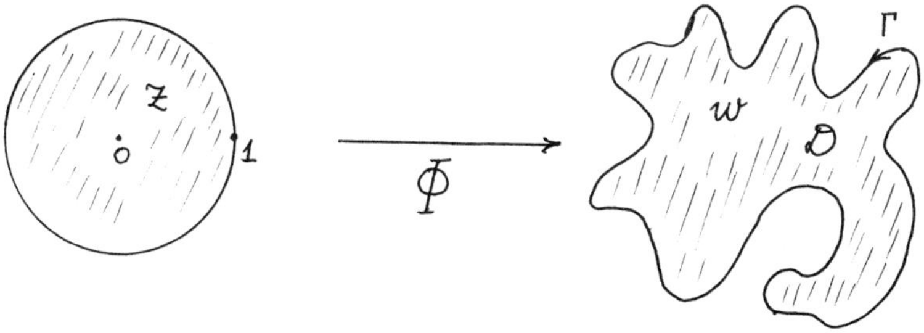

By Caratheodory's theorem, proved in Section C, 1, Φ has a continuous one-one extension up to $|z| = 1$ and maps that circumference onto Γ. So surely, if $[e^{i\theta_0}, e^{i\theta_1}, \ldots, e^{i\theta_p}]$ is a partition of $\{|z| = 1\}$, $[\Phi(e^{i\theta_0}), \Phi(e^{i\theta_1}), \ldots, \Phi(e^{i\theta_p})]$ is a partition of Γ.

1. Theorem. $\Phi'(z) \in H_1$.

Proof (Beckenbach): Let $\varepsilon = e^{2\pi i/n}$. Then, for $|z| < 1$,

$$S(z) = |\Phi(\varepsilon z) - \Phi(z)| + |\Phi(\varepsilon^2 z) - \Phi(\varepsilon z)| + \ldots$$
$$\ldots + |\Phi(\varepsilon^n z) - \Phi(\varepsilon^{n-1} z)|$$

is <u>subharmonic</u>, and $S(z)$ is continuous for $|z| < 1$ because $\Phi(z)$ is. So by <u>principle of maximum</u>, if $|z| < 1$,

$$S(z) \leq \max_{|\zeta|=1} S(\zeta).$$

But if $|\zeta| = 1$, the points

$$[\Phi(\zeta),\Phi(\varepsilon\zeta),\Phi(\varepsilon^2\zeta),\ldots,\Phi(\varepsilon^n\zeta)]$$

form a partition of Γ, so by <u>definition of curve length</u> (!)

$$S(\zeta) \leq \text{length } \Gamma < \infty.$$

$\therefore$ If $|z| < 1$, $S(z) \leq$ length Γ. Now fix $r < 1$. We have

$$\sum_{k=1}^{n} |\Phi(\varepsilon^k r) - \Phi(\varepsilon^{k-1} r)| \leq \text{length } \Gamma.$$

Making $n \to \infty$ and using continuity of $\Phi'(re^{i\theta})$ <u>in</u> θ for $r < 1$, we get in limit

$$\int_0^{2\pi} |\Phi'(re^{i\theta})| \, r \, d\theta \leq \text{length } \Gamma,$$

and since this is valid for all $r < 1$, $\Phi'(z) \in H_1$.

Q.E.D.

2. By Section C, 1, if Λ is an arc of $\{|z| = 1\}$, Φ maps some arc J of $\{|z| = 1\}$, in one-one bicontinuous fashion onto Λ. The very <u>definition</u> of arc-length now gives us

$$\text{length } \Lambda = \int_J |d\Phi(e^{i\theta})|.$$

<u>Theorem</u> (F. and M. Riesz). If J is an arc of the unit circle and $\Lambda = \Phi(J)$,

$$\text{length } \Lambda = \int_J |\Phi'(e^{i\theta})|\,d\theta.$$

Proof. By the theorem of the preceding section, $\Phi'(z) \in H_1$, so by Section B.2,

$$\int_{-\pi}^{\pi} |\Phi'(e^{i\theta}) - \Phi'(re^{i\theta})|\,d\theta \to 0$$

for $r \to 1$.

If J is given, we can find a continuous function $T(\theta)$, $|T(\theta)| \leq 1$, vanishing identically outside J, such that

$$\left| \int_J T(\theta)\,d\Phi(e^{i\theta}) - \int_J |d\Phi(e^{i\theta})| \right| < \varepsilon,$$

$$\left| \int_J T(\theta) \cdot ie^{i\theta}\Phi'(e^{i\theta})\,d\theta - \int_J |\Phi'(e^{i\theta})|\,d\theta \right| < \varepsilon,$$

$\varepsilon > 0$ being arbitrary. Here, where J is an arc, we can even take $T(\theta)$ to be continuously differentiable. Then, integrating by parts,

$$\int_J T(\theta)\,d\Phi(e^{i\theta}) = -\int_J \Phi(e^{i\theta})T'(\theta)\,d\theta$$

$$= -\lim_{r\to 1} \int_J \Phi(re^{i\theta})T'(\theta)\,d\theta$$

by the theorem of Section C.1. But, for each $r < 1$,

$$-\int_J \Phi(re^{i\theta})T'(\theta)\,d\theta = \int_J ire^{i\theta}\Phi'(re^{i\theta})T(\theta)\,d\theta,$$

which, by the remark made at the beginning, must tend to

$$\int_J ie^{i\theta}\Phi'(e^{i\theta})T(\theta)\,d\theta$$

as $r \to 1$. This is, by choice of T, within ε of $\int_J |\Phi'(e^{i\theta})|\,d\theta$, whilst the quantity we started with, which is equal to it, is within ε

of $\int_J |d\Phi(e^{i\theta})|$. Since this last is the same as length Λ, we get the theorem on making $\varepsilon \to 0$.

Arc length on Γ can be used in an evident fashion to define linear measure on Γ. First if $\mathfrak{G} \subset \Gamma$ is (relatively) open on Γ, $\mathfrak{G}$ is a disjoint countable union of open arcs Λ_k of Γ, and we take $|\mathfrak{G}| = \sum_k$ length Λ_k. Then, for $E \subset \Gamma$ arbitrary, we take $|E|$ to be $\inf\{|\mathfrak{G}|;\ \mathfrak{G} \supset E,\ \mathfrak{G}$ open on $\Gamma\}$. Using the one-one bicontinuity of Φ as a mapping from $\{|z| = 1\}$ to Γ it is now easy to build upon the above theorem and see that

$$|\Phi(E)| = \int_E |\Phi'(e^{i\theta})|d\theta$$

for Borel sets E on the unit circumference.

We have especially the important

Theorem (F. and M. Riesz). If E is on unit circumference and $|E| = 0$, then $|\Phi(E)| = 0$.

Proof. Let the Ω_n be open sets on $\{|z| = 1\}$, $\Omega_n \supset \Omega_{n+1} \supset \cdots$, $\Omega_n \supset E$, such that $|\Omega_n| \underset{n}{\to} 0$. Then $|\Phi(E)| \le |\Phi(\Omega_n)|$ for all n. But by the previous theorem and the discussion immediately following it, $|\Phi(\Omega_n)| = \int_{\Omega_n} |\Phi'(e^{i\theta})|d\theta$, which goes to zero as $n \to \infty$ because $|\Omega_n| \underset{n}{\to} 0$ and $\Phi'(e^{i\theta}) \in L_1(-\pi,\pi)$.

Q.E.D.

3. Theorem. The power series of $\Phi(z)$ converges absolutely up to $|z| = 1$ (Hardy).

Proof. By 1, $\Phi'(z) \in H_1$. Also, since Φ is conformal, $\Phi'(z)$ never $= 0$ in $|z| < 1$. Therefore we can define an analytic $\Psi(z) = \sqrt{\Phi'(z)}$ in $|z| < 1$. Now write, for $|z| < 1$,

$$\Phi'(z) = \sum_0^\infty a_n z^n.$$

Then we'll have

$$\Psi(z) = \sum_0^\infty b_n z^n$$

with

$$b_0^2 = a_0, \quad b_1 b_0 + b_0 b_1 = a_1,$$

$$b_2 b_0 + b_1 b_1 + b_0 b_2 = a_2, \ \&\ c.$$

Since $\Phi'(z) \in H_1$,

$$\int_{-\pi}^{\pi} |\Psi(re^{i\theta})|^2 d\theta$$

is bounded for $r < 1$. By Parseval, this yields

$$\sum_0^\infty |b_n|^2 < \infty.$$

Now write

$$\psi(z) = \sum_0^\infty |b_n| z^n.$$

By Parseval,

$$\int_{-\pi}^{\pi} |\psi(re^{i\theta})|^2 d\theta$$

is bounded for $|z| < 1$.

Let $\Theta(z) = [\psi(z)]^2 =$ say $\sum_0^\infty A_n z^n$. Then, on the one hand, $\Theta(z) \in H_1$, and on the other,

$$A_n = \sum_0^n |b_k||b_{n-k}| \geq |\sum_0^n b_k b_{n-k}| = |a_n|.$$

We have

$$\Phi(z) = c_0 + \sum_0^\infty \frac{a_n}{n+1} z^{n+1},$$

so to prove absolute convergence of power series of $\Phi(z)$ up to $|z| = 1$, we need to show

$$\sum_0^\infty \frac{|a_n|}{n+1} < \infty.$$

The above computation guarantees this if $\sum_0^\infty A_n/(n+1) < \infty$.

Now for $|z| < 1$, using the principal determination of the logarithm,

$$-\frac{\pi}{2} < \Im \log(1 - z) < \frac{\pi}{2}$$

and

$$\log(1 - z) = -\sum_1^\infty \frac{z^n}{n},$$

so that

$$\Im \log(1 - z) = \frac{i}{2} \sum_{-\infty}^{\infty}{}' \frac{\operatorname{sgn} n}{|n|} r^{|n|} e^{in\theta}$$

for $z = re^{i\theta}$, $0 \leq r < 1$, whence, with $\Theta(z) = \sum_0^\infty A_n z^n$, using absolute convergence and orthogonality,

$$-\pi i \sum_1^\infty \frac{r^{2n} A_n}{n} = \int_0^{2\pi} \Theta(re^{i\theta}) \Im \log(1 - re^{i\theta}) d\theta$$

and this is in absolute value $\leq \frac{\pi}{2} \int_0^{2\pi} |\Theta(re^{i\theta})| d\theta$ which is $\leq M$ independently of $r < 1$. Since $A_n \geq 0$, we get, making $r \to 1$, $\sum_1^\infty A_n/n \leq M/\pi$, as required.

Problem 3. Let Γ be a rectifiable Jordan curve bounding a domain $\mathfrak{D}$. If $f(z)$ is analytic in $\mathfrak{D}$ and continuous on $\bar{\mathfrak{D}}$, prove that $\int_\Gamma f(z)dz = 0$. (Hint: Use conformal mapping.)

III. Elementary Boundary Behaviour Theory for Analytic Functions

A. Existence of nontangential boundary values a.e.

Recall, from Chapter II, Section B (or simply from the general results of Chapter I):

> If $F(z) \in H_1$,
>
> $$F(e^{i\theta}) = \lim_{z \underset{\measuredangle}{\longrightarrow} e^{i\theta}} F(z)$$
>
> exists and is finite for almost all θ.

In particular, we have the

Corollary (Fatou). If $F(z)$ is regular and bounded in $\{|z| < 1\}$, it has nontangential boundary values almost everywhere on the unit circumference.

B. Uniqueness theorem for H_1 functions.

Theorem. Let $F(z) \in H_1$ and suppose, for a set E of positive measure, that $F(e^{i\theta}) = 0$ for $\theta \in E$. Then $F(z) \equiv 0$.

Proof. Wlog, $0 < |E| < 2\pi$. Put, for $0 \le r < 1$,

$$U(re^{i\theta}) = \frac{1}{2\pi|E|} \int_E P_r(\theta - t)dt - \frac{1}{2\pi(2\pi - |E|)} \int_{[0,2\pi]\sim E} P_r(\theta - t)dt.$$

Then $U(z)$ is harmonic for $|z| < 1$ and

$$-\frac{1}{2\pi - |E|} \le U(z) \le \frac{1}{|E|}$$

there.

Besides, $U(0) = 0$ and especially

$$U(z) \longrightarrow -\frac{1}{2\pi-|E|} \quad \text{as} \quad z \underset{\measuredangle}{\longrightarrow} e^{i\theta}$$

for almost all θ in $[0,2\pi] \sim E$. Using $\tilde{U}(z)$ to denote the harmonic conjugate of $U(z)$, we see that

$$\varphi(z) = \exp[U(z) + i\tilde{U}(z)]$$

is bounded in $|z| < 1$, $|\varphi(0)| = 1$, and for almost all $\theta \in [0,2\pi] \sim E$,

$$|\varphi(e^{i\theta})| = \exp\left(-\frac{1}{2\pi-|E|}\right) .$$

(A.e. existence of $\varphi(e^{i\theta})$ is guaranteed by Section A above!)

Since $\varphi(z)$ is bounded, for each k, $[\varphi(z)]^k F(z) \in H_1$. Suppose $F(z) \not\equiv 0$; we may, wlog, suppose that $F(0) \neq 0$; otherwise we just work with the H_1-function $F(z)/z^m$ instead of $F(z)$, m being the order of the zero F has at the origin. By Chapter II, Section B we now get

$$[\varphi(0)]^k F(0) = \frac{1}{2\pi} \int_0^{2\pi} [\varphi(e^{i\theta})]^k F(e^{i\theta}) d\theta$$

$$= \frac{1}{2\pi} \int_{[0,2\pi]\sim E} [\varphi(e^{i\theta})]^k F(e^{i\theta}) d\theta,$$

since $F(e^{i\theta}) = 0$, $\theta \in E$. From this,

$$|F(0)| = |\varphi(0)|^k |F(0)| \leq \frac{1}{2\pi} \exp\left(-\frac{k}{2\pi-|E|}\right) \int_{[0,2\pi]\sim E} |F(e^{i\theta})| d\theta$$

which tends to zero as $k \to \infty$ ($F(e^{i\theta})$ is in L_1!). So $F(0) = 0$, a contradiction. We are done.

C. More existence theorems for limits.

1. Theorem. If $f(z)$ is regular in $|z| < 1$ and $\Re f(z) \geq 0$ there, $\lim_{z \underset{\measuredangle}{\to} e^{i\theta}} f(z)$ exists and is finite for almost all θ.

Proof (Titchmarsh). $F(z) = \frac{1}{1+f(z)}$ is regular and bounded in $|z| < 1$ so, by Section A, $F(z) \to$ a finite limit $F(e^{i\theta})$ for almost all θ as $z \underset{\measuredangle}{\to} e^{i\theta}$. Since $F(z) = \frac{1}{F(z)} - 1$, $f(z)$ tends to the finite limit $\frac{1}{F(e^{i\theta})} - 1$ as $z \underset{\measuredangle}{\to} e^{i\theta}$ unless $F(e^{i\theta}) = 0$. Since, for $F \not\equiv 0$, $|F(e^{i\theta})| > 0$ a.e. by Section B, we are done.

2. Recall the notation of Chapter I, Section E, where (Section E.4) we proved existence of $\tilde{F}(\theta)$ a.e. for $F \in L_2(-\pi,\pi)$.

Theorem. Let $F(\theta + 2\pi) = F(\theta)$ and $F \in L_1(-\pi,\pi)$. Then the principal value

$$\tilde{F}(\theta) = \frac{1}{\pi} \int_{-\pi}^{\pi} \frac{F(t)}{2 \tan(\theta-t)/2}\, dt$$

exists and is finite a.e.

Proof (Titchmarsh). Wlog, $F(\theta) \geq 0$. Then, put $U(re^{i\theta}) = \frac{1}{2\pi} \int_{-\pi}^{\pi} P_r(\theta-t)F(t)dt$, and let $\tilde{U}(z)$ be the harmonic conjugate of $U(z)$.

$f(z) = U(z) + i\tilde{U}(z)$ is regular in $\{|z| < 1\}$ and $\Re f(z) \geq 0$ there. So by 1. above, $f(z) \to f(e^{i\theta})$ finite for almost all θ as $z \underset{\measuredangle}{\to} e^{i\theta}$. In particular, for almost all θ,

$$\lim_{r \to 1} \tilde{U}(re^{i\theta}) = \Im f(e^{i\theta})$$

exists and is finite. By Chapter I, Section E.3,

$$\frac{1}{2\pi} \int_{1-r<|t-\theta|<\pi} \{F(t)/\tan(\tfrac{\theta-t}{2})\}dt - \tilde{U}(re^{i\theta}) \longrightarrow 0$$

as $r \to 1$ for almost all θ. This does it.

D. Privalov's uniqueness theorem.

1. Privalov's ice-cream cone construction.

Definition. If $|\zeta| = 1$,

$$S_\zeta = \{z;\ |z| > \frac{1}{\sqrt{2}};\ |\arg(\zeta - z)| < \frac{\pi}{4}\}.$$

S_1 is the dashed region shown in the following figure:

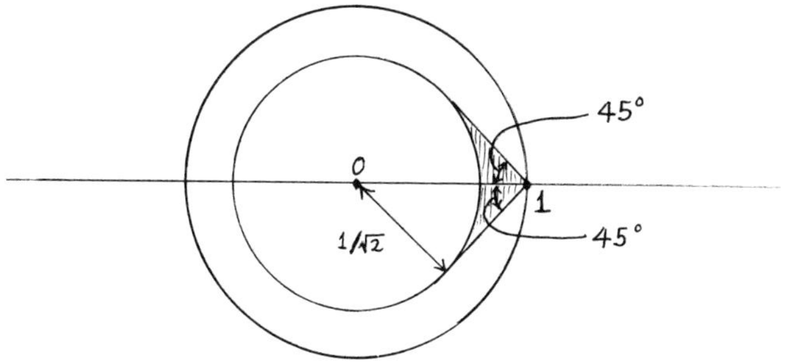

We make a series of obvious remarks:

a) $\bigcup_{|\zeta|=1} S_\zeta$ is all of $\{\frac{1}{\sqrt{2}} < |z| < 1\}$.

b) If $\frac{1}{\sqrt{2}} < |z| < 1$ for some z, $\{\zeta;\ |\zeta| = 1\ \&\ z \in S_\zeta\}$ is the (open) arc $\widehat{\zeta_1, \zeta_2}$ of the unit circle constructed as follows:

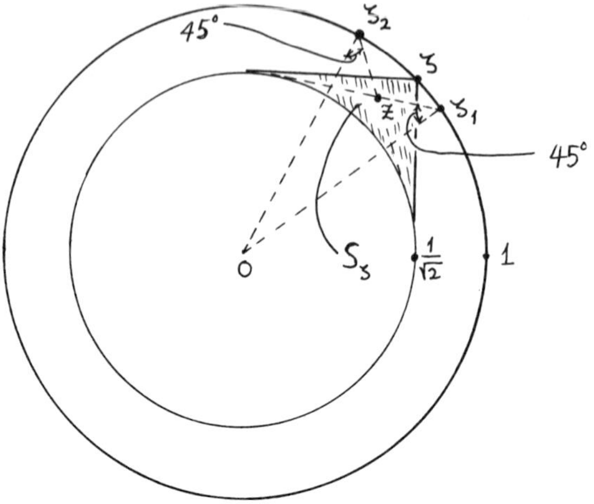

c) If $J = \widehat{\zeta_1, \zeta_2}$ is an arc of the unit circle with opening $\leq 90°$, the set of z, $\frac{1}{\sqrt{2}} < |z| < 1$, such that an S_ζ contains z only for some $\zeta \in J$, consists of the closed curvelinear triangle T constructed upon J in the following manner:

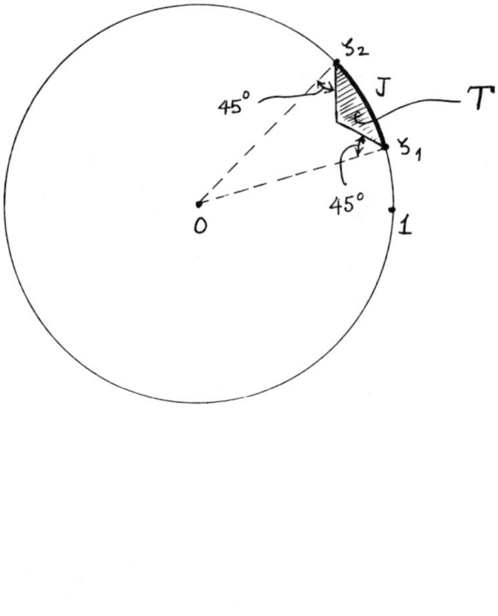

This follows very easily from a) and b).

If the arc J subtends <u>more</u> than $90°$, the set of z, $\frac{1}{\sqrt{2}} < |z| < 1$, with $z \in S_\zeta$ <u>only for</u> $\zeta \in J$ is the closed curvelinear <u>trapezoid</u> T constructed thus:

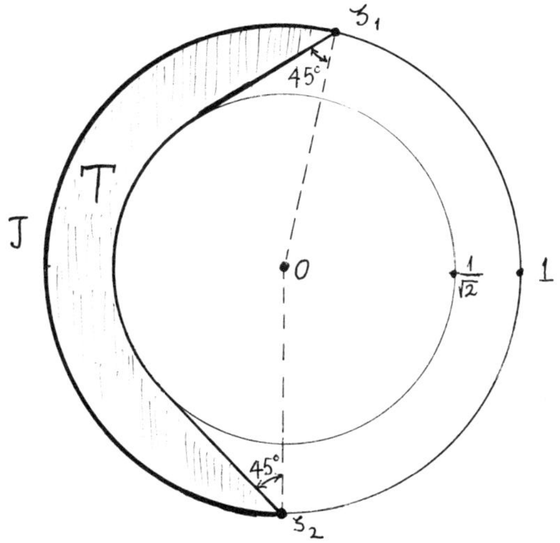

<u>Now we can describe Privalov's construction.</u>

Given a closed set E on the unit circumference, let $\{J_k\}$ be the (at most countable) set of arcs on that circumference <u>contiguous</u> (complementary) to E. Using each arc J_k as a <u>base</u> construct upon it the triangle or trapezoid T_k according to the recipe in c) above. Take the closed domain

$$\bar{\mathfrak{D}} = \{|z| \leq 1\} \sim \bigcup_k T_k^0 \sim \bigcup_k J_k.$$

(The supercript ^{0} denotes <u>interior</u> and the <u>bar</u> denotes <u>closure</u>.)

Our domain $\bar{\mathfrak{D}}$ has the following important property:

Every $z \in \overline{\mathcal{D}}$ of modulus $> \frac{1}{\sqrt{2}}$ is in an $\overline{S}_\zeta$ for some $\zeta \in E$.

This follows directly from remarks a) and c) above.

Here is a picture of $\overline{\mathcal{D}}$:

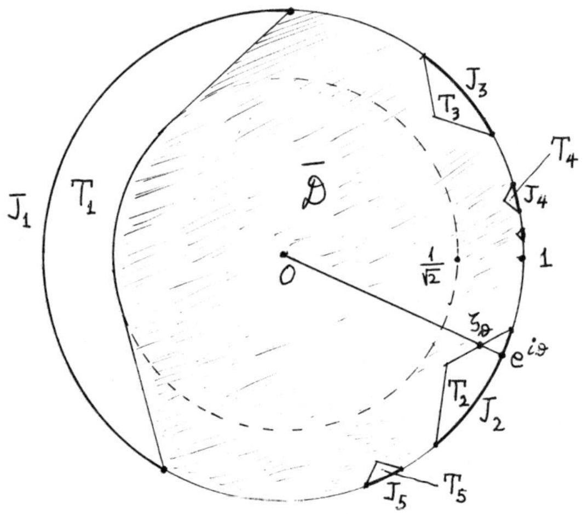

Note that $\partial\overline{\mathcal{D}}$ is a <u>Jordan curve</u>. Indeed, if ζ_θ denotes the single point where the ray from 0 to $e^{i\theta}$ hits $\partial\overline{\mathcal{D}}$, ζ_θ is a continuous one-one mapping from the unit circumference onto $\partial\overline{\mathcal{D}}$. This boundary of $\overline{\mathcal{D}}$ is even a <u>rectifiable</u> Jordan curve, because, for each k, perimeter $T_k \leq C|J_k|$, where C is a geometric constant whose value we need not calculate.

We denote by $\mathcal{S}$ the <u>interior</u> of $\bar{\mathcal{S}}$. <u>We see that the results of Chapter</u> II, <u>Sections</u> C, D <u>all apply to</u> $\mathcal{S}$.

<u>Remark</u>. Besides being used to prove Privalov's theorem below, $\mathcal{S}$ (or its analogue for the upper half plane) comes up in Chapter VIII in the study of maximal functions and of Carleson measures.

2. <u>Use of Egoroff's theorem</u>

<u>Theorem</u>. Let $f(z)$ be analytic in $|z| < 1$, and put, for $|\zeta| = 1$,

$$M_f(\zeta) = \sup\{|f(z)|;\ z \in S_\zeta\}.$$

Then $M_f(\zeta)$ is Lebesgue measurable.

<u>Proof</u>. For $n \geq 3$, take $r_n = 1 - \frac{1}{n}$ and put, for $|\zeta| = 1$,

$$M_n(\zeta) = \sup\{|f(z)|;\ \frac{1}{\sqrt{2}} \leq |z| \leq r_n\ \&\ |\arg(r_n\zeta - z)| \leq \frac{\pi}{4}\}.$$

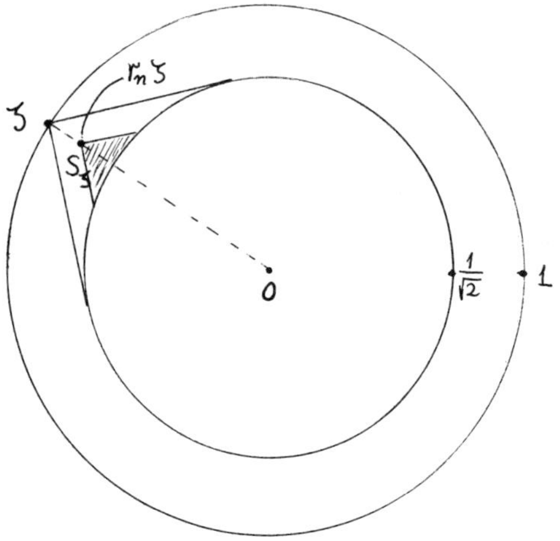

Because $f(z)$ is continuous for $|z| \leq r_n$, $M_n(\zeta)$ is <u>continuous</u>. Since clearly $M_n(\zeta) \underset{n}{\rightarrow} M_f(\zeta)$ pointwise in ζ, $M_f(\zeta)$ is Lebesgue measurable.

<u>Theorem</u>. Let $f(z)$ be regular in $|z| < 1$. Suppose there is a set G of positive measure on the unit circumference such that

$$\lim_{\substack{z \to \zeta \\ \angle}} f(z) = 0$$

for each $\zeta \in G$. Then there is a <u>closed</u> $E \subset G$, $|E| > 0$, such that $f(z) \to 0$ <u>uniformly for</u> $|z| \to 1$, z <u>in the union of the</u> S_ζ <u>with</u> $\zeta \in E$.

<u>Proof</u>. For $n \geq 3$ and $|\zeta| = 1$ put

$$P_n(\zeta) = \sup\{|f(z)|;\ z \in S_\zeta \ \&\ |z| \geq 1 - \tfrac{1}{n}\}.$$

The argument used in the proof of the previous theorem shows that each $P_n(\zeta)$ is Lebesgue measurable.

By hypothesis, $P_n(\zeta) \underset{n}{\rightarrow} 0$ for $\zeta \in G$, and $|G| > 0$. Egoroff's theorem now gives us an $E_0 \subseteq G$, $|E_0| > 0$, with $P_n(\zeta) \underset{n}{\rightarrow} 0$ <u>uniformly</u> for $\zeta \in E_0$. Taking a closed $E \subset E_0$ with $|E_0| > 0$, we have the theorem.

3. <u>Privalov's uniqueness theorem.</u>

<u>Theorem</u> (Privalov, ca. 1917). Let $f(z)$ be analytic in $|z| < 1$ and suppose G is a set of <u>positive measure</u> on the unit circumference with

$$\lim_{\substack{z \to \zeta \\ \angle}} f(z) = 0 \quad \text{for} \quad \zeta \in G.$$

<u>Then</u> f <u>vanishes identically</u>.

Proof. By 2° we can find a closed E, $|E| > 0$, on the unit circumference with $f(z) \to 0$ uniformly as $|z| \to 1$ if z is in the union of the S_ζ, $\zeta \in E$. This means that if we make Privalov's construction, described in 1°, starting from E, we will obtain a domain $\mathcal{D} \subset \{|z| < 1\}$ with $f(z) \to 0$ uniformly for $|z| \to 1$, $\zeta \in \mathcal{D}$.

Looking at the construction in 1°, we see that $\partial\mathcal{D}$ consists of some segments in $\{|z| < 1\}$ going out to points of E on the unit circumference, together with the set E thereupon. Therefore, if we define $f(z)$ to be zero on E, we get a function continuous on $\bar{\mathcal{D}}$ whose restriction to $\mathcal{D}$ is analytic.

As explained in 1°, $\partial\mathcal{D}$ is a rectifiable Jordan curve. Take any conformal mapping φ of $\{|w| < 1\}$ onto $\mathcal{D}$ and, for $|w| < 1$, put $F(w) = f(\varphi(w))$. By Carathéodory's theorem (Chapter II, Section C) φ actually extends continuously up to $\{|w| = 1\}$ and maps that circumference in continuous one-one fashion onto $\partial\mathcal{D}$. This means that $F(w)$ extends continuously up to $\{|w| = 1\}$ since $f(z)$ extends continuously up to $\partial\mathcal{D}$. Let $S = \varphi^{-1}(E)$. Then $F(w) = 0$ for $w \in S$. Since $|E| > 0$, E has positive linear measure as a subset of the rectifiable curve $\partial\mathcal{D}$. So by a theorem of F. and M. Riesz (Chapter II, Section D), $|S| > 0$. Since $F(w)$ is regular in $\{|w| > 1\}$, continuous on the closed unit circle, and zero on S, it therefore follows that $F \equiv 0$ by Section B. So $f(z) \equiv 0$.

Q.E.D.

Remark 1. Privalov's theorem is valid for functions $f(z)$ which are merely supposed meromorphic in $|z| < 1$. Indeed, the theorems of Section 2 hold (with the same proofs) for such functions, and we can proceed as above at the start of the proof of Privalov's theorem, getting a domain $\mathfrak{D}$ such that $f(z) \to 0$ uniformly for $|z| \to 1$, $z \in \mathfrak{D}$. That means we can certainly find an $r < 1$ such that $f(z)$ has no poles in $\mathfrak{D}$ for $|z| > r$. So the only poles of f in $\mathfrak{D}$ must be for $|z| \leq r$. But, f being meromorphic in the whole unit circle, its poles in $|z| \leq r$ are finite in number, say $z_1, \ldots, z_m$. Then

$$g(z) = (z - z_1)(z - z_2) \ldots (z - z_m) f(z)$$

is regular in $\mathfrak{D}$, continuous on $\bar{\mathfrak{D}}$, and zero on $E \subset \partial\mathfrak{D}$. The remaining part of the proof of Privalov's theorem is now applied to g instead of f, and one sees as before that $g \equiv 0$, hence $f \equiv 0$.

Remark 2. In the statement of Privalov's theorem (where no growth restrictions are imposed on $|f(z)|$ for $|z| \to 1$) it is crucial that nontangential boundary values (on the set of positive measure where these limits are supposed to vanish) and not merely radial ones be involved. One can, indeed, construct nonzero functions $f(z)$ regular for $|z| < 1$, for which

$$\lim_{r \to 1} f(re^{i\theta}) = 0, \qquad 0 \leq \theta \leq 2\pi.$$

We do not give such a construction here. One can be found in Privalov's book on boundary properties of analytic functions, whose second Russian edition came out around 1950. There is a German translation. More recently, the construction given in that book has been simplified, I believe, by Rudin.

E. Generalizations of the Schwarz reflection principle.

1. __Theorem.__ Let $F \in H_1$ and suppose that $\Im F(\zeta) = 0$ a.e. for ζ belonging to an __arc__ J of the unit circle. Then $F(z)$ can be continued analytically across J into $\{|z| > 1\}$ by putting $F(z^*) = \overline{F(z)}$ for $|z| < 1$, where z^* denotes $1/\bar{z}$.

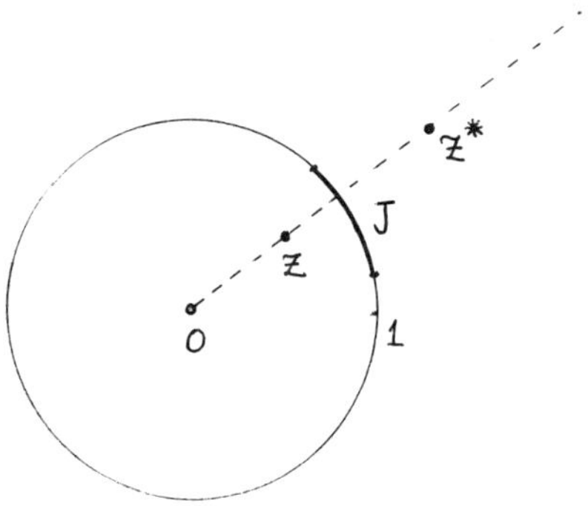

__Proof.__ By Chapter II, Section B, for $0 \leq r < 1$,

$$F(re^{i\theta}) = \frac{1}{2\pi} \int_{-\pi}^{\pi} P_r(\theta - t)F(e^{it})dt,$$

whence

$$\Im F(re^{i\theta}) = \frac{1}{2\pi} \int_{-\pi}^{\pi} P_r(\theta - t)\Im F(e^{it})dt,$$

which reduces (with slight abuse of notation!) to

$$\frac{1}{2\pi} \int_{[-\pi,\pi]\sim J} P_r(\theta - t)\Im F(e^{it})dt$$

by hypothesis. From Chapter I, Section D we now see that, as $r \to 1$,

$\Im F(re^{i\theta}) \to 0$ uniformly for $e^{i\theta}$ ranging over any proper closed subarc of J.

The result now follows from the classical Schwarz reflection principle.

2. Here is a result of Carleman on analytic continuation which has some applications in harmonic analysis.

Theorem. Let F and G be in H_1, let J be an arc of the unit circle, and suppose that $F(\zeta) = \overline{G(\zeta)}$ a.e. for $\zeta \in J$. Then $F(z)$ can be analytically continued across J by putting, for $|z| < 1$, $F(z^*) = \overline{G(z)}$, where $z^* = 1/\bar{z}$.

Proof. For $|z| < 1$, write $S(z) = F(z) + G(z)$ and $D(z) = F(z) - G(z)$, S and D are in H_1, and by hypothesis,

$$\Im S(\zeta) = 0 \quad \text{a.e.,} \quad \zeta \in J$$

$$\Re D(\zeta) = 0 \quad \text{a.e.,} \quad \zeta \in J.$$

Therefore by 1. the functions $S(z)$ and $D(z)$ can be analytically continued across J by putting, for $|z| < 1$ and $z^* = 1/\bar{z}$,

$$S(z^*) = \overline{S(z)},$$

$$D(z^*) = -\overline{D(z)}.$$

Therefore $F(z) = \frac{1}{2}(S(z) + D(z))$ also continues analytically across J by putting, for $|z| < 1$,

$$F(z^*) = \frac{1}{2}(\overline{S(z)} - \overline{D(z)}) = \overline{G(z)}.$$

Q.E.D.

<u>Problem</u> 4. Let $U(z)$ be harmonic in $|z| < 1$, and let $V(z)$ be a harmonic conjugate of $U(z)$ (in $|z| < 1$). Suppose there is a measurable subset E of the unit circumference such that $\lim_{z \not\to \zeta} U(z)$ exists and is finite for every $\zeta \in E$. Show that $\lim_{z \not\to \zeta} V(z)$ <u>then</u> exists and is finite for almost every $\zeta \in E$. <u>A fairly concise solution is required</u>.

IV. Application of Jensen's Formula.

Factorization into a Product of Inner and Outer Functions

A. If $0 < |z_n| < 1$, $n = 1,2,3,\ldots$, and the infinite product

$$\prod_1^\infty \frac{|z_n|}{z_n} \frac{z_n - z}{1 - \bar{z}_n z}$$

converges absolutely for $|z| < 1$, it represents a certain function, analytic there, called a Blaschke product. We can even allow a finite number of the z_n to be zero, in which case the factors corresponding to $\frac{|z_n|}{z_n} \frac{z_n - z}{1 - \bar{z}_n z}$ are simply replaced by z.

1°. We have

$$\frac{|z_n|}{z_n} \frac{z_n - z}{1 - \bar{z}_n z} = |z_n| \frac{1 - (z/z_n)}{1 - \bar{z}_n z} = |z_n| + \frac{(\bar{z}_n - (1/z_n))|z_n| z}{1 - \bar{z}_n z}$$

$$= |z_n| + \frac{|z_n|^2 - 1}{1 - \bar{z}_n z} \frac{|z_n| z}{z_n}$$

$$\therefore \quad \frac{|z_n|}{z_n} \frac{z_n - z}{1 - \bar{z}_n z} = 1 + \{|z_n| - 1\} \left\{ 1 + \frac{(|z_n| + 1)|z_n|}{z_n(1 - \bar{z}_n z)} \cdot z \right\}$$

$$= 1 + \{|z_n| - 1\} \left\{ 1 + \frac{(|z_n| + 1)|z_n|}{z_n(1 - \bar{z}_n z)} \cdot z \right\},$$

so the infinite product converges absolutely for $z = 0$ iff

$$\sum_n (1 - |z_n|) < \infty.$$

But if $\sum_n (1 - |z_n|) < \infty$, by the same formula just found,

$$\sum_n \left| 1 - \frac{|z_n|}{z_n} \cdot \frac{z_n - z}{1 - \bar{z}_n z} \right| = \sum_n [1 - |z_n|] \left| 1 + \frac{(|z_n| + 1)|z_n|}{z_n(1 - \bar{z}_n z)} \cdot z \right| < \infty$$

if $|z| < 1$,

so the infinite product converges absolutely in $\{|z| < 1\}$ if $\sum_n (1 - |z_n|) < \infty$. Thus:

$\prod_n \frac{|z_n|}{z_n} \frac{z_n - z}{1 - \bar{z}_n z}$ converges absolutely in $\{|z| < 1\}$ iff $\sum_n (1 - |z_n|) < \infty$.

2°. Let $\sum_n (1 - |z_n|) < \infty$, so that $\prod_n \frac{|z_n|}{z_n} \frac{z_n - z}{1 - \bar{z}_n z}$ converges absolutely in $\{|z| < 1\}$ and represents a function $B(z)$ analytic there. By elementary complex variable theory, each factor of the product is in modulus < 1 for $|z| < 1$, so $|B(z)| < 1$ for $\{|z| < 1\}$.

Therefore, for almost all ζ, $|\zeta| = 1$, $B(\zeta) = \lim B(z)$ for $z \underset{\not\, }{\longrightarrow} \zeta$ exists. (Fatou's Theorem; Chapter III, Section A.)

Theorem. $|B(e^{i\theta})| = 1$ a.e.

Proof. Wlog, all $|z_n| > 0$ (otherwise work with $B(z)/z^k$ instead of $B(z)$.) Then $\log|B(0)| = \sum_n \log|z_n|$. Now because $\sum_n (1 - |z_n|) < \infty$, $\sum_n \log|z_n| > -\infty$ (N.B. $\log|z_n| < 0$ for each n!) Let $0 < r < 1$, with r not equal to any $|z_n|$. Then, by the most elementary form of Jensen's formula,

$$\log|B(0)| = \sum_{|z_n|<r} \log\left(\frac{|z_n|}{r}\right) + \frac{1}{2\pi}\int_{-\pi}^{\pi} \log|B(re^{i\theta})|\,d\theta,$$

i.e.,

$$\sum_n \log|z_n| = \sum_{|z_n|<r} \log\frac{|z_n|}{r} + \frac{1}{2\pi}\int_{-\pi}^{\pi} \log|B(re^{i\theta})|\,d\theta.$$

That is,

$$\frac{1}{2\pi}\int_{-\pi}^{\pi} \log|B(re^{i\theta})|\,d\theta = \sum_{|z_n|<r} \log\left(\frac{r}{|z_n|}\right) - \sum_n \log\frac{1}{|z_n|}\,.$$

Pick any fixed p so that $\sum_{n>p} \log(1/|z_n|) < \varepsilon$, and take $r < 1$ so close to 1 that $|z_n| < r$ if $n = 1,2,\ldots\ p$. Then, the previous relation yields

$$\frac{1}{2\pi}\int_{-\pi}^{\pi} \log|B(re^{i\theta})|\,d\theta \geq \sum_1^p \log\frac{r}{|z_n|} - \sum_1^p \log\frac{1}{|z_n|} - \varepsilon,$$

or, making $r < 1$ <u>close enough</u> to 1,

$$\frac{1}{2\pi}\int_{-\pi}^{\pi} \log|B(re^{i\theta})|\,d\theta > -2\varepsilon.$$

That means

$$\limsup_{r\to 1-} \frac{1}{2\pi}\int_{-\pi}^{\pi} \log|B(re^{i\theta})|\,d\theta \geq 0$$

since $\varepsilon > 0$ is arbitrary.

But $B(re^{i\theta}) \to B(e^{i\theta})$ a.e. as $r \to 1$ and

$$\log|B(re^{i\theta})| \leq 0.$$

Therefore, by <u>Fatou's lemma</u> (use a <u>sequence</u> of r's tending to 1)

$$\frac{1}{2\pi}\int_{-\pi}^{\pi} \log|B(e^{i\theta})|\,d\theta \geq 0.$$

<u>Since</u> $|B(e^{i\theta})| \leq 1$, we have $\log|B(e^{i\theta})| = 0$ a.e. Q.E.D.

B. Factorizing out Blaschke products.

1°. <u>Theorem</u>. Let $F(z)$ be regular in $\{|z| < 1\}$, and let the z_n be its zeros there, $|z_n| < 1$. Suppose that

$$\int_{-\pi}^{\pi} \log|F(re^{i\theta})|\,d\theta$$

is bounded above for $r < 1$. Then

$$\sum_n (1 - |z_n|) < \infty,$$

so that

$$B(z) = \prod_n \frac{|z_n|}{z_n} \frac{z_n - z}{1 - \bar{z}_n z}$$

converges absolutely for $\{|z| < 1\}$, and we have $F(z) = B(z)G(z)$ where $G(z)$ is regular and has no zeros in $\{|z| < 1\}$.

Proof. Wlog, $F(0) \neq 0$ - if not, work with $F(z)/z^k$ instead of $F(z)$. Then, if $0 < r < 1$, and no $|z_n| = r$, by Jensen's formula:

$$\log|F(0)| = \sum_{|z_n| < r} \log\left|\frac{z_n}{r}\right| + \frac{1}{2\pi} \int_{-\pi}^{\pi} \log|F(re^{i\theta})|\,d\theta,$$

i.e., by hypothesis,

$$\sum_{|z_n| < r} \log\left|\frac{r}{z_n}\right| \leq M - \log|F(0)|,$$

where M is independent of r. Making $r \to 1$, we get, for any fixed p,

$$\sum_1^p \log\frac{1}{|z_n|} \leq M - \log|F(0)|,$$

so

$$\sum_1^\infty \log \frac{1}{|z_n|} < \infty.$$

The existence of $B(z)$ now follows by Section A,1. Defining, in $\{|z| < 1\}$, $G(z) = F(z)/B(z)$, we're done.

2°. Definition. If $p > 0$ (sic!), H_p is the set of $F(z)$ analytic in $\{|z| < 1\}$ with

$$\sup_{0 \leq r < 1} \int_{-\pi}^{\pi} |F(re^{i\theta})|^p \, d\theta < \infty .$$

(H_1 is a special case of this Definition.).

Theorem. Let $F(z) \in H_p$. Then there is a Blaschke product $B(z)$ and a $G(z) \in H_p$ with $F(z) = B(z)G(z)$, $G(z)$ without zeros in $\{|z| < 1\}$.

Proof. If $r < 1$, by the inequality between arithmetic and geometric means,

$$\frac{1}{2\pi} \int_{-\pi}^{\pi} p \log|F(re^{i\theta})| \, d\theta \leq \log \frac{1}{2\pi} \int_{-\pi}^{\pi} |F(re^{i\theta})|^p d\theta$$

which by hypothesis is $\leq \log C$ independent of r. So, by 1°, if the z_n are the zeros of $F(z)$ in $\{|z| < 1\}$, $\sum_n (1 - |z_n|) < \infty$ and the Blaschke product $B(z)$ can be formed. If $G(z) = \frac{F(z)}{B(z)}$, $G(z)$ has no zeros in $\{|z| < 1\}$.

Write

$$B_N(z) = \prod_1^N \frac{|z_n|}{z_n} \frac{z_n - z}{1 - \bar{z}_n z} .$$

Since the product for $B(z)$ converges absolutely in $\{|z| < 1\}$, if $r < 1$, $B_N(z) \xrightarrow[N]{} B(z)$ uniformly on $|z| \leq r$. Pick such an r for which no $|z_n| = r$.

Then

$$\int_{-\pi}^{\pi} |G(re^{i\theta})|^p d\theta = \lim_{N \to \infty} \int_{-\pi}^{\pi} \left| \frac{F(re^{i\theta})}{B_N(re^{i\theta})} \right|^p d\theta .$$

But for each N,

$$G_N(z) = \frac{F(z)}{B_N(z)}$$

is regular in $|z| < 1$, so $|G_N(z)|^p$ is subharmonic there, so for fixed $r < 1$,

$$\int_{-\pi}^{\pi} |G_N(re^{i\theta})|^p \, d\theta \leq \limsup_{R \to 1} \int_{-\pi}^{\pi} |G_N(Re^{i\theta})|^p \, d\theta.$$

However, for <u>fixed</u> N, as is easily seen, $|B_N(Re^{i\theta})| \to 1$ <u>uniformly</u> as $R \to 1$, so

$$\limsup_{R \to 1} \int_{-\pi}^{\pi} |G_N(Re^{i\theta})|^p d\theta = \limsup_{R \to 1} \int_{-\pi}^{\pi} |F(Re^{i\theta})|^p \, d\theta$$

which by hypothesis is $\leq 2\pi C$, say. Thus, for each $r < 1$,

$$\int_{-\pi}^{\pi} |G_N(re^{i\theta})|^p d\theta \leq 2\pi C$$

for <u>all</u> N, and finally, by the previous,

$$\int_{-\pi}^{\pi} |G(re^{i\theta})|^p d\theta \leq 2\pi C.$$

Since $r < 1$ is <u>arbitrary</u>, $G \in H_p$. Q.E.D.

<u>Scholium</u>. We see that, in the above factorization,

$$\sup_{r<1} \int_{-\pi}^{\pi} |G(re^{i\theta})|^p d\theta = \sup_{r<1} \int_{-\pi}^{\pi} |F(re^{i\theta})|^p \, d\theta.$$

For the argument at the end of the above proof yields

$$\sup_{r<1} \int_{-\pi}^{\pi} |G(re^{i\theta})|^p \, d\theta \leq \sup_{r<1} \int_{-\pi}^{\pi} |F(re^{i\theta})|^p \, d\theta.$$

<u>Since, however,</u> $F = BG$ and $|B(z)| \leq 1$ in $\{|z| < 1\}$, the reverse inequality must also hold.

Corollary. If $f \in H_1$, we can find g and $h \in H_1$, $g(z)$ and $h(z)$ without zeros in $\{|z| < 1\}$,

$$\int_{-\pi}^{\pi} |g(e^{i\theta})|\,d\theta \le \int_{-\pi}^{\pi} |f(e^{i\theta})|\,d\theta$$

$$\int_{-\pi}^{\pi} |h(e^{i\theta})|\,d\theta \le \int_{-\pi}^{\pi} |f(e^{i\theta})|\,d\theta,$$

such that $f = g + h$.

Remark. This is a useful technical device, because many inequalities for H_1 are easier to prove for functions without zeros in $\{|z| < 1\}$.

Proof. Let $B(z)$ be the Blaschke product formed from the zeros of $f(z)$, then, by Theorem, $f = BF$ where $F(z)$ has no zeros in $\{|z| < 1\}$, $F \in H_1$ and

$$\int_{-\pi}^{\pi} |F(e^{i\theta})|\,d\theta = \int_{-\pi}^{\pi} |f(e^{i\theta})|\,d\theta.$$

Now the result follows with

$$g(z) = \frac{1}{2}(1 + B(z))F(z)$$

$$h(z) = -\frac{1}{2}(1 - B(z))F(z),$$

because, by direct inspection (or the strong maximum principle),

$$|B(z)| < 1 \quad \text{for} \quad |z| < 1.$$

Corollary. Let $f \in H_p$. Then we can write

$$f(z) = B(z)[g(z)]^{1/p}$$

where $B(z)$ is a Blaschke product and $g \in H_1$ has no zeros in $\{|z| < 1\}$.

Proof. $f = BF$ with $F \in H_p$ having no zeros in $\{|z| < 1\}$. Take $g(z) = [F(z)]^p$, $|z| < 1$; the pth power is defined and regular in $\{|z| < 1\}$ because F never $= 0$ there.

C. Functions in H_p.

Definition. If $F \in H_p$ and $p \geq 1$, we write

$$\|F\|_p = \sup_{r<1} \sqrt[p]{\int_{-\pi}^{\pi} |F(re^{i\theta})|^p \, d\theta} .$$

If $0 < p < 1$, we write

$$\|F\|_p = \sup_{r<1} \int_{-\pi}^{\pi} |F(re^{i\theta})|^p \, d\theta,$$

(without the pth root!)

Then $\| \ \|_p$ satisfies the triangle inequality (hence yields a metric for H_p) in all cases, but is positive homogeneous (i.e., a norm) only for $p \geq 1$.

1°. Theorem. If $p > 0$ and $F \in H_p$, for almost all $e^{i\theta}$, $\lim F(z)$ for $z \underset{\measuredangle}{\longrightarrow} e^{i\theta}$ exists and is finite, and if we call that limit $F(e^{i\theta})$,

$$\int_{-\pi}^{\pi} |F(e^{i\theta})|^p \, d\theta = \|F\|_p, \qquad 0 < p < 1$$

$$\int_{-\pi}^{\pi} |F(e^{i\theta})|^p \, d\theta = \|F\|_p^p, \qquad p \geq 1.$$

Proof. By the scholium of Section B.2, $F = BG$ where B is a Blaschke product, $G(z)$ has no zeros in $\{|z| < 1\}$, and

$$\|F\|_p = \|G\|_p.$$

Now write $G(z) = [h(z)]^{1/p}$, as in the second corollary of Section B.2.

Then $h \in H_1$, and by Chapter II, Section C.2,

$$\|h\|_1 = \limsup_{r \to 1} \int_{-\pi}^{\pi} |h(re^{i\theta})|\,d\theta = \int_{-\pi}^{\pi} |h(e^{i\theta})|\,d\theta,$$

where the $\measuredangle$ boundary value $h(e^{i\theta})$ of $h(z)$ exists a.e. (Chapter II, Section C.1). Now we are of course assuming that $F \not\equiv 0$, so by Chapter III, Section B, for almost all θ, $h(e^{i\theta}) \neq 0$.

$\therefore$ For almost all θ, $\lim G(z) = \lim[h(z)]^{1/p}$ exists as $z \underset{\measuredangle}{\to} e^{i\theta}$. Call this limit $G(e^{i\theta})$. Then, since $B(z) \to B(e^{i\theta})$ as $z \underset{\measuredangle}{\to} e^{i\theta}$ for almost all θ, $F(z) \to B(e^{i\theta})G(e^{i\theta})$, which call $F(e^{i\theta})$, for almost all θ as $z \underset{\measuredangle}{\to} e^{i\theta}$.

Finally, if $0 < p < 1$,

$$\|F\|_p = \|G\|_p = \|h^{1/p}\|_p = (\text{clearly}) = \|h\|_1 = \int_{-\pi}^{\pi} |h(e^{i\theta})|\,d\theta$$
$$= \int_{-\pi}^{\pi} |G(e^{i\theta})|^p d\theta = \int_{-\pi}^{\pi} |F(e^{i\theta})|^p d\theta$$

because $B(e^{i\theta}) = 1$ a.e.

If $p \geq 1$, we get the same result, but the integrals in the above chain of equalities, and $\|h\|_1$, are affected with p^{th} roots. We're done.

2°. Theorem. If $f \in H_p$,

$$\int_{-\pi}^{\pi} |f(re^{i\theta}) - f(e^{i\theta})|^p \,d\theta \to 0$$

as $r \to 1$, where $f(e^{i\theta})$ is the $\measuredangle$ boundary value of $f(z)$ which, by the previous Theorem, exists a.e.

Proof. If $B(z)$ is the Blaschke product formed from the zeros of $f(z)$, $f(z) = B(z)F(z)$, where $F \in H_p$ and $F(z)$ is without zeros in $\{|z| < 1\}$. If $r < 1$, and $0 < p < 1$,

$$\int_{-\pi}^{\pi} |f(e^{i\theta}) - f(re^{i\theta})|^p d\theta \le \int_{-\pi}^{\pi} |B(re^{i\theta})|^p |F(e^{i\theta}) - F(re^{i\theta})|^p d\theta$$

$$+ \int_{-\pi}^{\pi} |B(e^{i\theta}) - B(re^{i\theta})|^p |F(e^{i\theta})|^p d\theta$$

$$\le \int_{-\pi}^{\pi} |F(e^{i\theta}) - F(re^{i\theta})|^p d\theta + \int_{-\pi}^{\pi} |B(re^{i\theta}) - B(e^{i\theta})|^p |F(e^{i\theta})|^p d\theta.$$

For $p \ge 1$, similar inequalities hold, save that the integrals are affected with p^{th} roots.

Now $|B(z)| \le 1$, $B(re^{i\theta}) \to B(e^{i\theta})$ a.e. as $r \to 1$, and $|F(e^{i\theta})|^p \in L_1$.

Therefore $\int_{-\pi}^{\pi} |B(re^{i\theta}) - B(e^{i\theta})|^p |F(e^{i\theta})|^p d\theta \to 0$ as $r \to 1$ by dominated convergence.

It is thus enough to prove

$$\int_{-\pi}^{\pi} |F(e^{i\theta}) - F(re^{i\theta})|^p \, d\theta \to 0$$

as $r \to 1$ if $F \in H_p$ is <u>without zeros</u> in $\{|z| < 1\}$.

Now, if $p \ge 1$, this is easy, because, by Chapter II, Section B.1,

$$F(re^{i\theta}) = \frac{1}{2\pi} \int_{-\pi}^{\pi} P_r(\theta - t) F(e^{it}) dt,$$

since, in particular, $F \in H_1$. By 1°, $F(e^{it}) \in L_p$ if $F \in H_p$, so, by the approximate identity property of P_r,

$$\int_{-\pi}^{\pi} |F(re^{i\theta}) - F(e^{i\theta})|^p \, d\theta \to 0$$

as $r \to 1$.

Suppose $p \ge \frac{1}{2}$. Using a clever trick of Zygmund, we put $G(z) = \sqrt{F(z)}$; the square root is well defined here because $F(z)$ never $= 0$ in $\{|z| < 1\}$. Then $G \in H_{2p}$ with $2p \ge 1$. We have:

$$\int_{-\pi}^{\pi} |F(re^{i\theta}) - F(e^{i\theta})|^p d\theta = \int_{-\pi}^{\pi} |[G(re^{i\theta}) - G(e^{i\theta})][G(re^{i\theta}) + G(e^{i\theta})]|^p d\theta$$

which, by Schwarz, is $\leq$

$$\leq \sqrt{\int_{-\pi}^{\pi} |G(re^{i\theta}) - G(e^{i\theta})|^{2p} d\theta} \sqrt{\int_{-\pi}^{\pi} |G(re^{i\theta}) + G(e^{i\theta})|^{2p} d\theta}$$

$$\leq C \sqrt{\int_{-\pi}^{\pi} |G(re^{i\theta}) - G(e^{i\theta})|^{2p} d\theta}$$

with C independent of r since $G \in H_{2p}$. But now, since $2p \geq 1$, this last expression $\to 0$ as $r \to 1$ by what has been proved above. So the result holds for $p \geq 1/2$.

If, now, $p \geq 1/4$, we make again the substitution $G = \sqrt{F}$ and argue as above, using the result just proven.

In this way we prove the Theorem succesively for $p \geq 1/4$, $p \geq 1/8$, $p \geq 1/16$, & c, and we GET **DONE**.

Scholium. The ideas of the above proof can be modified so as to obtain a new proof of the celebrated Theorem of the Brothers Riesz (Chapter II, Section A).

First of all, the theorem just given is elementary for $p = 2$. Indeed, if $f(z) = \sum_0^\infty A_n z^n$ is in H_2, we have

$$\int_{-\pi}^{\pi} |f(re^{i\theta})|^2 d\theta = \pi \sum_0^\infty |A_n|^2 r^{2n}$$

by absolute convergence and orthogonality, so $\sum_0^\infty |A_n|^2 < \infty$.

By the Riesz-Fischer theorem, there is a function $f(e^{i\theta}) \in L_2(-\pi,\pi)$ having the Fourier series $\sum_0^\infty A_n e^{in\theta}$. It is now easy to check by direct calculation that, for $r < 1$,

$$f(re^{i\theta}) = \frac{1}{2\pi}\int_{-\pi}^{\pi} P_r(\theta - t)f(e^{it})dt$$

(work with the series!), so that, <u>in fact</u>,

$$f(z) \to f(e^{i\theta}) \quad \text{a.e.} \quad \text{for} \quad z \underset{\not\,}{\longrightarrow} e^{i\theta}$$

and

$$\int_{-\pi}^{\pi} \left| f(re^{i\theta}) - f(e^{i\theta}) \right|^2 d\theta \to 0$$

as $r \to 1$, by the basic material in Chapter I, Section D.

Granted this result, we can prove the F. and M. Riesz theorem as follows. Take any measure μ on $[-\pi,\pi]$ with

$$\int_{-\pi}^{\pi} e^{in\theta}\, d\mu(\theta) = 0, \qquad n = 1,2,3,\ldots,$$

and put

$$F(re^{i\theta}) = \frac{1}{2\pi}\int_{-\pi}^{\pi} P_r(\theta - t)d\mu(t)$$

for $0 \le r < 1$. As we easily see,

$$F(re^{i\theta}) = \sum_0^\infty a_n\, r^n\, e^{in\theta},$$

where

$$a_n = \frac{1}{2\pi}\int_{-\pi}^{\pi} e^{-in\theta}\, d\mu(\theta),$$

so $F(z)$ is <u>regular</u> in $\{|z| < 1\}$ - by Chapter I, Section C, $F \in H_1$. As at the beginning of the proof of the theorem given above, we can now write $F(z) = B(z)G(z)$ where $B(z)$ is a Blaschke product and $G \in H_1$, $G(z)$ <u>without zeros</u> in $\{|z| < 1\}$. <u>Therefore we can write</u> $G = f^2$ for an $f(z)$ analytic in $\{|z| < 1\}$; it is practically evident that $f \in H_2$.

So use the special case of the theorem just proved; put $F(e^{i\theta}) = B(e^{i\theta})[f(e^{i\theta})]^2$; then $F(e^{i\theta}) \in L_1(-\pi,\pi)$. Now

$$\int_{-\pi}^{\pi} |F(re^{i\theta}) - F(e^{i\theta})|d\theta \le \int_{-\pi}^{\pi} |B(re^{i\theta}) - B(e^{i\theta})| \, |f(e^{i\theta})|^2 d\theta$$

$$+ \int_{-\pi}^{\pi} |B(re^{i\theta})| \, |(f(re^{i\theta}))^2 - (f(e^{i\theta}))^2|d\theta.$$

As $r \to 1$, the <u>first</u> integral on the right goes to zero by dominated convergence. The <u>second</u> is

$$\le \int_{-\pi}^{\pi} \left|[f(re^{i\theta})]^2 - [f(e^{i\theta})]^2\right| d\theta;$$

<u>this</u> is now shown to $\to 0$ as $r \to 1$ by the argument of Zygmund's trick, using the already known fact that

$$\int_{-\pi}^{\pi} |f(re^{i\theta}) - f(e^{i\theta})|^2 d\theta \to 0, \qquad r \to 1.$$

<u>In fine</u>,

$$\int_{-\pi}^{\pi} |F(re^{i\theta}) - F(e^{i\theta})|d\theta \to 0, \qquad r \to 1.$$

But by Chapter I, Section D, as $r \to 1$,

$$F(re^{i\theta})d\theta \to d\mu(\theta) \;\; w^*.$$

So $d\mu(\theta) = F(e^{i\theta})d\theta$ and μ is absolutely continuous, proving the F. and M. Riesz theorem.

3°. <u>Theorem</u> (Smirnov). Let $f \in H_p$. If $p' > p$ and $f(e^{i\theta}) \in L_{p'}$, then $f \in H_{p'}$.

<u>Proof</u>. We have $f = BF$ where B is a Blaschke product and $F(z)$ without zeros in $\{|z| < 1\}$. Since $|B(z)| \le 1$, it is enough to prove that $F \in H_{p'}$ if $F(e^{i\theta}) \in L_{p'}$. F being without zeros, write

$$G(z) = [F(z)]^p.$$

Then $G \in H_1$ and $G(e^{i\theta}) \in L_{p'/p}$ with $p'/p > 1$. It is enough to prove that $G(z) \in H_{p'/p}$. But now use the result of Chapter II, Section B.1:

$$G(re^{i\theta}) = \frac{1}{2\pi}\int_{-\pi}^{\pi} P_r(\theta - t)G(e^{it})dt,$$

and the fact that $G(e^{it}) \in L_{p'/p}$ with $p'/p > 1$ to show that

$$\int_{-\pi}^{\pi} |G(re^{i\theta})|^{p'/p}\, d\theta \leq \int_{-\pi}^{\pi} |G(e^{i\theta})|^{p'/p}\, d\theta$$

for $r > 1$. We're done.

D. Inner and Outer Factors

1°. Lemma. Let $\Phi(x)/x \to \infty$, $x \to \infty$, $\Phi(x) \geq 0$. Let $f_n(t) \geq 0$,

$$\int_{-\pi}^{\pi} \Phi(f_n(t))dt \leq C,$$

and $f_n(t)dt \to d\mu(t)$ w^* where μ is a measure on $[-\pi,\pi]$. Then μ is absolutely continuous.

Proof. For $K > 0$, denote by η_k the supremum of $\frac{x}{\Phi(x)}$ for $x \geq K$. Then $\eta_k \to 0$ as $K \to \infty$. Let $E \subseteq [-\pi,\pi]$ be compact and $|E| = 0$; to show that $\mu(E) = 0$. With $\varepsilon > 0$ arbitrary, take $\mathcal{O}$ open $\supset E$ with $|\mathcal{O}| < \varepsilon$, and let $\psi(t)$ be continuous on $[-\pi,\pi]$, $0 \leq \psi(t) \leq 1$, $\psi(t) \equiv 0$ outside $\mathcal{O}$, and $\psi(t) \equiv 1$ on E. Since $f_n(t)dt \underset{n}{\longrightarrow} d\mu(t)$ w^*,

$$\mu(E) \leq \int_{-\pi}^{\pi} \psi(t)d\mu(t) = \lim_{n\to\infty} \int_{-\pi}^{\pi} \psi(t)f_n(t)dt$$

$$\leq \limsup_{n\to\infty} \int_{\mathcal{O}} f_n(t)dt.$$

Let, for given large K, and any n,

$$\mathfrak{S}(K,n) = \{t \in \mathfrak{S} \mid f_n(t) \leq K\}.$$

Then

$$\int_{\mathfrak{S}} f_n(t)dt \leq K|\mathfrak{S}(k,n)| + \int_{\mathfrak{S}\sim\mathfrak{S}(K,n)} f_n(t)dt \leq K|\mathfrak{S}| +$$
$$+ \eta_k \int_{-\pi}^{\pi} \Phi(f_n(t))dt \leq K|\mathfrak{S}| + \eta_k C \leq K\varepsilon + C\eta_k.$$

Given $\delta > 0$ choose first K so large that $C\eta_k < \delta/2$, then take $\varepsilon > 0$ so small that $K\varepsilon < \delta/2$. One finds $\mu(E) < \delta$. Since $\delta > 0$ is arbitrary, $\mu(E) = 0$. Q.E.D.

2°. <u>Lemma</u>. If $F(z)$ <u>is regular in</u> $\{|z| < 1\}$ <u>and</u>

$$\int_{-\pi}^{\pi} \log^{+}|F(re^{i\theta})|d\theta$$

<u>is bounded for</u> $0 \leq r < 1$, <u>then, if</u> $F \not\equiv 0$,

$$\int_{-\pi}^{\pi} |\log|F(re^{i\theta})||d\theta$$

<u>is bounded for</u> $0 \leq r < 1$.

<u>Remark</u>. This is an important result in the theory of functions.

<u>Proof</u>. Wlog, $F(0) \neq 0$ (otherwise work with $F(z)/z^k$ instead of $F(z)$). Then, by Jensen's inequality,

$$- \infty < \log|F(0)| \leq \frac{1}{2\pi}\int_{-\pi}^{\pi} \log|F(re^{i\theta})|d\theta$$
$$\leq \frac{1}{2\pi}\int_{-\pi}^{\pi} 2 \log^{+}|F(re^{i\theta})|d\theta - \frac{1}{2\pi}\int_{-\pi}^{\pi} |\log|F(re^{i\theta})||d\theta$$

for $0 \leq r < 1$. Therefore

$$\int_{-\pi}^{\pi} |\log|F(re^{i\theta})||d\theta \le 2\int_{-\pi}^{\pi} \log^+|F(re^{i\theta})|d\theta - 2\pi \log|F(0)|.$$

The result follows.

3°. First, an identity:

$$\frac{1-r^2}{1+r^2-2r\cos\theta} + \frac{2ir\sin\theta}{1+r^2-2r\cos\theta} = \frac{(1+re^{i\theta})(1-re^{-i\theta})}{(1-re^{i\theta})(1-re^{-i\theta})} = \frac{1+re^{i\theta}}{1-re^{i\theta}}.$$

From Chapter I, if $F(z)$ is regular in a region including $\{|z|<1\}$ in its interior and $0 \le r < 1$,

$$F(re^{i\theta}) = \frac{1}{2\pi}\int_{-\pi}^{\pi} \frac{1+re^{i(\theta-t)}}{1-re^{i(\theta-t)}} \cdot \Re F(e^{it})dt + ic,$$

where c is a real constant $(= \Im F(0)!)$.

We usually write:

$$\boxed{F(re^{i\theta}) = \frac{1}{2\pi}\int_{-\pi}^{\pi} \frac{e^{it}+re^{i\theta}}{e^{it}-re^{i\theta}} \Re F(e^{it})dt + i\Im F(0).}$$

4°. Theorem. Let $F(z) \in H_p$, $p > 0$. Let $B(z)$ be the Blaschke product consisting of the zeros of $F(z)$. Then there is a singular measure $\sigma \ge 0$ on $[-\pi,\pi]$ with

$$\boxed{F(z) = B(z)e^{-\frac{1}{2\pi}\int_{-\pi}^{\pi}\frac{e^{it}+z}{e^{it}-z}d\sigma(t)} e^{ic} \cdot e^{\frac{1}{2\pi}\int_{-\pi}^{\pi}\frac{e^{it}+z}{e^{it}-z}\log|F(e^{it})|dt}}$$

where c is a real constant.

Proof. By Section B.2, we can write $F(z) = B(z)G(z)$ where $G \in H_p$ and $G(z)$ has no zeros in $\{|z| < 1\}$.

Thus, we can define $\varphi(z) = \log G(z)$ so as to be <u>analytic</u> in $\{|z| < 1\}$. Now for $0 \le r < 1$, by the inequality between arithmetic and geometric means,

$$\frac{1}{2\pi}\int_{-\pi}^{\pi} p \log^+|G(re^{i\theta})|d\theta \le \frac{1}{2\pi}\int_{-\pi}^{\pi} \log\{|G(re^{i\theta})|^p + 1\}d\theta$$

$$\le \log \frac{1}{2\pi}\int_{-\pi}^{\pi} (|G(re^{i\theta})|^p + 1)d\theta \quad \le C$$

because $G \in H_p$, so $\int_{-\pi}^{\pi} \log^+|G(re^{i\theta})|d\theta$ is bounded above for $0 \le r < 1$. Therefore by 2°, $\int_{-\pi}^{\pi} |\log|G(re^{i\theta})||$ is bounded above for $0 \le r < 1$.

<u>NOTATION</u>. Write

$$\log^-|G| = |\log|G|| - \log^+|G|.$$

($\therefore \log^-|G|$ is ≥ 0 !)

<u>Therefore</u>: <u>We have a sequence of</u> $r_\nu \to 1$, $r_\nu < 1$, <u>and measures</u> μ_+, μ_- on $[-\pi,\pi]$, $d\mu_+ \ge 0$, $d\mu_- \ge 0$, <u>with</u>

$$\log^+|G(r_\nu e^{i\theta})|d\theta \xrightarrow[\nu]{} d\mu_+(\theta) \quad w^*$$

$$\log^-|G(r_\nu e^{i\theta})|d\theta \xrightarrow[\nu]{} d\mu_-(\theta) \quad w^*$$

Now by the formulas in 3° with a change of variable, if z, $|z| < 1$, if <u>fixed</u>,

$$\varphi(r_\nu z) = i\,\Im\varphi(0) + \frac{1}{2\pi}\int_{-\pi}^{\pi} \frac{e^{it} + z}{e^{it} - z}\,\Re\,\varphi(r_\nu e^{it})dt$$

$$= i\,\Im\varphi(0) + \frac{1}{2\pi}\int_{-\pi}^{\pi} \frac{e^{it} + z}{e^{it} - z}\,\log|G(r_\nu e^{it})|dt.$$

By the aforementioned w^* convergence, the right side tends to

$$i\,\mathfrak{J}\varphi(0) + \frac{1}{2\pi}\int_{-\pi}^{\pi} \frac{e^{it}+z}{e^{it}-z}\,[d\mu_+(t) - d\mu_-(t)],$$

and the left side tends to $\varphi(z)$. So

$$\varphi(z) = i\,\mathfrak{J}\varphi(\nu) + \frac{1}{2\pi}\int_{-\pi}^{\pi} \frac{e^{it}+z}{e^{it}-z}\,(d\mu_+(t) - d\mu_-(t)).$$

<u>Now</u> μ_+ <u>is absolutely continuous</u>. Indeed, since $G \in H_p$, we can apply 1^0 - we have

$$\log^+|G(r_\nu e^{i\theta})|\,d\theta \xrightarrow[\nu]{} d\mu_+(\theta) \ \ w^*$$

whilst

$$\int_{-\pi}^{\pi} |G(r_\nu e^{i\theta})|^p\,d\theta \le c\,,$$

so surely,

$$\int_{-\pi}^{\pi} \exp(p\,\log^+|G(r_\nu e^{i\theta})|)\,d\theta \le c + 2\pi,$$

and $\frac{\exp x}{x} \to \infty$ for $x \to \infty$. We therefore have

$$d\mu_+(\theta) = h_+(\theta)d\theta, \qquad h_+ \in L_1.$$

<u>However, all we can say about</u> μ_- <u>is that</u>

$$d\mu_-(\theta) = h_-(\theta)d\theta + d\sigma(\theta)$$

where $h_- \in L_1$ and $\sigma \ge 0$ is <u>singular, not necessarily zero</u>.

Therefore, with $h = h_+ - h_- \in L_1$,

$$\varphi(z) = i\,\mathfrak{J}\varphi(0) + \frac{1}{2\pi}\int_{-\pi}^{\pi} \frac{e^{it}+z}{e^{it}-z}\,(h(t)dt - d\sigma(t)).$$

This yields, in particular (taking real parts - recall 3^0!)

$$\log|G(re^{i\theta})| = \Re\varphi(re^{i\theta}) = \frac{1}{2\pi}\int_{-\pi}^{\pi} \frac{1-r^2}{1+r^2-2r\cos(\theta-t)}[h(t)dt - d\sigma(t)].$$

Therefore, by Chapter I, Section D.2°, at any θ_0 for which $\sigma'(\theta_0)$ exists and the derivative of the indefinite integral of $h(t)$ equals $h(\theta_0)$,

$$\log|G(z)| \to h(e^{i\theta_0}) + \sigma'(\theta_0) \quad \text{as} \quad z \underset{\measuredangle}{\longrightarrow} e^{i\theta_0}.$$

But, σ being singular, $\sigma'(\theta) = 0$ a.e. Also, G being $\in H_p$, the $\measuredangle$ limit $G(e^{i\theta})$ exists a.e. and is $\neq 0$ a.e. (A.2, B.2, and Chapter III, Section B.) So it makes sense to talk about $\log|G(e^{i\theta})|$ a.e., and since $|G(e^{i\theta})| = |F(e^{i\theta})|$ a.e., we have the identification for the values of the real L_1 function h:

$$h(e^{i\theta}) = \log|F(e^{i\theta})| \quad \text{a.e.}$$

We've thus proven

$$\varphi(z) = i\,\Im\varphi(0) + \frac{1}{2\pi}\int_{-\pi}^{\pi}\frac{e^{it}+z}{e^{it}-z}[\log|F(e^{it})|dt - d\sigma(t)].$$

Finally,

$$F(z) = B(z)G(z) = B(z)e^{\varphi(z)},$$

and our desired representation is established. Q.E.D.

Scholium. As a by-product of the above proof we have $\log|F(e^{it})| \in L_1(-\pi,\pi)$ for non-zero $F \in H_p$. This is a quantitative form of Chapter III, Section B. The result also follows more quickly from 2° and Fatou's lemma.

Definition. The formula

$$F(z) = B(z) e^{-\frac{1}{2\pi}\int_{-\pi}^{\pi} \frac{e^{it}+z}{e^{it}-z}\, d\sigma(t)}\, e^{ic} e^{\frac{1}{2\pi}\int_{-\pi}^{\pi} \frac{e^{it}+z}{e^{it}-z} \log|F(e^{it})|\, dt}$$

is called the <u>canonical representation</u> of $F \in H_p$. We call (after Beurling)

$$I_F(z) = e^{ic} B(z) \exp\left\{-\frac{1}{2\pi}\int_{-\pi}^{\pi} \frac{e^{it}+z}{e^{it}-z}\, d\sigma(t)\right\}$$

the <u>inner factor</u> of $F(z)$, and

$$\Theta_F(z) = \exp\left(\frac{1}{2\pi}\int_{-\pi}^{\pi} \frac{e^{it}+z}{e^{it}-z} \log|F(e^{it})|\, dt\right)$$

the <u>outer factor</u> of $F(z)$.

Note that ($d\sigma \geq 0$!) $|I_F(z)| \leq 1$, $\{|z| < 1\}$, and by the above discussion, $|I_F(\zeta)| \equiv 1$ a.e., $|\zeta| = 1$. (Cf. argument showing that $\Re\varphi(z) \to h(e^{i\theta})$ a.e. for $z \xrightarrow{\measuredangle} e^{i\theta}$.)

E. <u>Beurling's Theorem</u>

1°. <u>Lemma</u>. Let $p > 0$. Then, if $F \in H_p$, there exist <u>polynomials</u> $P(z)$ with $\|F - P\|_p$ <u>arbitrarily small</u>.

<u>Proof</u>. By C.2, $\|F(z) - F(rz)\|_p < \varepsilon$ if $r < 1$ is sufficiently close to 1. But the Taylor series of $F(rz)$ converges <u>even uniformly</u> for $|z| \leq 1$; cutting it off far enough out, we get a polynomial $P(z)$ with $\|F - P\|_p < 2\varepsilon$.

2°. <u>Lemma</u>. Let F and G belong to (perhaps different) H_p-spaces. If <u>the ratio</u> of their <u>outer factors</u> $k(z) = \Theta_F(z)/\Theta_G(z)$ has $k(e^{i\theta}) \in L_{p'}$, say, then $\Theta_F(z)/\Theta_G(z) \in H_{p'}$. (Generalization of Smirnov's Theorem.)

Proof. For $r < 1$, by the inequality between arithmetic and geometric means

$$\left(\text{N.B. } \frac{1}{2\pi}\int_{-\pi}^{\pi} \frac{1-r^2}{1+r^2-2r\cos(\theta-t)}\,dt = 1!\right),$$

$$p\log|k(re^{i\theta})| = \frac{p}{2\pi}\int_{-\pi}^{\pi} \frac{1-r^2}{1+r^2-2r\cos(\theta-t)} \log\left|\frac{F(e^{it})}{G(e^{it})}\right| dt$$

$$\leq \log \frac{1}{2\pi}\int_{-\pi}^{\pi} \frac{1-r^2}{1+r^2-2r\cos(\theta-t)} \left|\frac{F(e^{it})}{G(e^{it})}\right|^p dt,$$

so

$$\int_{-\pi}^{\pi} |k(re^{i\theta})|^p d\theta \leq \frac{1}{2\pi}\int_{-\pi}^{\pi}\int_{-\pi}^{\pi} \frac{1-r^2}{1+r^2-2r\cos(\theta-t)} \left|\frac{F(e^{it})}{G(e^{it})}\right|^p dt\,d\theta$$

$$= \int_{-\pi}^{\pi} \left|\frac{F(e^{it})}{G(e^{it})}\right|^p dt,$$

which is enough.

3°. Theorem. (For case $p = 2$ this is due to Beurling.) Let $F = I_F \mathcal{O}_F \in H_p$, $p > 0$. The closure of $F(z)\cdot\{\text{polynomials in } z\}$ in H_p is precisely $I_F \cdot H_p$. (General case due to Srinivasan and Wang.)

Proof. a) The closure is not more than $I_F \cdot H_p$. Indeed, let $G \in H_p$, and let $P_n(z)$ be polynomials with $\|FP_n - G\|_p \xrightarrow[n]{} 0$. We may, wlog, assume that $\|FP_n - FP_{n+1}\|_p < 2^{-n}$. Then, since $|I_F(e^{i\theta})| = 1$ a.e., by C.1, if $0 < p < 1$,

$$\int_{-\pi}^{\pi} |\mathcal{O}_F(e^{i\theta})P_n(e^{i\theta}) - \mathcal{O}_F(e^{i\theta})P_{n+1}(e^{i\theta})|^p \, d\theta < 2^{-n},$$

and if $p \geq 1$, a similar inequality holds with the integral affected by a pth root. Since $\mathcal{O}_F \in H_p$ if F is (cf. proof of 2° above!), $\mathcal{O}_F P_n \in H_p$ and by C.1, $\|\mathcal{O}_F P_n - \mathcal{O}_F P_{n+1}\|_p \leq 2^{-n}$. By inequalities like those used in 2^0, it is easy to see that

$$\mathcal{O}_F(z)P_1(z) + \sum_{n=1}^{\infty} [\mathcal{O}_F(z)P_{n+1}(z) - \mathcal{O}_F(z)P_n(z)]$$

converges uniformly in the interior of $\{|z| < 1\}$ to an analytic function $R(z)$, say, and by the triangle inequality for $\| \|_p$ (N.B. valid also for $0 < p < 1$!) in case $r < 1$, (assuming $p < 1$; if $p \geq 1$, integral has a pth root):

$$\int_{-\pi}^{\pi} |R(re^{i\theta})|^p d\theta \leq \|\mathcal{O}_F P_1\|_p + \sum_{n=1}^{\infty} \|\mathcal{O}_F P_{n+1} - \mathcal{O}_F P_n\|_p < \infty,$$

so $R \in H_p$.

Now

$$FP_n = I_F \mathcal{O}_F P_n = I_F\{\mathcal{O}_F P_1 + \sum_{m=1}^{n-1} (\mathcal{O}_F P_{m+1} - \mathcal{O}_F P_m)\} \xrightarrow[n]{} I_F R.$$

So, since $\|FP_n - G\|_p \xrightarrow[n]{} 0$, we have $G = I_F R$, a member of $I_F H_p$.

b) The closure <u>is at least</u> $I_F \cdot H_p$. Since $|I_F(z)| \leq 1$ for $|z| < 1$, it is enough to show that $\mathcal{O}_F \cdot \{\text{polynomials}\}$ is <u>dense</u> in H_p.

<u>First</u>, suppose $p \geq 1$. Then we can argue by <u>duality</u>. Let $\frac{1}{q} + \frac{1}{p} = 1$, assume $k \in L_q$ and

$$\int_{-\pi}^{\pi} e^{in\theta}\mathcal{O}_F(e^{i\theta})k(\theta)d\theta = 0 \quad \text{for} \quad n = 0,1,2,\ldots \quad .$$

It is enough to show that then $0 = \int_{-\pi}^{\pi} G(e^{i\theta})k(\theta)d\theta$ for any $G \in H_p$. Since $\mathcal{O}_F(e^{i\theta})k(\theta) \in L_1$, $\int_{-\pi}^{\pi} e^{in\theta}\mathcal{O}_F(e^{i\theta})k(\theta)d\theta = 0$ for $n = 0,1,2,\ldots$ makes $\mathcal{O}_F(e^{i\theta})k(\theta) = e^{i\theta}H(e^{i\theta})$ where $H \in H_1$. Therefore

$$k(\theta) = e^{i\theta}I_H(e^{i\theta}) \frac{\mathcal{O}_H(e^{i\theta})}{\mathcal{O}_F(e^{i\theta})},$$

and

$$\left|\frac{\mathcal{O}_H(e^{i\theta})}{\mathcal{O}_F(e^{i\theta})}\right| = |k(\theta)| \in L_q.$$

So by 2°, $\mathfrak{G}_H/\mathfrak{G}_F = R$ is in H_q, and $k(\theta) = e^{i\theta}I_H(e^{i\theta})R(e^{i\theta})$. Therefore, if $G \in H_p$, $G(e^{i\theta})k(\theta) = e^{i\theta}I_H(e^{i\theta})G(e^{i\theta})R(e^{i\theta})$ with $I_H GR \in H_1$, so $\int_{-\pi}^{\pi} e^{in\theta}k(\theta)G(e^{i\theta})d\theta = 0$ for $n = 0,1,2,\ldots$. We're done in case $p \geq 1$.

Now, using Zygmund's idea, suppose that $F \in H_p$ with $p \geq 1/2$. Clearly we can take $F_1(z) = \sqrt{\mathfrak{G}_F(z)}$ and $F_1(z) \in H_{2p}$, $\mathfrak{G}_{F_1} = F_1$. Therefore, if G is any element of H_{2p} we can find a polynomial P with $\|F_1P - G\|_{2p} < \varepsilon$ by the above. Using Schwarz, we then get $\|\mathfrak{G}_F P - F_1\|_p \leq$ $\leq \sqrt{\|\mathfrak{G}_F\|_p}\sqrt{\varepsilon}$ (or another variant, depending on whether $p > 1$ or not).

In particular, if Q is a polynomial we can find another polynomial P such that $\|\mathfrak{G}_F P - F_1 Q\|_p \leq \delta$, say, $\delta > 0$ being arbitrary. Now, given $H \in H_p$, use first the result of 1° to find a polynomial R with $\|H - R\|_p \leq \delta$. Then, as remarked above, we can find another polynomial Q making $\|F_1Q - R\|_{2p}$ as small as we like - using Schwarz, we see that we can choose Q with $\|F_1Q - R\|_p < \delta$. Finally, we have the polynomial P with $\|\mathfrak{G}_F P - F_1Q\|_p < \delta$, whence $\|H - \mathfrak{G}_F P\|_p \leq \|H - R\|_p + \|R - F_1Q\|_p +$ $+ \|F_1Q - \mathfrak{G}_F P\| \leq 3\delta$. This extends the result we are proving to all values of $p \geq 1/2$.

The same argument may now be used to extend it to values of $p \geq 1/4$, then to values $\geq 1/8$, & c.

We're done.

F. Invariant Subspaces

Let $\mathfrak{H}$ be a Hilbert space with orthonormal basis $\{e_n;\ n = 0, \pm 1, \pm 2, \pm 3, \ldots\}$, and consider the unitary transformation V defined on $\mathfrak{H}$ by putting

$$Ve_n = e_{n+1}.$$

The problem is to study the invariant subspaces of V.

We map $\mathfrak{H}$ isometrically onto $L_2(-\pi,\pi)$ by making e_n correspond to $\frac{1}{\sqrt{2\pi}}\, e^{in\theta}$; then V corresponds to multiplication by $e^{i\theta}$.

Theorem. Let a closed subspace E of $L_2(-\pi,\pi)$ be such that $e^{i\theta}E = E$. Then $E = \chi_A L_2$ where χ_A is the characteristic function of some measurable set $A \subset (-\pi,\pi)$.

Proof. If $E \subset L_2(-\pi,\pi)$ properly, then $1 \notin E$. Otherwise $e^{i\theta}$ and therefore $e^{2i\theta}$, $e^{3i\theta}$, &c $\in E$. Also, from $e^{i\theta}E = E$ we get $e^{-i\theta}E = E$, so from $1 \in E$ we get $e^{-i\theta} \in E$, $e^{-2i\theta} \in E$, & c, and finally E would $= L_2$.

So let φ be the closest element of E to 1. The, arguing as above,

$$e^{in\theta}\varphi \in E, \qquad n = 0, \pm 1, \pm 2, \ldots,$$

so

$$1 - \varphi \perp e^{in\theta}\varphi, \qquad n \in \mathbb{Z}.$$

Therefore

$$\int_{-\pi}^{\pi} e^{in\theta}(\varphi(\theta) - |\varphi(\theta)|^2)\,d\theta = 0, \qquad n \in \mathbb{Z},$$

so $\varphi = |\varphi|^2$ a.e., and, finally, $\varphi(\theta) = 1$ or 0 a.e., so $\varphi = \chi_A$ for some measurable A. Therefore $\chi_A \in E$ so $\chi_A L_2 \subseteq E$. $\chi_A L_2$ is clearly a closed subspace of L_2. If $\chi_A L_2$ is not all of E, there is a $\psi \in E$ which is orthogonal to $\chi_A L_2$.

In particular,

$$\int_{-\pi}^{\pi} \chi_A(\theta)\psi(\theta)e^{in\theta}\,d\theta = 0, \qquad n \in \mathbb{Z},$$

so $\psi\chi_A \equiv 0$, so $\psi \equiv 0$ a.e. on A. But at the same time

$e^{in\theta}\psi \in E$ for $n \in \mathbb{Z}$, so since $1 - \chi_A = 1 - \varphi \perp E$,

$$\int_{-\pi}^{\pi} (1 - \chi_A(\theta))\psi(\theta)e^{in\theta}d\theta = 0, \qquad n \in \mathbb{Z},$$

i.e.,

$$\int_{-\pi}^{\pi} \chi_{\sim A}(\theta)\psi(\theta)e^{in\theta}d\theta = 0, \qquad n \in \mathbb{Z}.$$

So $\psi \equiv 0$ a.e. on $\sim A$. So $\psi \equiv 0$ a.e., contradiction. So $\chi_A L_2 = E$,

Q.E.D.

Theorem. Let E be a closed subspace of L_2 and suppose that $e^{i\theta}E \subset E$ properly. Then $E = \omega H_2$, where $|\omega(\theta)| \equiv 1$ a.e.

Proof. $e^{i\theta}E$ is also closed. Let $\omega \in E$ be $\neq 0$, $\omega \perp e^{i\theta}E$. Then in particular,

$$\omega \perp e^{in\theta}\omega, \qquad n = 1,2,\ldots,$$

so

$$\int_{-\pi}^{\pi} |\omega(\theta)|^2 e^{in\theta}\, d\theta = 0, \qquad n = 1,2,\ldots,$$

and by conjugation

$$\int_{-\pi}^{\pi} |\omega(\theta)|^2 e^{-in\theta}\, d\theta = 0, \qquad n = 1,2,\ldots,$$

so finally $|\omega(\theta)| = \text{const.}$ a.e. Wlog, $|\omega(\theta)| \equiv 1$ a.e.

Now ω, $e^{i\theta}\omega$, $e^{2i\theta}\omega,\ldots$ & c are all in E, so $\omega H_2 \subset E$. I claim $\omega H_2 = E$. ωH_2 is closed in L_2, because $|\omega(\theta)| = 1$, and the orthogonal complement of ωH_2 in L_2 is $e^{-i\theta}\omega(\theta)\overline{H}_2$, as is easily seen by direct calculation.

But if $F \in E$, $g \in H_2$, then $gF \in E$ so $e^{i\theta}gF \in e^{i\theta}E$, so by choice of ω,

$$\omega \perp e^{i\theta}gF.$$

That is,

$$\int_{-\pi}^{\pi} e^{-i\theta}\overline{g(\theta)}\omega(\theta)\overline{F(\theta)}\, d\theta = 0,$$

i.e.,

$$e^{-i\theta}\omega(\theta)\overline{H}_2 \perp E.$$

So $p \in L_2$ and $P \perp \omega H_2$ imply $P \perp E$. Therefore $E \subseteq \omega H_2$. So $E = \omega H_2$.

Corollary. Let E be a subspace of H_2 and satisfy $e^{i\theta}E \subseteq E$, $E \neq \{0\}$. Then $E = \omega H_2$ where ω is an inner function*.

Proof. If $e^{i\theta}E = E$, then $e^{in\theta}E = E$ for $n = 1,2,\ldots$, so $E \subseteq \bigcap_1^\infty e^{in\theta}H_2$. But this intersection consists only of 0. So $e^{i\theta}E \subset E$ properly. Therefore $E = \omega H_2$ where $|\omega(\theta)| \equiv 1$ a.e. $\omega \in E \subseteq H_2$. So ω is an inner function.

Q.E.D.

Corollary. Let $f_\alpha = I_\alpha \mathcal{O}_\alpha$, each I_α inner, and each $\mathcal{O}_\alpha$ outer in H_2, and write

$$I_\alpha(z) = B_\alpha(z)\exp\left\{-\frac{1}{2\pi}\int_{-\pi}^{\pi}\frac{e^{it}+z}{e^{it}-z}\, d\sigma_\alpha(t)\right\},$$

with Blaschke products B_α and singular $d\sigma_\alpha \geq 0$. Let E be the invariant subspace of H_2 generated by the f_α; in other words, let E be the smallest closed subspace of H_2 containing all the f_α, such that $e^{i\theta}E \subset E$.

Then $E = \omega H_2$ where

$$\omega(z) = B(z)\exp\left\{-\frac{1}{2\pi}\int_{-\pi}^{\pi}\frac{e^{it}+z}{e^{it}-z}\, d\sigma(t)\right\}.$$

---

*I.e., one whose outer factor is identically 1.

Here, $B(z)$ is the greatest common divisor of the $B_\alpha(z)$, and $d\sigma \geq 0$ is the largest measure $\leq$ all of the $d\sigma_\alpha$.

Proof. E is by the preceding Corollary of the form ωH_2, where ω is an inner function.

Each $f_\alpha \in E$, so $f_\alpha = I_\alpha \theta_\alpha$ is of the form ωg, $g \in H_2$. Therefore if we write

$$\omega(z) = B(z)\exp\left(-\frac{1}{2\pi}\int_{-\pi}^{\pi} \frac{e^{it}+z}{e^{it}-z}\, d\sigma(t)\right) ,$$

we certainly see that $B \mid B_\alpha$ and $d\sigma \leq d\sigma_\alpha$ by taking into account the inner factor of g.

Now suppose that there is a Blaschke product $\mathfrak{B}$ and a singular measure σ' with $\mathfrak{B} \mid$ every B_α and $d\sigma' \leq$ every $d\sigma_\alpha$. If also $B \mid \mathfrak{B}$ and $d\sigma \leq d\sigma'$, and one of these relations is proper, then, putting

$$\Omega(z) = \mathfrak{B}(z)\exp\left(-\frac{1}{2\pi}\int_{-\pi}^{\pi} \frac{e^{it}+z}{e^{it}-z}\, d\sigma'(t)\right) ,$$

we know that

$$\omega H_2 \supset \Omega H_2$$

properly. But each $f_\alpha \in \Omega H_2$ by the choice of $\mathfrak{B}$ & σ'. In this case we would have $E \subseteq \Omega H_2$, so then E couldn't equal ωH_2. But it does.

Therefore ω is as specified in the statement of the Corollary.

Q.E.D.

Problem 5.

1. Let the $f_n(z)$ be outer functions in some H_p, $p > 0$, and suppose that $|f_1(z)| \geq |f_2(z)| \geq |f_3(z)| \geq \dots$ for $|z| < 1$. Suppose that

$$f_n(z) \underset{n}{\longrightarrow} f(z)$$

uniformly in the interior of $\{|z| < 1\}$. Prove that if $f(z)$ is not identically zero, it is outer.

2. If $f(z) \in H_1$ and $\Re f(z) > 0$ for $|z| < 1$, $f(z)$ is outer. (Hint: Consider the functions $\frac{1}{n} + f(z)$.)

3. Let $\omega(z)$ be a non-constant inner function. For which complex α is $\omega(z) - \alpha$ an outer function? (Hint: In case $|\alpha| < 1$ look at $\dfrac{\omega(z)-\alpha}{1-\bar{\alpha}\omega(z)}$.)

G. Approximation of inner functions by Blaschke products.

Definition. H_∞ denotes the set of functions analytic and bounded in $\{|z| < 1\}$. If $F \in H_\infty$ we put

$$\|F\|_\infty = \sup\{|F(z)|;\ |z| < 1\}.$$

Most of the results of Sections A-D hold for H_∞ if $\| \ \|_p$ is replaced by $\| \ \|_\infty$.

A notable exception is the theorem of Section C.2 saying that $\|F(re^{i\theta}) - F(e^{i\theta})\|_p \to 0$ as $r \to 1$ if $F \in H_p$. This is in general FALSE for $F \in H_\infty$ if we use the norm $\| \ \|_\infty$!

Indeed, if $F \in H_\infty$ and we have $\|F(e^{i\theta}) - F(re^{i\theta})\|_\infty \to 0$ as $r \to \infty$, then $F(e^{i\theta})$ is continuous. And there are plenty of $F \in H_\infty$ with $F(e^{i\theta})$ not continuous!

However, we do have, for $F \in H_\infty$, $\|F\|_\infty = \text{ess}_\theta \sup |F(e^{i\theta})|$. (If $F \in H_\infty$, F is in every H_p, so the ∤ boundary value $F(e^{i\theta})$ exists a.e.!) The material in Section D also applies.

Proofs that all this work carries over to H_∞ are left to the reader - they follow easily from the results presented thus far in this course.

A special class of functions in H_∞ is constituted by the inner functions, i.e., those whose outer factor is equal to 1. An inner function $\omega(z)$ has the property that $|\omega(e^{i\theta})| \equiv 1$ a.e. The importance of these functions is already seen in Sections E, F above.

As we see from Section D.4, an inner function $\omega(z)$ has the representation

$$\omega(z) = e^{ic}B(z)\exp\left\{-\frac{1}{2\pi}\int_{-\pi}^{\pi}\frac{e^{it}+z}{e^{it}-z}\,d\sigma(t)\right\}$$

where $c \in \mathbb{R}$, $B(z)$ is a Blaschke product, and $d\sigma \geq 0$ is a singular measure on $[-\pi,\pi]$.

Lemma. An inner function $\omega(z)$ is a constant multiple of a Blaschke product iff $\int_{-\pi}^{\pi}\log|\omega(re^{i\theta})|\,d\theta \to 0$ as $r \to 1$.

Proof. By the work of Section A.2,

$$\int_{-\pi}^{\pi}\log|B(re^{i\theta})|\,d\theta \to 0 \quad \text{as} \quad r \to 1$$

for the Blaschke product factor $B(z)$ of $\omega(z)$. As for the "singular" factor,

$$S(z) = \exp\left\{-\frac{1}{2\pi}\int_{-\pi}^{\pi} \frac{e^{it}+z}{e^{it}-z}\, d\sigma(t)\right\} ;$$

$$\log|S(re^{i\theta})| = -\frac{1}{2\pi}\int_{-\pi}^{\pi} \frac{1-r^2}{1+r^2-2r\cos(\theta-t)}\, d\sigma(t),$$

so

$$\int_{-\pi}^{\pi} \log|S(re^{i\theta})|\, d\theta = -\int_{-\pi}^{\pi} d\sigma(t)$$

as $r \to 1$, and since $d\sigma \geq 0$, this last expression is zero iff $\sigma \equiv 0$, i.e., iff $S(z) \equiv 1$. Now there is the remarkable

Theorem (Frostman, rediscovered years later by D. J. Newman). Let $\omega(z)$ be any inner function. Then, given any $\varepsilon > 0$ there is a Blaschke product $B(z)$ and a real c with

$$\|\omega - e^{ic}B\|_\infty < \varepsilon.$$

Remark. Thus, the measurable function $e^{ic}B(e^{i\theta})$ is (a.e.) uniformly within ε of $\omega(e^{i\theta})$.

Proof. Let the number ω satisfy $|\omega| \leq 1$, and let $0 < \rho < 1$. Then we have the elementary formula

$$\frac{1}{2\pi}\int_{-\pi}^{\pi} \log\left|\frac{\omega - \rho e^{i\varphi}}{1-\rho e^{-i\varphi}\omega}\right| d\varphi = \max(\log\rho, \log|\omega|).$$

Putting $\omega = \omega(re^{it})$ and integrating again,

$$\frac{1}{2\pi}\int_{-\pi}^{\pi}\int_{-\pi}^{\pi} \log\left|\frac{\omega(re^{it}) - \rho e^{i\varphi}}{1-\rho e^{-i\varphi}\omega(re^{it})}\right| d\varphi dt$$

$$= \int_{-\pi}^{\pi} \max(\log\rho, \log|\omega(re^{it})|)dt.$$

But $|\omega(re^{it})| \leq 1$ and $\log|\omega(re^{it})| \to \log|\omega(e^{it})| = 0$ a.e.

as $r \to 1$. So $(\log \rho > -\infty!)$, $\int_{-\pi}^{\pi} \max(\log \rho, \log|\omega(re^{it})|)dt \to 0$ as $r \to 1$ by bounded convergence.

Therefore, by Fubini's theorem,

$$\frac{1}{2\pi}\int_{-\pi}^{\pi}\left\{\int_{-\pi}^{\pi} \log\left|\frac{\omega(re^{it}) - \rho e^{i\varphi}}{1 - \rho e^{-i\varphi}\omega(re^{it})}\right| dt\right\} d\varphi \to 0 \quad \text{as } r \to 1.$$

Since

$$\log\left|\frac{\omega(z) - \rho e^{i\varphi}}{1 - \rho e^{-i\varphi}\omega(z)}\right| \leq 0,$$

the inner integral in the above expression is ≤ 0, and, by Fatou's lemma applied to $\frac{1}{2\pi}\int_{-\pi}^{\pi}(\quad)d\varphi$,

$$\frac{1}{2\pi}\int_{-\pi}^{\pi}\left\{\limsup_{r\to 1}\int_{-\pi}^{\pi} \log\left|\frac{\omega(re^{it}) - \rho e^{i\varphi}}{1 - \rho e^{-i\varphi}\omega(re^{it})}\right| dt\right\} d\varphi = 0.$$

Now, since $0 < \rho < 1$, for each φ,

$$\omega_\varphi(z) = \frac{\omega(z) - \rho e^{i\varphi}}{1 - \rho e^{-i\varphi}\omega(z)}$$

is regular and bounded in $\{|z| < 1\}$ ($|\omega(z)| \leq 1$ there) and is indeed an inner function, because $|\omega_\varphi(e^{it})| \equiv 1$ a.e., due to the fact that $|\omega(e^{it})| \equiv 1$ a.e.

The lemma (and its proof) shows that

$$\lim_{r\to 1}\int_{-\pi}^{\pi} \log|\omega_\varphi(re^{it})|dt$$

exists, and that if it's zero, $\omega_\varphi(z)$ is a constant multiple of a Blaschke product. By the calculation made above, this limit must be zero for almost every φ. So almost every ω_φ is a constant multiple of a Blaschke product.

Finally, if $|z| < 1$,

$$|\omega(z) - \omega_\varphi(z)| = \left|\frac{\rho e^{i\varphi} - \rho e^{-i\varphi}(\omega(z))^2}{1 - \rho e^{-i\varphi}\omega(z)}\right| \leq \frac{2\rho}{1-\rho},$$

and this is $< \varepsilon$ (for all z, $\{|z| < 1\}$) if $\rho > 0$ is sufficiently small.

We are done.

Scholium. This result has several applications in the deeper study of H_∞ (e.g., Carleson's proof of the corona conjecture, not given in this course, but to be found in Duren's book).

V. Norm Inequalities for Harmonic Conjugation

A. Review. Hilbert transforms of L_2-functions. Recall, from Chapter I, Section E.4, the

Theorem. Let $f(\theta) \in L_2(-\pi,\pi)$ and write, for $0 \le r < 1$,

$$U(re^{i\theta}) = \frac{1}{2\pi}\int_{-\pi}^{\pi} \frac{1-r^2}{1+r^2-2r\cos(\theta-t)} f(t)dt,$$

$$\tilde{U}(re^{i\theta}) = \frac{1}{2\pi}\int_{-\pi}^{\pi} \frac{2r\sin(\theta-t)}{1+r^2-2r\cos(\theta-t)} f(t)dt,$$

so that $\tilde{U}(z)$ is a harmonic conjugate of $U(z)$, and $U(0) = 0$.

Then (taking $f(\theta+2\pi)$ equal to $f(\theta)$)

$$\tilde{f}(\theta) = \lim_{\varepsilon\to 0} \frac{1}{\pi} \int_{\varepsilon\le|t|\le\pi} \{f(\theta-t)/2\tan\tfrac{t}{2}\}\, dt$$

exists a.e., $\|\tilde{f}\|_2 \le \|f\|_2$, and

$$\tilde{U}(re^{i\theta}) = \frac{1}{2\pi}\int_{-\pi}^{\pi} \frac{1-r^2}{1+r^2-2r\cos(\theta-t)} \tilde{f}(t)dt,$$

so that $\tilde{U}(z) \to \tilde{f}(\theta)$ a.e. for $z \underset{\not\,}{\longrightarrow} e^{i\theta}$.

As mentioned in Chapter I, Section E.4, the Cauchy principal value defining $\tilde{f}(\theta)$ in the above theorem is usually written

$$\frac{1}{\pi}\int_{-\pi}^{\pi} \frac{f(\theta-t)}{2\tan t/2}\, dt.$$

Given $f(\theta) \in L_2(-\pi,\pi)$, it follows immediately from the theorem that $F(z) = U(z) + i\tilde{U}(z)$ is in H_2. The boundary values $F(e^{i\theta})$ satisfy $F(e^{i\theta}) = f(\theta) + i\tilde{f}(\theta)$ a.e., so if $f \in L_2$ is real, we have a way of constructing $F \in H_2$ with $\Re F(e^{i\theta}) = f(\theta)$ a.e.

Now we saw in Chapter III, Section C.2 that as long as $f \in L_1(-\pi,\pi)$ and $f(\theta + 2\pi) = f(\theta)$,

$$\tilde{f}(\theta) = \frac{1}{\pi} \int_{-\pi}^{\pi} \frac{f(\theta - t)}{2 \tan t/2} \, dt$$

exists and is finite a.e.

Definition. $\tilde{f}(\theta)$ is called the Hilbert transform of $f(\theta)$. Sometimes, by abuse of language, we call it the harmonic conjugate of $f(\theta)$, on account of the theorem quoted at the beginning of this section. It would be better to call it the function conjugate to $f(\theta)$.

The inequality $\|\tilde{f}\|_2 \leq \|f\|_2$ has a valid analogue for other L_p-spaces.

B. Hilbert transforms of L_p-functions $1 < p < \infty$.

1°. Lemma. Let $U(z)$, $V(z)$ be harmonic in $\{|z| < 1\}$, and let $\tilde{U}(z)$, $\tilde{V}(z)$ be their harmonic conjugates with $\tilde{U}(0) = 0$, $\tilde{V}(0) = 0$. Then, if $0 \leq r < 1$,

$$\int_{-\pi}^{\pi} U(re^{i\theta})\tilde{V}(re^{i\theta})\,d\theta = - \int_{-\pi}^{\pi} \tilde{U}(re^{i\theta})V(re^{i\theta})\,d\theta.$$

Proof. If

$$U(re^{i\theta}) = \sum_{-\infty}^{\infty} A_n r^{|n|} e^{in\theta}$$

$$V(re^{i\theta}) = \sum_{-\infty}^{\infty} B_n r^{|n|} e^{in\theta},$$

we have

$$\tilde{U}(re^{i\theta}) = \sum_{-\infty}^{\infty} (-i \operatorname{sgn} n)A_n r^{|n|} e^{in\theta},$$

$$\tilde{V}(re^{i\theta}) = \sum_{-\infty}^{\infty} (-i \operatorname{sgn} n)B_n r^{|n|} e^{in\theta}.$$

All the series here are absolutely convergent if $r < 1$, so that the desired relation may be verified directly by termwise integration.

2°. Theorem (Riesz). Let $f(\theta) \in L_p$, $1 < p < \infty$, (sic!) and let

$$U(re^{i\theta}) = \frac{1}{2\pi}\int_{-\pi}^{\pi} \frac{(1-r^2)f(t)dt}{1+r^2-2r\cos(\theta-t)},$$

with $\tilde{U}$ the harmonic conjugate of U satisfying $\tilde{U}(0) = 0$. Then (taking always f as 2π-periodic)

$$\tilde{f}(\theta) = \frac{1}{\pi}\lim_{\varepsilon\to 0}\int_{\varepsilon}^{\pi} \frac{f(\theta-t) - f(\theta+t)}{2\tan t/2}\,dt$$

exists a.e., $\tilde{f} \in L_p$, and

$$\tilde{U}(re^{i\theta}) = \frac{1}{2\pi}\int_{-\pi}^{\pi} \frac{(1-r)\tilde{f}(t)dt}{1+r^2-2r\cos(\theta-t)}$$

for $r < 1$. There is a constant K_p depending only on p with $\|\tilde{f}\|_p \le K_p\|f\|_p$.

Proof. For $p = 2$, this is the result quoted in Section A, so we may suppose $1 < p < 2$ or $2 < p < \infty$.

As remarked in Section A, existence of $\tilde{f}(\theta)$ a.e. was already proved in Chapter III, Section C.2 under the assumption that $f \in L_1$; this is certainly fulfilled for $f \in L_p$, $p > 1$. It is thus enough to show that

$$\int_{-\pi}^{\pi} |\tilde{U}(re^{i\theta})|^p \, d\theta \leq (K_p \|f\|_p)^p$$

for $0 \leq r < 1$, for then, by Chapter 1, Section C there is a $g \in L_p$, $\|g\|_p \leq K_p \|f\|_p$, with

$$\tilde{U}(re^{i\theta}) = \frac{1}{2\pi} \int_{-\pi}^{\pi} \frac{(1 - r^2) g(t) dt}{1 + r^2 - 2r \cos(\theta - t)},$$

and we can argue as in Chapter I, Section D.4 to show that $g(\theta) = \tilde{f}(\theta)$ a.e. (The argument of Chapter I, Section D.4 also furnishes a proof, valid under the present circumstances, that $\tilde{f}(\theta)$ *exists* a.e. - and is equal to $g(\theta)$. So in fact, we do not really *require* the theorem of Chapter III, Section C.2 at this point!)

We first prove that

$$\int_{-\pi}^{\pi} |\tilde{U}(re^{i\theta})|^p \, d\theta \leq (K_p \|f\|_p)^p, \qquad 0 \leq r < 1,$$

in the case that $1 < p < 2$. Afterwards, we'll use the lemma of 1° and a duality argument to obtain the same result for $2 < p < \infty$.

The following argument is due to Katznelson. (It is, however, essentially an elegant manner of presenting reasoning already found in the second edition of Zygmund's book.) *It is enough* to prove the desired inequality for the case $f(\theta) \geq 0$, for in the general case we can break up a real $f \in L_p$ into the *difference* of *two positive ones*, and use the triangle inequality for L_p-norms, together with the fact that

$$\|f\|_p^p = \|f_+\|_p^p + \|f_-\|_p^p \geq C_p(\|f_-\|_p + \|f_-\|_p)^p$$

for $f = f_+ - f_-$, if $f_+ \cdot f_- \equiv 0$ a.e.

Assuming $f(\theta) \geq 0$, and $f \not\equiv 0$!, write $F(z) = U(z) + i\tilde{U}(z)$ for $|z| < 1$. $\Re F(z) = U(z) > 0$ for $|z| < 1$, so $F(z)$ is free of zeros there, and we can take the analytic function $G(z) = (F(z))^p$. Since $F(0) = U(0) \geq 0$, we have, for $0 \leq r < 1$, by Cauchy's theorem

$$\int_{-\pi}^{\pi} \Re G(re^{i\theta})d\theta = 2\pi \Re G(0) = 2\pi(U(0))^p \geq 0.$$

Now, for given $r < 1$, we break up $[-\pi,\pi]$ into two complementary sets, E_1 and E_2. Take a γ, $0 < \gamma < \frac{\pi}{2}$, such that $\frac{\pi}{2} < p\gamma < \pi$. There is such a γ, because $1 < p < 2$. Then, let

$$E_1 = \{\theta;\ -\gamma \leq \arg F(re^{i\theta}) \leq \gamma\}$$

and

$$E_2 = \{\theta;\ \gamma < |\arg F(re^{i\theta})| \leq \tfrac{\pi}{2}\}.$$

Since $\Re F(z) > 0$, $|\arg F(z)| < \frac{\pi}{2}$, so $E_1 \cup E_2 = [-\pi,\pi]$.

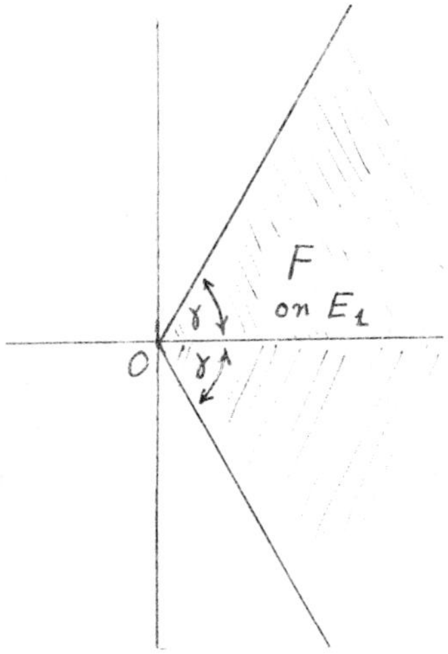

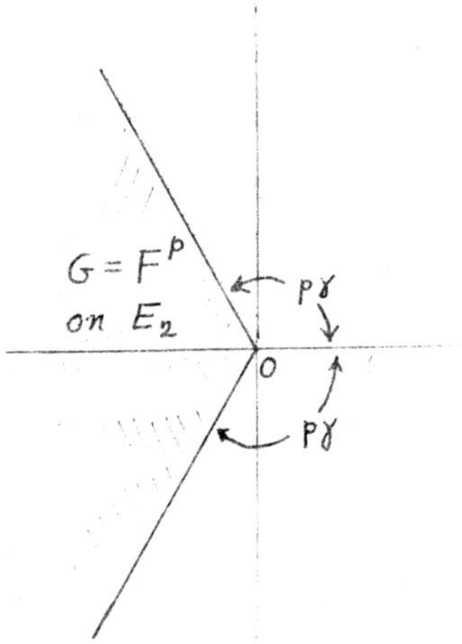

We now have

$$(*)\qquad \int_{E_1} \Re G(re^{i\theta})d\theta + \int_{E_2} \Re G(re^{i\theta})d\theta = \int_{-\pi}^{\pi} \Re G(re^{i\theta})d\theta \geq 0.$$

For $\theta \in E_2$, $\Re G(re^{i\theta})$ is <u>negative</u> and $\leq -|G(re^{i\theta})||\cos p\gamma|$ - look at the <u>second</u> of the above two diagrams. So by (*),

$$|\cos p\gamma| \int_{E_2} |G(re^{i\theta})|d\theta \leq \int_{E_1} \Re G(re^{i\theta})d\theta.$$

For $\theta \in E_1$ (<u>first</u> of above two diagrams!)

$$|F(re^{i\theta})| \leq |U(re^{i\theta})| / \cos \gamma,$$

so

$$|G(re^{i\theta})| \leq \cos^{-p}\gamma |U(re^{i\theta})|^p,$$

which, substituted in the previous, yields

$$\int_{E_2} |G(re^{i\theta})|d\theta \leq |\sec p\gamma| \int_{E_1} |G(re^{i\theta})|d\theta$$

$$\leq (\cos^{-p}\gamma)|\sec p\gamma| \int_{E_1} |U(re^{i\theta})|^p d\theta.$$

We also see that

$$\int_{E_1} |G(re^{i\theta})|d\theta \leq \cos^{-p}\gamma \int_{E_1} |U(re^{i\theta})|^p\, d\theta.$$

Since $E_1 \cup E_2 = [-\pi,\pi]$ and $E_1 \cap E_2 = \emptyset$, we have, adding the above two inequalities,

$$\int_{-\pi}^{\pi} |G(re^{i\theta})|d\theta \leq \frac{1+|\sec p\gamma|}{\cos^p\gamma} \int_{E_1} |U(re^{i\theta})|^p\, d\theta,$$

whence, a fortiori,

$$\int_{-\pi}^{\pi} |\tilde{U}(re^{i\theta})|^p \, d\theta \le \frac{1+|\sec p\gamma|}{\cos^p\gamma} \int_{-\pi}^{\pi} |U(re^{i\theta})|^p d\theta,$$

since $G(z) = [U(z) + i\tilde{U}(z)]^p$. Finally,

$$\int_{-\pi}^{\pi} |\tilde{U}(re^{i\theta})|^p d\theta \le \frac{1+|\sec p\gamma|}{\cos^p\gamma} \|f\|_p^p ,$$

by the properties of $P_r(\theta)$ (Chapter I, Section D.1), and we are done in case $1 < p < 2$.

Remark. This very clever idea is like the one used in Chapter IV, Section D.2 to show that $\int_{-\pi}^{\pi} |\log|F(re^{i\theta})||\,d\theta$ is bounded if $\int_{-\pi}^{\pi} \log^+|F(re^{i\theta})|\,d\theta$ is.

In case $2 < p < \infty$, we prove

$$\int_{-\pi}^{\pi} |\tilde{U}(re^{i\theta})|^p d\theta \le C_p \int_{-\pi}^{\pi} |U(re^{i\theta})|^p \, d\theta$$

in the following manner. Let $\frac{1}{q} = 1 - \frac{1}{p}$, then $1 < q < 2$, and by Hölder,

$$\|\tilde{U}(re^{i\theta})\|_p = \sqrt[p]{\int_{-\pi}^{\pi} |\tilde{U}(re^{i\theta})|^p \, d\theta}$$

is the supremum of $\left|\int_{-\pi}^{\pi} \tilde{U}(re^{i\theta})T(re^{i\theta})d\theta\right|$ taken over all finite sums

$$T(re^{i\theta}) = \sum_n B_n r^{|n|} e^{in\theta}$$

with

$$\|T(re^{i\theta})\|_q = \sqrt[q]{\int_{-\pi}^{\pi} |T(re^{i\theta})|^q d\theta} \le 1.$$

However, since $1 < q < 2$, for any such T, by what has just been proven,

$$\|\tilde{T}(re^{i\theta})\|_q \le K_q \|T(re^{i\theta})\|_q,$$

therefore, by 1°,

$$\left|\int_{-\pi}^{\pi} \tilde{U}(re^{i\theta})T(re^{i\theta})d\theta\right| = \left|\int_{-\pi}^{\pi} U(re^{i\theta})\tilde{T}(re^{i\theta})d\theta\right|$$

$$\le \|\tilde{T}(re^{i\theta})\|_q \ \|U(re^{i\theta})\|_p \le K_q \|T(re^{i\theta})\|_q \|U(re^{i\theta})\|_p$$

$$\le K_q \|U(re^{i\theta})\|_p,$$

and

$$\|\tilde{U}(re^{i\theta}\|_p \le K_q \|U(re^{i\theta})\|_p$$

for $2 < p < \infty$, as required. We are done.

C. $\tilde{f}(\theta)$ for $f \in L_1$. Let $f(\theta + 2\pi) = f(\theta)$, $f \in L_1(-\pi,\pi)$. By Chapter III, Section C.2, $\tilde{f}(\theta) = \frac{1}{2\pi}\int_{-\pi}^{\pi} \frac{f(\theta - t)}{\tan t/2}\, dt$ exists and is finite a.e.

1°. Definition. If h is a measurable function defined on $[-\pi,\pi]$, we write, for $\lambda \ge 0$,

$$m_h(\lambda) = |\{\theta \in [-\pi,\pi]; \ |h(\theta)| \ge \lambda\}|$$

Clearly $m_h(0) = 2\pi$, $m_h(\lambda)$ decreases, and $m_h(\lambda) \to 0$ as $\lambda \to \infty$ if $h(\theta)$ is finite a.e. m_h is called the distribution function of h.

Theorem (Kolmogorov). If $f \in L_1$ and $\lambda > 0$, $m_{\tilde{f}}(\lambda) \le \frac{K}{\lambda}\|f\|_1$, where K is a constant independent of f & λ.

Proof (Carleson, via Katznelson). Suppose $\tilde{f}(\theta) = g(\theta) + h(\theta)$. Then, if $|\tilde{f}(\theta)| \ge \lambda$ we must have $|g(\theta)| \ge \frac{\lambda}{2}$ or $|h(\theta)| \ge \frac{\lambda}{2}$, so

$m_{\tilde{f}}(\lambda) \leq m_g(\frac{\lambda}{2}) + m_h(\frac{\lambda}{2})$. Therefore it is enough to prove the theorem for the case where $f(\theta) \geq 0$, since, by the above observation, we can extend it therefrom to general f by taking a larger constant K.

So assume $f(\theta) \geq 0$; since harmonic conjugation is linear, we may, wlog, take $\|f\|_1$ to be 2π, then we're done if we show $m_{\tilde{f}}(\lambda) \leq \frac{4\pi}{\lambda}$.

For $|z| < 1$, write

$$F(z) = \frac{1}{2\pi}\int_{-\pi}^{\pi} \frac{e^{it} + z}{e^{it} - z} f(t)dt;$$

then $\Re F(z) > 0$, and, from Chapter III, Section C.2, for almost all θ,

$$F(z) \longrightarrow f(\theta) + i\tilde{f}(\theta) \quad \text{as} \quad z \underset{\measuredangle}{\longrightarrow} e^{i\theta}.$$

Given $\lambda > 0$, let

$$\varphi(z) = 1 + \frac{F(z) - \lambda}{F(z) + \lambda}.$$

Observe that $w \to 1 + \frac{w-\lambda}{w+\lambda}$ is the following conformal mapping:

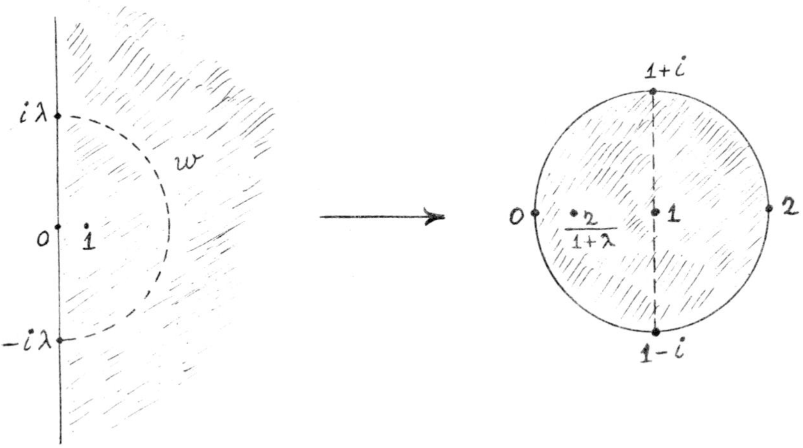

In this mapping, $0 \to 0$, $1 \to \frac{2}{1+\lambda}$, and the semi-circle joining $i\lambda$ to $-i\lambda$ goes onto the diameter from $1+i$ to $1-i$. We surely have $|\varphi(z)| \leq 2$, $|z| < 1$, so by Chapter II, Section B.1,

$$2\pi\varphi(0) = \int_{-\pi}^{\pi} \varphi(e^{i\theta})d\theta,$$

i.e.,

$$2\pi \Re\varphi(0) = \int_{-\pi}^{\pi} \Re\varphi(e^{i\theta})d\theta.$$

Now

$$F(0) = \frac{1}{2\pi}\int_{-\pi}^{\pi} f(t)dt = \frac{\|f\|_1}{2\pi} = 1$$

by assumption, since $f(t) \geq 0$, therefore, $\varphi(0) = \frac{2}{1+\lambda}$. Also $\Re\varphi(e^{i\theta}) \geq 0$. Finally, if $|\tilde{f}(\theta)| \geq \lambda$ we have $|F(e^{i\theta})| \geq \lambda$ so $\Re\varphi(e^{i\theta}) \geq 1$. Thus, from

$$\int_{-\pi}^{\pi} \Re\varphi(e^{i\theta})d\theta = \frac{4\pi}{1+\lambda}$$

we get $|\{\theta;\ \Re\varphi(e^{i\theta}) \geq 1\}| \leq \frac{4\pi}{1+\lambda}$, i.e. $m_{\tilde{f}}(\lambda) \leq \frac{4\pi}{1+\lambda}$, more than what was needed.

We are done.

<u>Scholium</u>. If $f \geq 0$ but $\|f\|_1 \neq 2\pi$, we can make a change of variable in the more precise result found at the end of the above proof to obtain

$$m_{\tilde{f}}(\lambda) \leq \frac{4\pi\|f\|_1}{\|f\|_1 + 2\pi\lambda}.$$

Finally, if f is merely <u>real</u>, $f = f_+ - f_-$ with f_+, $f_- \geq 0$, $\|f\|_1 = \|f_+\|_1 + \|f_-\|_1$, we use $m_{\tilde{f}}(\lambda) \leq m_{\tilde{f}_+}(\frac{\lambda}{2})$ and the result just found to get

$$m_{\tilde{f}}(\lambda) \le \frac{4\pi\|f_+\|_1}{\|f_+\|_1 + \pi\lambda} + \frac{4\pi\|f_-\|_1}{\|f_-\|_1 + \pi\lambda} .$$

Observe that for fixed λ, $\frac{x}{x+\pi\lambda}$ is an increasing function of $x \ge 0$, so, for f real in L_1,

$$\boxed{m_{\tilde{f}}(\lambda) \le \frac{8\pi\|f\|_1}{\|f\|_1 + \pi\lambda}} .$$

Sometimes this more accurate version of the above theorem is useful.

2°. We use the result in 1° together with that quoted in Section A to give a new proof of Riesz' Theorem (Section B.2) in order to show an example of Marcinkiewicz interpolation for linear operations. As the duality argument at the end of Section B.2 shows, the version proved below actually covers all values of p, $1 < p < \infty$, if we throw in the easy case $p = 2$, given in Section A.

Theorem (Riesz). If $1 < p < 2$ there is a constant K_p depending on p with

$$\|\tilde{U}(re^{i\theta})\|_p \le K_p \|U(re^{i\theta})\|_p$$

if $r < 1$ and $U(z)$ is harmonic in $\{|z| < 1\}$.

Proof. Write $f(\theta) = U(re^{i\theta})$; then $\tilde{f}(\theta) = \tilde{U}(re^{i\theta})$. Here, f and $\tilde{f}$ are both continuous, so by the definition of integrals,

$$\int_{-\pi}^{\pi} |f(\theta)|^p d\theta = -\int_0^{\infty} \lambda^p dm_f(\lambda),$$

$$\int_{-\pi}^{\pi} |\tilde{f}(\theta)|^p d\theta = -\int_0^{\infty} \lambda^p dm_{\tilde{f}}(\lambda).$$

Our job is to estimate the <u>second</u> integral in terms of the <u>first</u>. Since $\tilde{f}(\theta)$ is here <u>continuous</u>, $m_{\tilde{f}}(\lambda) = 0$ for λ large enough, and integration by parts gives

$$\int_{-\pi}^{\pi} |\tilde{f}(\theta)|^p d\theta = p \int_0^\infty \lambda^{p-1} m_{\tilde{f}}(\lambda) d\lambda.$$

<u>Suppose</u> $f = g + h$. Then, as we argued in 1°, $\tilde{f} = \tilde{g} + \tilde{h}$, so

$$m_{\tilde{f}}(\lambda) \le m_{\tilde{g}}\left(\frac{\lambda}{2}\right) + m_{\tilde{h}}\left(\frac{\lambda}{2}\right).$$

<u>Marcinkiewicz had the crazy idea of using this inequality to estimate</u> $m_{\tilde{f}}(\lambda)$ <u>by breaking up</u> f <u>as</u> $g + h$ <u>in a way depending on</u> λ !

<u>Given</u> $\lambda > 0$, <u>take</u>:

$$g(\theta) = \begin{cases} f(\theta) & \text{if } |f(\theta)| < \lambda \\ 0 & \text{otherwise}. \end{cases}$$

$$h(\theta) = \begin{cases} 0 & \text{if } |f(\theta)| < \lambda \\ f(\theta) & \text{if } |f(\theta)| \ge \lambda. \end{cases}$$

The idea here is that g is <u>not large</u>, so $g \in L_2$.

Indeed, by Section A,

$$\|\tilde{g}\|_2^2 \le \|g\|_2^2 = \int_{|f(\theta)| < \lambda} |f(\theta)|^2 d\theta = -\int_0^\lambda s^2 dm_f(s),$$

so

$$m_{\tilde{g}}\left(\frac{\lambda}{2}\right) \le \frac{4}{\lambda^2} \|\tilde{g}\|_2^2 \le -\frac{4}{\lambda^2} \int_0^\lambda s^2 dm_f(s).$$

For h, we use the result of 1°, which says that

$$m_{\tilde{h}}\left(\frac{\lambda}{2}\right) \le \frac{2K}{\lambda} \|h\|_1 = \frac{2K}{\lambda} \int_{|f(\theta)| \ge \lambda} |f(\theta)| d\theta = -\frac{2K}{\lambda} \int_\lambda^\infty s \, dm_f(s).$$

Therefore, finally,

$$\boxed{m_{\tilde f}(\lambda) \le -\frac{4}{\lambda^2}\int_0^\lambda s^2 dm_f(s) - \frac{2K}{\lambda}\int_\lambda^\infty s\, dm_f(s).}$$

Now we have:

$$\int_{-\pi}^{\pi} |\tilde f(\theta)|^p d\theta = p\int_0^\infty \lambda^{p-1} m_{\tilde f}(\lambda)d\lambda$$

$$\le -4p\int_0^\infty \lambda^{p-3}\int_0^\lambda s^2 dm_f(s)d\lambda - 2K_p\int_0^\infty \lambda^{p-2}\int_\lambda^\infty s\, dm_f(s)d\lambda .$$

By Fubini's theorem:

$$-\int_0^\infty \lambda^{p-3}\int_0^\lambda s^2 dm_f(s)d\lambda = -\int_0^\infty\int_s^\infty \lambda^{p-3}s^2 d\lambda dm_f(s)$$

$$= -\frac{1}{2-p}\int_0^\infty s^{p-2}s^2 dm_f(s)$$

$$= -\frac{1}{2-p}\int_0^\infty s^p dm_f(s) = \frac{\|f\|_p^p}{2-p},$$

because $-2 < p-3 < -1$.

Similarly,

$$-\int_0^\infty \lambda^{p-2}\int_\lambda^\infty s\, dm_f(s)d\lambda = -\int_0^\infty\int_0^s \lambda^{p-2} s\, d\lambda\, dm_f(s)$$

$$= -\frac{1}{p-1}\int_0^\infty s^p dm_f(s) = \frac{\|f\|_p^p}{p-1},$$

because $-1 < p-2 < 0$.

So finally,

$$\|\tilde f\|_p^p \le 2p\left[\frac{2}{2-p} + \frac{K}{p-1}\right] \|f\|_p^p.$$

Q.E.D.

3°. It turns out that $f \in L_1$ is <u>not enough</u> to guarantee $\tilde{f} \in L_1$. The theorem which applies here is due to <u>Zygmund</u>:

<u>Theorem</u>. There is a constant c with

$$\int_{-\pi}^{\pi} |\tilde{f}(\theta)|\,d\theta \le c \int_{-\pi}^{\pi} |f(\theta)|(1 + \log^+|f(\theta)|)\,d\theta + 2\pi,$$

and a similar inequality holds with $f(\theta)$ replaced by $U(re^{i\theta})$ & $\tilde{f}(\theta)$ replaced by $\tilde{U}(re^{i\theta})$.

<u>Remark</u>. $x(1+\log^+ x)$ is <u>convex</u> for $x > 0$ so, since $P_r \ge 0$ & $\frac{1}{2\pi}\int_{-\pi}^{\pi} P_r(t)\,dt = 1$, if $U(re^{i\theta}) = \frac{1}{2\pi}\int_{-\pi}^{\pi} P_r(\theta - t)f(t)\,dt$, we have, for $r < 1$, by Jensen's inequality,

$$\int_{-\pi}^{\pi} |U(re^{i\theta})|(1 + \log^+|U(re^{i\theta})|)\,d\theta$$

$$\le \int_{-\pi}^{\pi} |f(\theta)|(1 + \log^+|f(\theta)|)\,d\theta.$$

Therefore $|f(\theta)|\log^+|f(\theta)| \in L_1(-\pi,\pi)$ <u>guarantees that</u>

$$F(z) = \frac{1}{2\pi}\int_{-\pi}^{\pi} \frac{e^{i\theta} + z}{e^{i\theta} - z} f(\theta)\,d\theta \in H_1.$$

<u>Proof of Theorem</u>. We use the estimate for $m_{\tilde{f}}(\lambda)$ obtained in 2° (boxed formula) to estimate

$$\int_{-\pi}^{\pi} |\tilde{f}(\theta)|\,d\theta = -\int_0^\infty \lambda\, dm_{\tilde{f}}(\lambda) = \int_0^\infty m_{\tilde{f}}(\lambda)\,d\lambda \le 2\pi + \int_1^\infty m_{\tilde{f}}(\lambda)\,d\lambda.$$

($m_{\tilde{f}}(\lambda) \le 2\pi$ for all $\lambda \ge 0$!). We have, by the formula in 2°,

$$\int_1^\infty m_{\tilde{f}}(\lambda)\,d\lambda \le -4\int_1^\infty \lambda^{-2}\int_0^\lambda s^2\,dm_f(s)\,d\lambda - 2K\int_\lambda^\infty \lambda^{-1}\int_\lambda^\infty s\,dm_f(s)\,d\lambda.$$

Since $dm_f(s) \leq 0$, this is

$$\leq -4\int_0^\infty \int_0^\lambda \frac{s^2}{\lambda^2}\, dm_f(s)d\lambda - 2K\int_1^\infty \int_1^s \frac{s}{\lambda}\, d\lambda\, dm_f(s)$$

$$= -4\int_0^\infty \int_s^\infty \frac{s^2}{\lambda^2}\, d\lambda\, dm_f(s) - 2K\int_1^\infty s\log s\, dm_f(s)$$

$$= -\int_0^\infty (4s + 2Ks\ \log^+ s)dm_f(s)$$

$$= -\int_{-\pi}^{\pi} |f(\theta)|(4 + 2K\log^+|f(\theta)|)d\theta.$$

This does it.

4°. In case $f(\theta) \geq 0$, the above theorem has a converse!

Theorem (Zygmund). Let $f(\theta) \geq 0$, let $F(z) = \frac{1}{2\pi}\int_{-\pi}^{\pi}\frac{e^{it}+z}{e^{it}-z}f(t)dt$, and suppose $F \in H_1$. Then

$$|f(\theta)|\log^+|f(\theta)| \in L_1(-\pi,\pi).$$

Proof. Let $G(z) = 1 + F(z)$, then $\Re G(z) > 1$, $|z| < 1$ and $G(0) = 1 + F(0)$ is real. So we can define a regular branch of $\log G(z)$ in $\{|z| < 1\}$ with $\log G(0)$ real and > 0.

By Cauchy's theorem, for $r < 1$,

$$G(0)\log G(0) = \frac{1}{2\pi}\int_{-\pi}^{\pi} G(re^{i\theta})\log G(re^{i\theta})\, d\theta,$$

so,

$$\Re G(0)\log G(0) = \frac{1}{2\pi}\int_{-\pi}^{\pi} [\Re G(re^{i\theta})\log|G(re^{i\theta})| - \Im G(re^{i\theta})\arg G(re^{i\theta})]d\theta.$$

In the present case, $\Re G(z) > 0$, so $-\frac{\pi}{2} < \arg G(z) < \frac{\pi}{2}$, and

$$\int_{-\pi}^{\pi} \Re G(re^{i\theta})\log|G(re^{i\theta})|\,d\theta \le 2\pi\,\Re G(0)\log G(0) + \frac{\pi}{2}\int_{-\pi}^{\pi} |\Im G(re^{i\theta})|\,d\theta.$$

As $r \to 1$,

$$\Re G(re^{i\theta})\log|G(re^{i\theta})| \longrightarrow [1 + f(\theta)]\log|1 + f(\theta) + i\tilde{f}(\theta)|$$

a.e., whilst

$$\int_{-\pi}^{\pi} |\Im G(re^{i\theta})|\,d\theta \le 2\pi + \|F\|_1.$$

So by Fatou's lemma,

$$\int_{-\pi}^{\pi} (1 + f(\theta))\log|1 + f(\theta) + i\tilde{f}(\theta)|\,d\theta$$

$$\le 2\pi[1 + \|F\|_1]\log[1 + \|F\|_1] + 2\pi + \|F\|_1,$$

using a trivial estimate for $G(0) = 1 + F(0)$. Since $f(\theta) > 0$, this is enough.

5°. Using 4°, it is <u>easy</u> to construct examples of $f \in L_1$ <u>with</u> $\tilde{f}(\theta)$ <u>not in</u> L_1 <u>on any interval</u>. However,

<u>Theorem</u> (Kolmogorov). If $f \in L_1$, $\int_{-\pi}^{\pi} |\tilde{f}(\theta)|^p d\theta < \infty$ whenever $0 < p < 1$, so $F(z) = \frac{1}{2\pi}\int_{-\pi}^{\pi} \frac{e^{it}+z}{e^{it}-z} f(t)\,dt$ belongs to H_p for $0 < p < 1$.

<u>Proof</u>. By the result in 1°, if $0 < p < 1$,

$$\int_{-\pi}^{\pi} |\tilde{f}(\theta)|^p d\theta = -\int_0^\infty \lambda^p dm_{\tilde{f}}(\lambda) = p\int_0^\infty \lambda^{p-1} m_{\tilde{f}}(\lambda)\,d\lambda \le 2\pi p \;+$$

$$+\; p\int_1^\infty \lambda^{p-1} m_{\tilde{f}}(\lambda)\,d\lambda \le 2\pi p + K_p\|f\|_1 \int_1^\infty \lambda^{p-2}\,d\lambda =$$

$$= 2\pi p + K_p\|f\|_1 \cdot \frac{1}{1-p}\,.$$

Q.E.D.

Remark. Using the inequality actually proved in 1°, and given in the Scholium of that subsection,

$$m_{\tilde{f}}(\lambda) \le \frac{8\pi\|f\|_1}{\|f\|_1 + \pi\lambda}$$

(valid for real f), we obtain a better inequality of the form

$$\int_{-\pi}^{\pi} |\tilde{f}(\theta)|^p d\theta \le C_p \|f\|_1^p, \qquad 0 < p < 1.$$

Indeed,

$$\int_{-\pi}^{\pi} |\tilde{f}(\theta)|^p d\theta = p\int_0^\infty \lambda^{p-1} m_{\tilde{f}}(\lambda)d\lambda \le 8\pi p \int_0^\infty \frac{\lambda^{p-1}\|f\|_1 d\lambda}{\|f\|_1 + \pi\lambda}$$

$$= 8\pi p\|f\|_1^p \int_0^\infty \frac{s^{p-1}\,ds}{1+\pi s},$$

the integral being finite for $0 < p < 1$.

There is, however, a much easier way to establish the same result. Taking, wlog and as usual, $f(\theta) \ge 0$, form the analytic function

$$F(z) = \frac{1}{2\pi}\int_{-\pi}^{\pi} \frac{e^{it} + z}{e^{it} - z} f(t)dt,$$

then apply Cauchy's theorem to $(F(re^{i\theta}))^p$, $0 < p < 1$, observing that $|\arg F(z)| \le \frac{\pi}{2}$. The desired inequality comes out quite easily.

D. $\tilde{f}(\theta)$ for bounded f and for continuous periodic $f(\theta)$.

1°. Theorem. If $f(\theta)$ is real and $\|f\|_\infty = \operatorname{ess\,sup}_{-\pi\le\theta\le\pi} |f(\theta)|$ is $\le \frac{\pi}{2}$, then $\int_{-\pi}^{\pi} e^{\lambda|\tilde{f}(\theta)|} d\theta \le C_\lambda$, a constant depending on λ alone, for each $\lambda < 1$.

Remark. Thus, although $f \in L_\infty$ does not imply $\tilde{f} \in L_\infty$, $\tilde{f}$ still has very strong boundedness properties.

Remark. The theorem cannot be improved to cover the case $\lambda = 1$, as is seen by considering the example where

$$f(\theta) = \frac{\pi}{2} \times \text{the characteristic function of an interval.}$$

Proof. Write

$$F(z) = \frac{1}{2\pi}\int_{-\pi}^{\pi} \frac{e^{it}+z}{e^{it}-z} f(t)dt,$$

then, if f is real & $|f(t)| \leq \frac{\pi}{2}$, $|\Re F(z)| \leq \frac{\pi}{2}$ and $\Im F(0) = 0$, hence, if $0 \leq \lambda < 1$, by Cauchy,

$$\int_{-\pi}^{\pi} e^{-i\lambda F(re^{i\theta})} d\theta = 2\pi e^{-i\lambda F(0)}.$$

Taking real parts,

$$\int_{-\pi}^{\pi} \Re e^{-i\lambda F(re^{i\theta})} d\theta \leq 2\pi |e^{-i\lambda F(0)}| = 2\pi.$$

But

$$|\arg e^{-i\lambda F(re^{i\theta})}| \leq \frac{\pi}{2}\lambda,$$

so

$$\int_{-\pi}^{\pi} |e^{-i\lambda F(re^{i\theta})}| \cos\frac{\pi}{2}\lambda\, d\theta = \int_{-\pi}^{\pi} e^{\lambda \Im F(re^{i\theta})} \cos\frac{\pi}{2}\lambda\, d\theta$$

$$\leq \int_{-\pi}^{\pi} |e^{-i\lambda F(re^{i\theta})}| d\theta \leq 2\pi,$$

and finally,

$$\int_{-\pi}^{\pi} \exp(\lambda \Im F(re^{i\theta}))d\theta \leq 2\pi/\cos\frac{\pi}{2}\lambda.$$

Similarly, using $\exp(i\lambda F(z))$;

$$\int_{-\pi}^{\pi} \exp(-\lambda \Im F(re^{i\theta}))d\theta \leq 2\pi/\cos\frac{\pi}{2}\lambda.$$

Since $\Im F(re^{i\theta}) \to \tilde{f}(\theta)$ a.e. as $r \to 1$, the theorem follows by Fatou's lemma.

2°. Corollary. If $f(\theta)$ is continuous of period 2π, and real, $\exp \lambda|\tilde{f}(\theta)|$ is in L_1 for all λ.

Proof. Given $\varepsilon > 0$, we can find a finite sum $S(\theta) = \sum_n A_n e^{in\theta}$ with $g(\theta) = f(\theta) - S(\theta)$ satisfying $\|g(\theta)\|_\infty \le \varepsilon$. Then $\exp \lambda|\tilde{g}(\theta)|$ is in L_1 for $\lambda < \frac{\pi}{2\varepsilon}$. But $\tilde{f}(\theta) = \tilde{g}(\theta) + \tilde{S}(\theta)$, and $\tilde{S}(\theta)$, being a finite sum of the same form as $S(\theta)$, is bounded.

Q.E.D.

E. Lipschitz classes

It is useful to know:

Theorem. Let $f(\theta)$ be 2π-periodic and Lip α, $0 < \alpha < 1$. Then $\tilde{f}(\theta)$ is Lip α.

Proof. First of all, if $h > 0$,

$$\left|\tilde{f}(\theta) - \frac{1}{\pi}\int_h^\pi \frac{f(\theta-t) - f(\theta+t)}{\tan t/2}\,dt\right| \le \frac{2}{\pi}\int_0^h \frac{|f(\theta-t) - f(\theta+t)|}{t}\,dt$$
$$\le \frac{2}{\pi}\int_0^h \frac{ct^\alpha}{t}\,dt = O(h^\alpha)$$

for $0 < \alpha < 1$. So we may work with

$$\frac{1}{\pi}\int_{2h}^\infty \frac{f(\theta - t) - f(\theta + t)}{\tan t/2}\,dt,$$

say, instead of $\tilde{f}(\theta)$, if we wish to estimate $\tilde{f}(\theta + h) - \tilde{f}(\theta)$.

We may also assume $f(\theta) = 0$, since $\tilde{f}(\varphi)$ and

$$\frac{1}{\pi}\int_{2h}^\pi [f(\varphi - t) - f(\varphi + t)]\cot\frac{t}{2}\,dt$$

don't change if $f(s)$ is replaced by $f(s) - f(\theta)$. Thus, we may assume $|f(\theta - t)| \le Ct^\alpha$. Now

$$\int_{2h}^{\pi} \frac{f(\theta+h-t) - f(\theta+h+t)}{\tan t/2}\, dt - \int_{2h}^{\infty} \frac{f(\theta-t) - f(\theta+t)}{\tan t/2}\, dt$$

$$= \int_{2h}^{\pi} \left[\frac{f(\theta+h-t)}{\tan t/2} - \frac{f(\theta-t)}{\tan t/2}\right] dt + \int_{2h}^{\pi} \frac{f(\theta+t) - f(\theta+h+t)}{\tan t/2}\, dt.$$

We evaluate the first integral on the right, the second being handled similarly. The first integral is

$$\int_{h}^{2h} \frac{f(\theta-s)}{\tan\frac{1}{2}(s+h)}\, ds + \int_{2h}^{\pi-h} \frac{f(\theta-s)}{\tan\frac{1}{2}(s+h)}\, ds - \int_{2h}^{\pi-h} \frac{f(\theta-t)}{\tan t/2}\, dt$$

$$- \int_{\pi-h}^{\pi} \frac{f(\theta-t)}{\tan t/2}\, dt =$$

$$= \Theta(h^{\alpha}) + \int_{2h}^{\pi-h} f(\theta-t) \frac{\tan\frac{1}{2}t - \tan\frac{1}{2}(t+h)}{(\tan\frac{1}{2}t)(\tan\frac{1}{2}(t+h))}\, dt + \Theta(h)$$

since $|f(\theta-t)| \leq \Theta(t^{\alpha})$. The integral is

$$\leq \text{const} \cdot \int_{2h}^{\pi} t^{\alpha} \cdot \frac{h}{t(t+h)}\, dt \leq \text{const} \cdot \int_{2h}^{\infty} \frac{ht^{\alpha} dt}{t^2} = \text{const} \cdot h^{\alpha}$$

since $0 < \alpha < 1$.

We are done.

VI. H_p Spaces for the Upper Half Plane

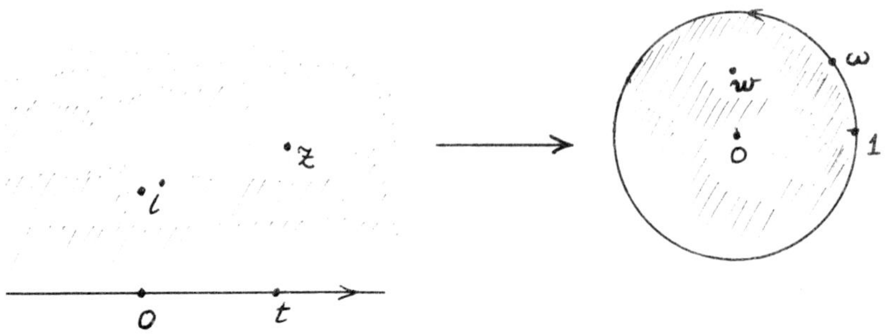

In order to study certain classes of functions analytic or harmonic in $\Im z > 0$ we make the conformal mapping

$$z \longrightarrow w = \frac{i - z}{i + z}$$

onto the unit circle $\{|w| < 1\}$ and apply the results already obtained in previous lectures.

A. Poisson's formula for the half-plane. If $\Im z > 0$ and t is real, z corresponds in the above conformal mapping to

$$w = \frac{i - z}{i + z} = re^{i\theta},$$

say, and t corresponds to

$$\omega = \frac{i - t}{i + t} = e^{i\tau},$$

say.

We want to work out the Poisson kernel

$$\frac{1-r^2}{1+r^2-2r\cos(\theta-\tau)}\,d\tau = \frac{1-|w|^2}{|w-\omega|^2}\,\frac{d\omega}{i\omega}.$$

$$\omega = \frac{2i}{i+t} - 1, \qquad \therefore\ d\omega = -\frac{2i\ dt}{(i+t)^2},$$

$$\& \qquad \frac{d\omega}{i\omega} = -\frac{2\ dt}{(i+t)(i-t)} = \frac{2\ dt}{t^2+1}.$$

Thus,

$$\frac{1-|w|^2}{|w-\omega|^2}\cdot\frac{d\omega}{i\omega} = \frac{1-\left|\frac{i-z}{i+z}\right|^2}{\left|\frac{i-z}{i+z}-\frac{i-t}{i+t}\right|^2}\cdot\frac{2\,dt}{t^2+1} =$$

$$= \frac{\dfrac{|z+i|^2-|z-i|^2}{|z+i|^2}}{\dfrac{|(i+t)(i-z)-(i-t)(i+z)|^2}{|z+i|^2\,|t+i|^2}}\cdot\frac{2\,dt}{t^2+1} =$$

$$= \frac{4y}{|2i(z-t)|^2}\cdot 2\ dt = \frac{2y\ dt}{|z-t|^2}.$$

Then by theorems of Chapter I

Theorem. Let $V(z)$ be harmonic and bounded for $\Im z > 0$. Then the limit $V(t) = \lim V(z)$ for $z \xrightarrow[\not\ ]{} t$ exists a.e. for t on $\mathbb{R}$, and for $\Im z > 0$

$$V(z) = \frac{1}{\pi}\int_{-\infty}^{\infty}\frac{y}{(x-t)^2+y^2}\,V(t)dt.$$

Theorem. Let $V(z)$ be harmonic and ≥ 0 for $\Im z > 0$. Then there is a constant $\alpha \geq 0$ and a measure $\mu \geq 0$ on $\mathbb{R}$ with $\int_{-\infty}^{\infty}\frac{d\mu(t)}{1+t^2} < \infty$, such that, for $\Im z > 0$,

$$V(z) = \alpha y + \frac{1}{\pi}\int_{-\infty}^{\infty}\frac{y\ d\mu(t)}{(x-t)^2+y^2}.$$

Proof. If, in the correspondence $z \longrightarrow w = \frac{i-z}{i+z}$, we let $U(w) = V(z)$, then we have

$$U(re^{i\theta}) = \frac{1}{2\pi}\int_{-\pi}^{\pi} \frac{1 - r^2}{1 + r^2 - 2r\cos(\theta-\tau)}\, d\nu(\tau)$$

where ν is a (finite) positive measure on $(-\pi,\pi)$. For

$$e^{i\tau} = \frac{i - t}{i + t},$$

$d\mu(t)$ is given by

$$d\mu(t) = \frac{1}{2}(1 + t^2)\, d\nu(\tau).$$

α is simply the point mass $\frac{1}{2\pi}\{\nu(-\pi) + \nu(\pi)\}$, if there be any.

Remark. It is frequent to think of α as "the mass at ∞".

What is the harmonic conjugate of

$$V(z) = \frac{1}{\pi}\int_{-\infty}^{\infty} \frac{y}{(x-t)^2 + y^2}\, d\mu(t) \ ?$$

Observe that

$$\frac{y}{(x-t)^2 + y^2} + \frac{i(x-t)}{(x-t)^2 + y^2} = \frac{i}{z - t}$$

is, for each $t \in \mathbb{R}$, analytic in $\Im z > 0$.

So one choice for the harmonic conjugate of $V(z)$ would be

$$\frac{1}{\pi}\int_{-\infty}^{\infty} \frac{x - t}{(x-t)^2 + y^2}\, d\mu(t).$$

This particular choice is frequently used.

The above formula is applicable as long as

$$\int_{-\infty}^{\infty} \frac{|d\mu(t)|}{1+|t|} < \infty .$$

But in the most general situations, <u>all we know is that</u>

$$\int_{-\infty}^{\infty} \frac{|d\mu(t)|}{1+t^2} < \infty .$$

(Cf. above Theorem for the Poisson representation of <u>positive harmonic functions</u>.)

<u>In that case we use the harmonic conjugate</u>

$$\frac{1}{\pi}\int_{-\infty}^{\infty}\left(\frac{x-t}{(x-t)^2+y^2}+\frac{t}{t^2+1}\right)d\mu(t).$$

<u>Here, the integral is absolutely convergent as long as</u>

$$\int_{-\infty}^{\infty} \frac{|d\mu(t)|}{1+t^2} < \infty .$$

<u>So, for</u> $V(z) = \frac{1}{\pi}\int_{-\infty}^{\infty} \frac{y}{(x-t)^2+y^2}\, d\mu(t)$ <u>we have</u>

<u>the following two choices for the harmonic conjugate</u>

$$\frac{1}{\pi}\int_{-\infty}^{\infty} \frac{x-t}{(x-t)^2+y^2}\, d\mu(t)$$

& $\frac{1}{\pi}\int_{-\infty}^{\infty}\left(\frac{x-t}{(x-t)^2+y^2}+\frac{t}{t^2+1}\right)d\mu(t) .$

B. Boundary behaviour.

Theorem. Let $V(z) = \frac{1}{\pi}\int_{-\infty}^{\infty} \frac{y}{(x-t)^2+y^2}\,d\mu(t)$, μ a signed measure on $\mathbb{R}$ with $\int_{-\infty}^{\infty} \frac{|d\mu(t)|}{1+t^2} < \infty$. Then, at any t_0 where $\mu'(t_0)$ exists and is finite, $V(z) \longrightarrow \mu'(t_0)$ as $z \underset{\measuredangle}{\longrightarrow} t_0$.

Proof. If $\frac{i-z}{i+z} = re^{i\theta}$, $\frac{i-t}{i+t} = e^{i\tau}$, then

$$V(z) = u(re^{i\theta}) = \frac{1}{2\pi}\int_{-\pi}^{\pi} \frac{1-r^2}{1+r^2-2r\cos(\theta-\tau)}\,d\nu(\tau),$$

where the relation between ν & μ is given by

$$d\nu(\tau) = \frac{2\,d\mu(t)}{1+t^2}.$$

We also have $d\tau = \frac{2\,dt}{1+t^2}$, so $\frac{d\nu(\tau)}{d\tau} = \frac{d\mu(t)}{dt}$. By Chapter I, wherever $\nu'(\tau_0)$ exists and is finite,

$$u(w) \longrightarrow \nu'(\tau_0) \quad \text{for } w \underset{\measuredangle}{\longrightarrow} e^{i\tau_0}.$$

So our result follows by direct carrying-over.

Remark. It is also instructive to go through a direct proof of this statement, keeping the kernel $y/[(x-t)^2+y^2]$ but following the procedure of Chapter I (integration by parts). It will be found that the computational details are simpler here!

In fact, this is generally the case - so much so that frequently results for $\{|w| < 1\}$ are proved by first going over to the half-plane $\Im z > 0$.

$$\frac{y}{(x-t)^2 + y^2} \quad \& \quad \frac{x-t}{(x-t)^2 + y^2}$$

are just easier to handle than

$$\frac{1 - r^2}{1 + r^2 - 2r\cos(\theta - \tau)} \quad \& \quad \frac{2r\sin(\theta - \tau)}{1 + r^2 - 2r\cos(\theta - \tau)}$$

respectively!

Similarly, by direct translation from Chapter I:

Theorem. Let $V(z) = \frac{1}{\pi}\int_{-\infty}^{\infty} \frac{y}{(x-t)^2 + y^2}\, d\mu(t)$, where $\int_{-\infty}^{\infty} \frac{|d\mu(t)|}{1 + t^2} < \infty$. If $\mu'(t_0)$ exists and is infinite,

$$V(t_0 + iy) \longrightarrow \mu'(t_0) \quad \text{as} \quad y \longrightarrow 0+.$$

Theorem. Let $\int_{-\infty}^{\infty} \frac{|F(t)|\,dt}{1 + t^2} < \infty$ and let

$$\tilde{V}(z) = \frac{1}{\pi}\int_{-\infty}^{\infty} \left(\frac{x - t}{(x-t)^2 + y^2} + \frac{t}{t^2 + 1} \right) F(t)\,dt.$$

Then, for almost all $x \in \mathbb{R}$,

$$\frac{1}{\pi}\int_{|t-x| \geq y} \left(\frac{1}{x - t} + \frac{t}{t^2 + 1} \right) F(t)\,dt - \tilde{V}(x + iy) \longrightarrow 0$$

as $y \longrightarrow 0$.

Remark. In case $\int_{-\infty}^{\infty} \frac{|t\,F(t)|}{t^2 + 1}\, dt < \infty$ we have simply

$$\frac{1}{\pi}\int_{-\infty}^{\infty} \frac{x-t}{(x-t)^2+y^2} F(t)dt - \frac{1}{\pi}\int_{|t-x|\geq y} \frac{F(t)}{x-t}\, dt \longrightarrow 0 \quad \text{a.e.}$$

as $y \to 0$.

Thus, existence of the principal value

$$\frac{1}{\pi}\int_{-\infty}^{\infty} \frac{F(t)}{x-t}\, dt = \lim_{y\to 0} \frac{1}{\pi}\int_{|t-x|\geq y} \frac{F(t)}{x-t}\, dt$$

is equivalent almost everywhere to existence of

$$\lim_{y\to 0} \frac{1}{\pi}\int_{-\infty}^{\infty} \frac{x-t}{(x-t)^2+y^2} F(t)dt.$$

There is a coarser kind of boundary behaviour which is easy to verify here.

Note that

$$\frac{1}{\pi}\int_{-\infty}^{\infty} \frac{y}{(x-t)^2+y^2}\, dt = 1,$$

that

$$\frac{y}{(x-t)^2+y^2} \geq 0,$$

and that

$$\frac{1}{\pi}\int_{|t-x|\geq\delta} \frac{y}{(x-t)^2+y^2}\, dt \longrightarrow 0 \quad \text{as} \quad y \longrightarrow 0$$

for any $\delta > 0$.

From this we get easily, by the usual arguments:

Theorem. If $\varphi(t)$ is bounded on $\mathbb{R}$,

$$\frac{1}{\pi}\int_{-\infty}^{\infty} \frac{y}{(x-t)^2+y^2} \varphi(t)dt \longrightarrow \varphi(x_0)$$

for $z \to x_0$, x_0 being any point of continuity of φ. If $\varphi(t)$ is uniformly continuous, the convergence is uniform.

Theorem. Let μ be a totally finite measure on $\mathbb{R}$ and write

$$d\mu_y(x) = \left\{\frac{1}{\pi}\int_{-\infty}^{\infty} \frac{y}{(x-t)^2+y^2}\, d\mu(t)\right\} dx,$$

$y > 0$. Then $d\mu_y(x) \longrightarrow d\mu(x)$ w^* as $y \longrightarrow 0$.

(Proof. By preceding and duality!)

Theorem. Let $F \in L_p(-\infty, \infty)$, $1 \leq p < \infty$. Let

$$F_y(x) = \frac{1}{\pi}\int_{-\infty}^{\infty} \frac{y}{(x-t)^2+y^2}\, F(t)dt.$$

Then $\|F_y - F\|_p \longrightarrow 0$ as $y \longrightarrow 0$.

C. The H_p-spaces for $\Im z > 0$.

Definition. $F(z)$, analytic for $\Im z > 0$, is said to belong to $H_p(\Im z > 0)$ or, just H_p if we know we are dealing with the upper half plane, provided that there is a constant $C < \infty$ with

$$\int_{-\infty}^{\infty} |F(x+iy)|^p dx \leq C$$

for all $y > 0$. We use this definition for all $p > 0$.

Remark. $H_p(\Im z > 0)$ is NOT just obtained from H_p for $\{|w| < 1\}$ by conformal mapping! There is a factor (depending on p) which also comes in!

Lemma. If $F \in H_p$, $|F(z)| \leq \dfrac{C}{y^{1/p}}$ with a constant C depending on F.

Proof.

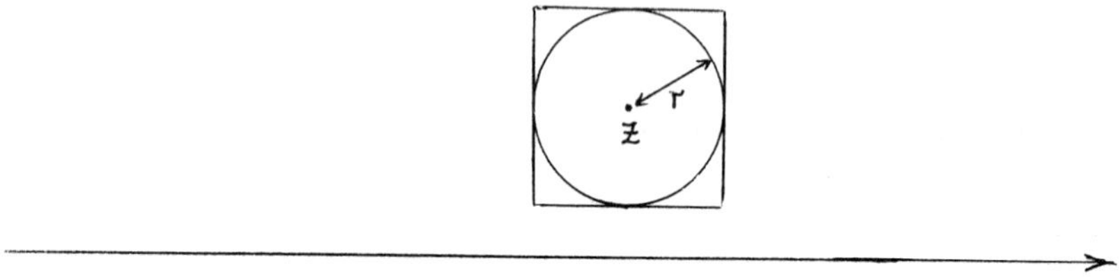

If $\Im z > 0$, by subharmonicity of $|F(x)|^p$ for $p > 0$,

$$|F(z)|^p \leq \frac{1}{2\pi} \int_0^{2\pi} |F(z + \rho e^{i\varphi})|^p \, d\varphi$$

for $0 < \rho < y$. Therefore, integrating ρ from 0 to $r < y$,

$$\frac{r^2}{2} |F(z)|^p \leq \frac{1}{2\pi} \int_0^r \int_0^{2\pi} |F(z + \rho e^{i\varphi})|^p \rho \, d\varphi d\rho$$

$$\leq \frac{1}{2\pi} \int_{x-r}^{x+r} \int_{y-r}^{y+r} |F(\xi + i\eta)|^p \, d\eta \, d\xi \quad \text{(picture)},$$

which, by hypothesis, is (changing the order of integration):

$$\leq \frac{1}{2\pi} \cdot C \cdot 2r = \frac{Cr}{\pi},$$

so $|F(z)|^p \leq \frac{2C}{\pi r}$. Make $r \longrightarrow y$. Done.

Lemma. Let $F \in H_p$, $p \geq 1$, and let $h > 0$. Then

$$F(z + ih) = \frac{1}{\pi} \int_{-\infty}^{\infty} \frac{y \, F(t + ih)}{(x - t)^2 + y^2} \, dt.$$

Proof. By Cauchy's theorem.

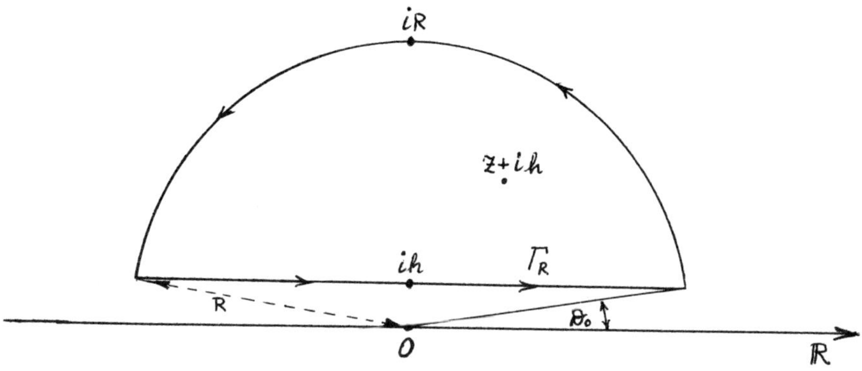

Let Γ_R be the contour shown, with R very large, and let $R\sin\theta_0 = h$. Then

$$F(z+ih) = \frac{1}{2\pi i}\int_{\Gamma_R} \frac{F(\zeta)}{\zeta-(z+ih)}\,d\zeta = \frac{1}{2\pi i}\int_{\Gamma_R} \frac{F(\zeta)}{(\zeta-ih)-z}\,d\zeta$$

$$= \frac{1}{2\pi i}\int_{-R\cos\theta_0}^{R\cos\theta_0} \frac{F(t+ih)dt}{t-z} + \frac{1}{2\pi i}\int_{\theta_0}^{\pi-\theta_0} \frac{F(re^{i\theta})}{Re^{i\theta}-ih-z}\, iRe^{i\theta}d\theta.$$

For very large R, the second integral is bounded by

$$J = \frac{1}{2\pi}\int_{\theta_0}^{\pi-\theta_0} |F(Re^{i\theta})|\ (1 + \mathcal{O}(\tfrac{1}{R}))d\theta.$$

Now use $|F(Re^{i\theta})| \leq k(R\sin\theta)^{-1/p}$, proved in the above lemma. If $p > 1$, we get immediately $J \leq k'R^{-1/p}$ since $\int_0^\pi d\theta/\sqrt[p]{\sin\theta} < \infty$ for $p > 1$. If $p = 1$, use $\sin\theta \geq \text{const.}\theta$ on $(\theta_0, \pi/2]$ and $\sin\theta \geq \text{const.}(\pi-\theta)$ on $[\pi/2, \pi-\theta_0)$ to get

$$J \leq \frac{k''}{R}\int_{\theta_0}^{\pi/2}\frac{d\theta}{\theta} = \frac{k''}{R}\log\frac{\pi}{2\theta_0} \leq \frac{k''}{R}\log\frac{\pi}{2\sin\theta_0} = \frac{k''}{R}\log\frac{\pi R}{2h}.$$

This also $\to 0$ as $R \to \infty$.

So

$$F(z + ih) = \lim_{R' \to \infty} \frac{1}{2\pi i} \int_{-R'}^{R'} \frac{F(t+ih)dt}{t - z} .$$

Now $\bar{z} + ih$ is outside Γ_R, so, again by Cauchy's theorem,

$$0 = \frac{1}{2\pi i} \int_{\Gamma_R} \frac{F(\zeta)d\zeta}{\zeta - ih - \bar{z}} = \frac{1}{2\pi i} \int_{-R\cos\theta_0}^{R\cos\theta_0} \frac{F(t+ih)dt}{t - \bar{z}} +$$

$$+ \frac{1}{2\pi i} \int_{\theta_0}^{\pi-\theta_0} \frac{F(Re^{i\theta})}{Re^{i\theta} - ih - \bar{z}} iRe^{i\theta} d\theta .$$

By the argument given above the second integral $\to 0$ as $R \to \infty$. So finally

$$0 = \lim_{R' \to \infty} \frac{1}{2\pi i} \int_{-R'}^{R'} \frac{F(t+ih)dt}{t - \bar{z}} .$$

Subtract this from the previous boxed relation and note the important identity

$$\frac{1}{t - z} - \frac{1}{t - \bar{z}} = \frac{2iy}{|t - z|^2} = \frac{2iy}{(x - t)^2 + y^2} .$$

We get

$$F(z + ih) = \frac{1}{\pi} \lim_{R' \to \infty} \int_{-R'}^{R'} \frac{yF(t+ih)dt}{(x-t)^2 + y^2} .$$

Q.E.D.

Remark. We define H_∞ to be the set of functions analytic and bounded in $\Im z >$. Then the above lemma still holds for $p = \infty$.

Note that here we cannot prove that

$$\int_{\theta_0}^{\pi-\theta_0} \frac{F(Re^{i\theta})}{Re^{i\theta} - ih - z} Re^{i\theta}\, d\theta \quad \& \quad \int_{\theta_0}^{\pi-\theta_0} \frac{F(Re^{i\theta})Re^{i\theta} d\theta}{Re^{i\theta} - ih - \bar{z}}$$

separately go to zero as $R \to \infty$. But their difference does, because $\frac{1}{Re^{i\theta} - ih - z} - \frac{1}{Re^{i\theta} - ih - \bar{z}}$ is $\mathcal{O}(R^{-2})$ for large R!

Remark. If, in the above proof (for $1 \le p < \infty$) we add instead of subtracting, we get

$$\boxed{F(z + ih) = \frac{i}{\pi} \int_{-\infty}^{\infty} \frac{(x - t)F(t + ih)dt}{(x - t)^2 + y^2}} ,$$

the integral being absolutely convergent for $1 \le p < \infty$. This fails if $p = \infty$.

Theorem. Let $F(z) \in H_p(\Im z > 0)$, $p \ge 1$. Then, for almost all $t \in \mathbb{R}$,

$$\lim_{z \not\to t} F(z) = F(t) \quad \text{exists,}$$

$$F(t) \in L_p(-\infty, \infty), \quad \text{and}$$

$$\boxed{F(z) = \frac{1}{\pi} \int_{-\infty}^{\infty} \frac{y}{(x - t)^2 + y^2} F(t)dt, \qquad \Im z > 0.}$$

Proof. If $p > 1$, we can get a sequence of $h > 0$ tending to zero such that the

$$F_h(t) = F(t + ih)$$

tend weakly to some $F \in L_p(-\infty, \infty)$ - for $\|F_h\|_p \leq C$, $h > 0$, by hypothesis. Substitute in the formula derived in the above lemma: for fixed z, $\Im z > 0$, we have

$$F(z+ih) = \frac{1}{\pi}\int_{-\infty}^{\infty} \frac{y}{(x-t)^2+y^2} F_h(t)dt \longrightarrow \frac{1}{\pi}\int_{-\infty}^{\infty} \frac{y}{(x-t)^2+y^2} F(t)dt$$

as $h \to 0$ through the particular sequence of values given. Since $F(z+ih) \to F(z)$ also, we get the above boxed formula.

Now $F(t) \in L_p(-\infty, \infty)$, $p > 1$, so if

$$\Phi(x) = \int_0^x F(t)dt,$$

$\Phi'(x)$ exists and equals $F(x)$ a.e. Observe that $d\Phi(t)/(1+t^2) = F(t)dt/(1+t^2)$ is absolutely integrable, so a theorem of Section B teaches us that $F(z) \longrightarrow F(x_0)$ for $z \underset{\not\,}{\longrightarrow} x_0$ a.e. in $x_0 \in \mathbb{R}$. We are done in case $p > 1$. (N.B. Including for the case $p = \infty$, by a preceding remark!)

Now for $p = 1$ all we know is that

$$F_h(t)dt \longrightarrow d\mu(t) \quad w^*$$

as $h \to 0$ through its sequence, where μ is some measure on $\mathbb{R}$ with $\int_{-\infty}^{\infty} |d\mu(t)| < \infty$. The above argument then yields

$$F(z) = \frac{1}{\pi}\int_{-\infty}^{\infty} \frac{y}{(x-t)^2+y^2}\, d\mu(t).$$

But now we can apply the theorem of F. and M. Riesz.

If $\Im z > 0$, the second boxed relation in the proof of the above lemma reads, for $h > 0$,

$$\int_{-\infty}^{\infty} \frac{F_h(t)dt}{t - \bar{z}} = 0.$$

Therefore, since $F_h(t)dt \longrightarrow d\mu(t)\quad w^*$,

$$\int_{-\infty}^{\infty} \frac{d\mu(t)}{t-\bar{z}} = 0, \qquad \Im z > 0.$$

Write $z = ik$, $k > 0$, we see that

$$\int_{-\infty}^{\infty} \frac{d\mu(t)}{t+ik} = 0, \qquad k > 0.$$

Differentiating successively with respect to k we get

$$\int_{-\infty}^{\infty} \frac{d\mu(t)}{(t+ik)^n} = 0, \qquad n = 1,2,\ldots \; .$$

Finally, we have

$$\int_{-\infty}^{\infty} \frac{1}{(i+t)^n}\, d\mu(t) = 0, \qquad n = 1,2,\ldots \; .$$

Now, in the conformal mapping

$$z \longrightarrow \frac{i-z}{i+z}$$

$$t \longrightarrow \frac{i-t}{i+t} = e^{i\tau},$$

define a measure ν on $[-\pi,\pi]$ by $d\nu(\tau) = \dfrac{d\mu(t)}{i-t}$. Then, for $n = 1,2,\ldots,$

$$\int_{-\pi}^{\pi} e^{in\tau}\, d\nu(\tau) = \int_{-\infty}^{\infty} \frac{(i-t)^{n-1}}{(i-t)^n}\, d\mu(t)$$

which can be rewritten in the form

$$\sum_{k=1}^{n} a_k \int_{-\infty}^{\infty} \frac{d\mu(t)}{(i+t)^k}$$

and is hence $= 0$. So

$$\int_{-\pi}^{\pi} e^{in\tau}\, d\nu(\tau) = 0, \qquad n = 1,2,3,$$

and by F. and M. Riez Theorem, $d\nu(\tau)$ is absolutely continuous. So $d\mu(t)/(i - t)$, and hence $d\mu(t)$ is; $d\mu(t) = F(t)dt$ for some $F \in L_1(-\infty,\infty)$. We now conclude the proof as in the case $p > 1$.

Q.E.D.

Theorem. If $F \in H_p$, $1 \leq p < \infty$, (sic!) and $\Im z > 0$, then

$$\frac{1}{2\pi i}\int_{-\infty}^{\infty} \frac{F(t)}{t - z}\, dt = F(z)$$

and

$$\int_{-\infty}^{\infty} \frac{F(t)dt}{t - \bar{z}} = 0.$$

Proof. By the two boxed formulas in the proof of a preceding lemma,

$$F(z + ih) = \frac{1}{2\pi i}\int_{-\infty}^{\infty} \frac{F_h(t)dt}{t - z},$$

$$0 = \int_{-\infty}^{\infty} \frac{F_h(t)dt}{t - \bar{z}}, \qquad \text{where} \quad F_h(t) = F(t + ih).$$

But by the above theorem

$$F_h(x) = \frac{1}{\pi}\int_{-\infty}^{\infty} \frac{hF(t)dt}{(x - t)^2 + h^2}$$

where $F \in L_p(-\infty, \infty)$. So by a result of Section B, $\|F_h - F\|_p \to 0$ as $h \to 0$. Substitution into the formulas on the preceding page gives the desired result.

The first of the above two theorems has its analogue for harmonic functions.

Theorem. Let $U(z)$ be harmonic in $\Im z > 0$ and suppose that, for some $p \geq 1$,

$$\int_{-\infty}^{\infty} |U(t+iy)|^p \, dt \leq C < \infty$$

independently of y. If $p > 1$, then

$$U(z) = \frac{1}{\pi} \int_{-\infty}^{\infty} \frac{y}{(x-t)^2 + y^2} u(t)dt, \qquad \Im z > 0,$$

where $u(t) \in L_p(-\infty,\infty)$.

If $p = 1$,

$$U(z) = \frac{1}{\pi} \int_{-\infty}^{\infty} \frac{y}{(x-t)^2 + y^2} d\mu(t), \qquad \Im z > 0,$$

where

$$\int_{-\infty}^{\infty} |d\mu(t)| < \infty.$$

Proof. $|U(z)|^p$ is subharmonic, as is easily verified from the mean value formula of Gauss and Hölder's inequality - therefore the argument of the first lemma of this section can be applied to it, and we see that $|U(z)| \leq C/y^{1/p}$ for $\Im z = y > 0$.

In particular each function $U_h(z) = U(z+ih)$, $h > 0$, is bounded (and harmonic!) for $\Im z > 0$. Now we can use the conformal mapping $\frac{i-z}{i+z} = w$, put $U_h(z) = \check{U}_h(w)$, and use the known representation (Poisson's formula) for functions $\check{U}_h(w)$ harmonic and BOUNDED in $\{|w| < 1\}$. If we go back to the z-plane, the Poisson formula for $\check{U}_h(re^{i\theta})$ in terms of $\check{U}_h(e^{i\tau})$ goes over into one for $U_h(z)$ in terms of $U_h(t)$, $t \in \mathbb{R}$. The calculation has already been carried out in Section A, and we find

$$U_h(z) = \frac{1}{\pi}\int_{-\infty}^{\infty} \frac{y\, U_h(t)}{(x-t)^2+y^2}\,dt,$$

i.e.,

$$U(z+ih) = \frac{1}{\pi}\int_{-\infty}^{\infty} \frac{y\, U_h(t)}{(x-t)^2+y^2}\,dt.$$

The rest of the argument consists in passing to the limit $h \to 0$ and is exactly like what was done in proving Poisson's formula for H_p-functions above. We're done.

Remark. The above proof works also when $p = \infty$ ($U(z)$ bounded in $\Im z > 0$).

Theorem. If $F(z) \in H_p(\Im z > 0)$, $p \geq 1$, and, for $w = \frac{i-z}{i+z}$, $f(w) = F(z)$, then $f(w) \in H_p\{|w| < 1\}$.

Proof. By the first of the above theorems,

$$F(z) = \frac{1}{\pi}\int_{-\infty}^{\infty} \frac{y}{|z-t|^2} F(t)\,dt,$$

where $F(t) \in L_p(-\infty,\infty)$. The change of variable $z \to w$ was carried out in Section A where we found

$$f(re^{i\varphi}) = \frac{1}{2\pi}\int_{-\pi}^{\pi} \frac{1-r^2}{1+r^2-2r\cos(\varphi-\tau)} f(e^{i\tau})\,d\tau,$$

putting

$$F(t) = f\left(\frac{i-t}{i+t}\right).$$

Now

$$\int_{-\pi}^{\pi} |f(e^{i\tau})|^p d\tau = \int_{-\infty}^{\infty} \frac{2|F(t)|^p dt}{1+t^2} \leq 2\int_{-\infty}^{\infty} |F(t)|^p dt < \infty,$$

so, by Chapter I, the means $\int_{-\pi}^{\pi} |f(re^{i\varphi})|^p d\varphi$ are bounded. So $f \in H_p\{|w| < 1\}$.

Q.E.D.

> Remark. The computation used in proving the above theorem shows that the CONVERSE
>
> $$f(w) \in H_p(\{|w| < 1\}) \Rightarrow F(z) \in H_p(\Im z > 0)$$
>
> IS FALSE!

The correct equivalence is obviously, for

$$f\left(\frac{i - z}{i + z}\right) = F(z),$$

> $f(w) \in H_p(\{|w| < 1\})$ iff
>
> $$\frac{F(z)}{(z + i)^{2/p}} \in H_p(\Im z > 0).$$

Theorem. If $0 < p < 1$ and $F(z) \in H_p$, $F(w) \in H_p(\{|w| < 1\})$ for

$$f\left(\frac{i - z}{i + z}\right) = F(z).$$

Proof. $|F(z)|^p$ is still subharmonic, so we still have $|F(z)| \leq K/y^{1/p}$, and in particular $|F(z)|$, and hence $|F(z)|^p$, is still bounded in each half-plane $\Im z \geq h > 0$. Therefore $|F_h(z)|^p = |F(z + ih)|^p$, being bounded and subharmonic in $\Im z > 0$, is there majorized by the bounded harmonic function taking the same boundary values on $\mathbb{R}$, i.e.,

$$|F(z + ih)|^p \leq \frac{1}{\pi} \int_{-\infty}^{\infty} \frac{y|F_h(t)|^p dt}{(x - t)^2 + y^2} .$$

Now

$$\int_{-\infty}^{\infty} |F_h(t)|^p dt \leq C, \qquad h > 0,$$

so, making $h \to 0$ we find, in the usual way, a measure μ, $\int_{-\infty}^{\infty} |d\mu(t)| < \infty$, with

$$|F(z)|^p \leq \frac{1}{\pi} \int_{-\infty}^{\infty} \frac{y}{|z-t|^2} d\mu(t).$$

Going over to

$$f\left(\frac{i-z}{i+z}\right) = F(z),$$

we get

$$|f(re^{i\theta})|^p \leq \frac{1}{2\pi} \int_{-\pi}^{\pi} \frac{1-r^2}{1+r^2-2r\cos(\theta-\tau)} d\nu(\tau),$$

where, for $\frac{i-t}{i+t} = e^{i\tau}$,

$$d\nu(\tau) = \frac{2\, d\mu(t)}{1+t^2} .$$

Therefore

$$\int_{-\pi}^{\pi} |f(re^{i\varphi})|^p d\varphi \leq 2 \int_{-\infty}^{\infty} \frac{|d\mu(t)|}{1+t^2}$$

by Fubini's theorem, for any $r < 1$. But $\int_{-\infty}^{\infty} \frac{|d\mu(t)|}{1+t^2} < \infty$. So done.

Because of these theorems, functions in $H_p(\Im z > 0)$ can be decomposed into inner and outer factors.

Theorem. Let $F(z) \in H_p(\Im z > 0)$. Then for $\Im z > 0$,

$$F(z) = e^{i\gamma} I_F(z) \cdot \mathcal{O}_F(z)$$

where:

i) γ is real

ii) $I_F(z)$, the inner factor of F, is

$$I_F(z) = B(z)\exp\left(\frac{1}{\pi}\int_{-\infty}^{\infty}\left(\frac{i}{z-t}+\frac{it}{t^2+1}\right)d\sigma(t)\right)e^{i\alpha z}$$

with

a) $B(z)$ a Blaschke product for $\Im z > 0$,

$$B(z) = \prod_k \left(e^{i\alpha_k}\cdot\frac{z-z_k}{z-\bar{z}_k}\right),$$

where the z_k are the zeros of $F(z)$ in $\Im z > 0$ and the real α_k are so chosen that

$$e^{i\alpha_k}\cdot\frac{i-z_k}{i-\bar{z}_k} \geq 0,$$

b) $d\sigma(t) \leq 0$ is a singular measure on $\mathbb{R}$ with

$$\int_{-\infty}^{\infty}\frac{|d\sigma(t)|}{1+t^2} < \infty$$

c) $\alpha \geq 0$ (the "mass at ∞" is $-\alpha$)

iii) The outer factor of F, $\mathcal{O}_F$, is

$$\mathcal{O}_F(z) = \exp\left(\frac{i}{\pi}\int_{-\infty}^{\infty}\left(\frac{1}{z-t}+\frac{t}{t^2+1}\right)\log|F(t)|\,dt\right).$$

Proof. Take the corresponding factorization for functions $f(w)$ in $H_p(\{|w| < 1\})$ and change the variables in it:

$$w = \frac{i-z}{i+z}, \qquad f(w) = F(z),$$

$$e^{i\tau} = \frac{i-t}{i+t}.$$

The calculations have mostly been already made in Section A.

Scholium. In terms of the z_k, the necessary and sufficient condition for the convergence of the Blaschke product $B(z)$ is $\Pi_k |(i-z_k)/(i-\bar{z}_k)| > 0$. This can be written more easily as follows: Firstly,

$$1 - \left|\frac{i - z_k}{i - \bar{z}_k}\right|^2 = \frac{|z_k + i|^2 - |z_k - i|^2}{|z_k + i|^2} = \frac{4\Im z_k}{|z_k + i|^2} .$$

Therefore $\Pi_k |(i-z_k)/(i-\bar{z}_k)|$ converges to a number > 0 iff

$$\sum_k \frac{\Im z_k}{|z_k + i|^2} < \infty .$$

This is usually stated as follows:

The Blaschke product $\Pi_k \left(e^{i\alpha_k}(z-z_k)/(z-\bar{z}_k)\right)$ is convergent in $\Im z > 0$ iff $\sum_{|z_k| < 1} \Im z_k < \infty$ and $\sum_{|z_k| \geq 1} \Im z_k / |z_k|^2 < \infty$.

The following picture shows that it is much easier to visualize Blaschke products for the upper half plane than for the circle:

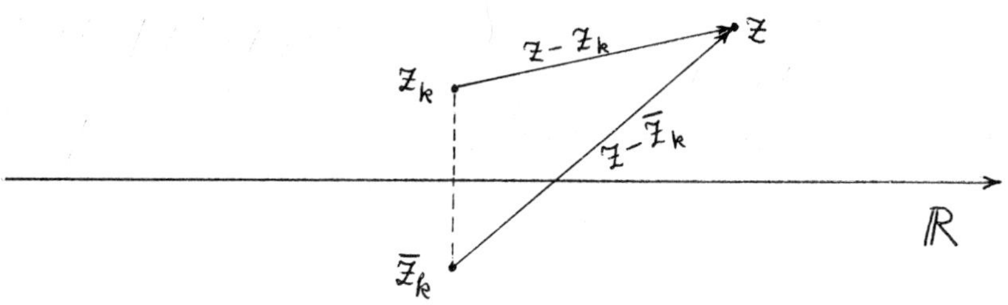

From this it is manifest why the individual factors have modulus < 1 precisely in $\Im z > 0$!

<u>Scholium</u>. Of course, if $F(z) \in H_p(\Im z > 0)$,

$$\int_{-\infty}^{\infty} \frac{|\log|F(t)||\,dt}{1 + t^2} < \infty.$$

This follows from the corresponding result for the circle (Chapter IV) by a change of variable.

<u>Remark</u>. The above given factorization for the <u>half-plane</u> is sometimes credited to Nevanlinna. Most of the obvious analogues of Beurling's theorem (Chapter IV, Section E) hold.

D. <u>Riesz' theorems for the Hilbert transform</u>. <u>If</u> $u(t) \in L_p(-\infty,\infty)$, $1 \le p \le \infty$, we can form the function, harmonic in $\Im z > 0$,

$$U(z) = \frac{1}{\pi}\int_{-\infty}^{\infty} \frac{y u(t)\,dt}{(x - t)^2 + y^2},$$

such that $U(z) \to u(t)$ for $z \not\to t$ a.e. in t (Section B).

<u>If, also</u>, $p < \infty$ (sic!), we can form the harmonic conjugate

$$\tilde{U}(z) = \frac{1}{\pi}\int_{-\infty}^{\infty} \frac{(x - t)u(t)\,dt}{(x - t)^2 + y^2}$$

(Section A).

$U(z) + i\tilde{U}(z)$ is ANALYTIC <u>in</u> $\Im z > 0$.

Lemma. If $F(z) \in H_1(\Im z > 0)$,

$$\int_{-\infty}^{\infty} F(t)dt = 0.$$

Proof. By a theorem in Section C,

$$\int_{-\infty}^{\infty} \frac{F(t)dt}{t + iN} = 0$$

for each $N > 0$. Since $F \in H_1$, by Section C, $F(t) \in L_1(-\infty,\infty)$ so

$$\int_{-\infty}^{\infty} F(t)dt = \lim_{N\to\infty} \int_{-\infty}^{\infty} \frac{iN}{t + iN} F(t)dt = 0.$$

Q.E.D.

Lemma. Let $1 < p \leq 2$, let $u \in L_p(-\infty,\infty)$, and let $U(z)$ and $\tilde{U}(z)$ be related to $u(t)$ by the above boxed formulas. There is a constant K_p depending only on p such that, for each $h > 0$,

$$\int_{-\infty}^{\infty} |\tilde{U}(x + ih)|^p dx \leq K_p \int_{-\infty}^{\infty} |U(x + ih)|^p dx.$$

Proof. Clearly there is no loss of generality in assuming $u(t)$ real, since the general case follows from this, with perhaps a worse value of K_p.

We can also assume that $u(t)$ has compact support. For, in general, we can find $u_n(t)$ of compact support with $\|u_n - u\|_p \xrightarrow[n]{} 0$, so that, if $U_n(z) = \frac{1}{\pi}\int_{-\infty}^{\infty} y\, u_n(t)dt/|z - t|^2$, and $\tilde{U}_n(z) = \frac{1}{\pi}\int_{-\infty}^{\infty} (x-t)u_n(t)dt/|z - t|^2$, we have, for each fixed h, $h > 0$,

$$\int_{-\infty}^{\infty} |U_n(x + ih) - U(x + ih)|^p dx \leq \int_{-\infty}^{\infty} |u_n(t) - u(t)|^p dt \xrightarrow[n]{} 0,$$

whilst $\tilde{U}_n(x + ih) \xrightarrow[n]{} \tilde{U}(x + ih)$ u.c.c. in x, so by Fatou's lemma,

$$\int_{-\infty}^{\infty} |\tilde{U}(x + ih)|^p dx \leq \liminf_{n\to\infty} \int_{-\infty}^{\infty} |\tilde{U}_n(x + ih)|^p\, dx,$$

and the validity of the inequality for each n would imply the same for $\tilde{U}$ and U.

So henceforth, without loss of generality, we assume that $u(t)$ is of compact support.

We now assume also that $u(t) \geq 0$ and $u \not\equiv 0$.

Let $F(z) = U(z) + i\tilde{U}(z)$. Then, for $h > 0$, $U(z) = \Re F(z)$ is > 0 for $\Im z \geq h$, so $F(z)$ has no zeros in any such half plane. Therefore

$$G(z) = [F(z + ih)]^p$$

is also analytic for $\Im z \geq 0$.

Now $G(z) \in H_1(\Im z > 0)$. Indeed,

$$\int_{-\infty}^{\infty} [U(x + iy + ih)]^p dx \leq C < \infty$$

for each $y > 0$, whilst

$\lvert\tilde{U}(z + ih)\rvert$ is bounded for $\Im z \geq 0$ AND HERE ALSO $O(\frac{1}{z})$ there because $u(t)$ is of compact support.

So, since $p > 1$,

$$\int_{-\infty}^{\infty} \lvert\tilde{U}(x + iy + ih)\rvert^p dx \leq \text{const.}$$

for $y \geq 0$. (N.B. The constant depends upon h, but we don't care about that now!)

So $\int_{-\infty}^{\infty} \lvert F(x + iy + ih)\rvert^p dx \leq c$, $y > 0$, and $G(z) \in H_1$.

<u>Therefore by the previous lemma</u>

$$\int_{-\infty}^{\infty} G(x)dx = 0.$$

Now we can use the method of Katznelson and Zygmund (Chapter V).

$$G(z) = (F_h(z))^p, \qquad 1 < p \le 2,$$

with $\Re F_h = U_h > 0$.

Choose a γ, $0 < \gamma < \frac{\pi}{2}$, <u>such that</u> $\frac{\pi}{2} < p\gamma < \pi$ - <u>because</u> $1 < p \le 2$ <u>we can do that</u>. Let

$$E = \{x \in \mathbb{R};\ |\arg F_h(x)| \le \gamma\}.$$

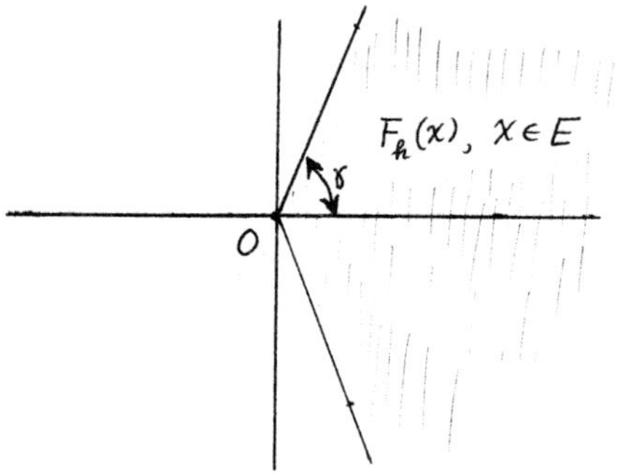

We play these two situations against each other.

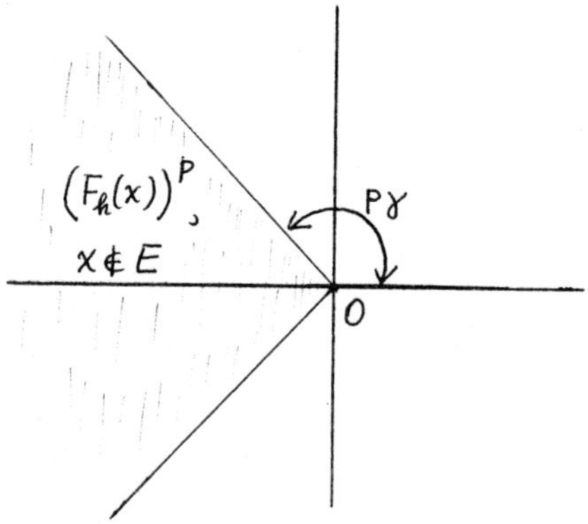

From the previous boxed relation,

$$\int_{\sim E} \Re(F_h(x))^p \, dx + \int_E \Re(F_h(x))^p \, dx = 0.$$

But for $x \in \sim E$,

$$\Re(F_h(x))^p \leq -|F_h(x)|^p \; |\cos p\gamma|$$

(see second picture!).

Therefore

$$\int_{\sim E} |F_h(x)|^p \; |\cos p\gamma| dx \leq \int_E \Re(F_h(x))^p dx \leq \int_E |F_h(x)|^p dx.$$

But for $x \in E$ (first picture!),

$$|F_h(x)| \leq \frac{U_h(x)}{\cos \gamma} ,$$

so

$$\int_E |F_h(x)|^p dx \leq \frac{1}{\cos^p \gamma} \int_E [U_h(x)]^p dx.$$

Finally we see that

$$\int_{-\infty}^{\infty} |F_h(x)|^p dx = \left(\int_{\sim E} + \int_E \right) |F_h(x)|^p dx$$

$$\leq (\cos \gamma)^{-p} \left(1 + \frac{1}{|\cos p\gamma|} \right) \int_E [U_h(x)]^p dx,$$

or, since $|\tilde{U}_h(x)| \leq |F_h(x)|$,

$$\int_{-\infty}^{\infty} |\tilde{U}_h(x)|^p dx \leq (\cos\gamma)^{-p}\left(1 + \frac{1}{|\cos p\gamma|}\right)\int_{-\infty}^{\infty} [U_h(x)]^p dx,$$

as desired, in case $u(t) \geq 0$.

Now we wish to remove that assumption. Observe that we certainly have

$$\int_{-\infty}^{\infty} |\tilde{U}_h(x)|^p dx \leq \check{K}_p \int_{-\infty}^{\infty} [u(t)]^p dt$$

with

$$\check{K}_p = (\cos\gamma)^{-p}\left(1 + \frac{1}{|\cos p\gamma|}\right)$$

in case $u \geq 0$. For a general real u of compact support, we write $u = u_+ - u_-$, where

$$\int_{-\infty}^{\infty} |u(t)|^p dt = \int_{-\infty}^{\infty} |u_+(t)|^p dt + \int_{-\infty}^{\infty} |u_-(t)|^p dt,$$

where u_+ and u_- have disjoint compact supports, and both are ≥ 0.

Then, with obvious notation,

$$\tilde{U}_h(x) = \tilde{U}_h^+(x) - \tilde{U}_h^-(x),$$

so $\|\tilde{U}_h\|_p \leq \|\tilde{U}_h^+\|_p + \|\tilde{U}_h^-\|_p$, and

$$\int_{-\infty}^{\infty} |\tilde{U}_h(x)|^p dx \leq 2^{p-1}\left(\int_{-\infty}^{\infty} |\tilde{U}_h^+(x)|^p dx + \int_{-\infty}^{\infty} |\tilde{U}_h^-(x)|^p dx\right)$$

which, by what has already been done, is

$$\leq \frac{2^{p-1}}{\cos^p\gamma}\left(1 + \frac{1}{|\cos p\gamma|}\right)\left[\int_{-\infty}^{\infty} |u_+(t)|^p dt + \int_{-\infty}^{\infty} |u_-(t)|^p dt\right]$$

$$= \frac{2^{p-1}}{\cos^p\gamma}\left(1 + \frac{1}{|\cos p\gamma|}\right)\int_{-\infty}^{\infty} |u(t)|^p dt.$$

We see that

$$\int_{-\infty}^{\infty} |\tilde{U}(x+ih)|^p dx \le K_p \int_{-\infty}^{\infty} |u(t)|^p dt$$

is now proven for general real u of variable sign and compact support (and hence for real u without compact support by the preliminary discussion given above). This would suffice for many purposes.

We wish, however, to prove the lemma as stated.

> This is based on a trick with the Poisson and conjugate Poisson kernels.

Observe that for fixed real x_0 and $h > 0$,

$$\Re \frac{1}{z - x_0 + ih}$$

is harmonic and bounded in $\Im z > 0$, so by Section C,

$$\Re \frac{1}{z - x_0 + ih} = \frac{1}{\pi} \int_{-\infty}^{\infty} \Re \frac{1}{\xi - x_0 + ih} \frac{y\, d\xi}{|z - \xi|^2},$$

i.e., changing sign,

$$\frac{x_0 - x}{(x_0 - x)^2 + (y+h)^2} = \frac{1}{\pi} \int_{-\infty}^{\infty} \frac{x_0 - \xi}{(x_0 - \xi)^2 + h^2} \cdot \frac{y}{(\xi - x)^2 + y^2}\, d\xi.$$

Therefore

$$\tilde{U}(x_0 + iy + ih) = \frac{1}{\pi} \int_{-\infty}^{\infty} \frac{x_0 - t}{(x_0 - t)^2 + (y+h)^2}\, u(t)dt =$$

$$= \left(\frac{1}{\pi}\right)^2 \int_{-\infty}^{\infty} \int_{-\infty}^{\infty} \frac{x_0 - \xi}{(x_0 - \xi)^2 + h^2} \cdot \frac{y}{(\xi - t)^2 + y^2}\, u(t) d\xi\, dt.$$

By Fubini's theorem $(u(t)$ can be assumed of compact support!), this works out to

$$\frac{1}{\pi}\int_{-\infty}^{\infty} \frac{x_0 - \xi}{(\xi - x_0)^2 + h^2} U(\xi + iy)d\xi.$$

Thus,

$$\tilde{U}(x + iy + ih) = \frac{1}{\pi}\int_{-\infty}^{\infty} \frac{x - t}{(x - t)^2 + h^2} U(t + iy)dt.$$

By what has just been proven,

$$\int_{-\infty}^{\infty} |\tilde{U}(x + iy + ih)|^p dx \le K_p \int_{-\infty}^{\infty} |U(t + iy)|^p \, dt$$

for all $h > 0$ with a K_p depending only on p. Now, keeping $y > 0$ fixed, make $h \to 0$ so that

$$\tilde{U}(x + iy + ih) \to \tilde{U}(x + iy)$$

u.c.c. in x and use Fatou's lemma. We find in the limit that

$$\int_{-\infty}^{\infty} |\tilde{U}(x + iy)|^p dx \le K_p \int_{-\infty}^{\infty} |U(t + iy)|^p dt,$$

the full strength of what was required.

Q.E.D.

Scholium. In case $p = 2$, we can say more. For then $(F(z))^2$ is regular in $\Im z > 0$ whether $U(z)$ is ≥ 0 or not, so

$$\int_{-\infty}^{\infty} (F(x + ih))^2 dx = 0$$

for each $h > 0$, F_h^2 being in H_1. Taking real parts, we have

$$\int_{-\infty}^{\infty} [U(x + ih))^2 - (\tilde{U}(x + ih))^2]dx = 0,$$

i.e.,

$$\int_{-\infty}^{\infty} [\tilde{U}(x+ih)]^2 dx = \int_{-\infty}^{\infty} [U(x+ih)]^2 dx.$$

For $p = 2$ we have equality!

Now we remove the restriction $p \le 2$ by a duality argument.

Theorem (Riesz). Let $1 < p < \infty$ and let $u(t) \in L_p(-\infty,\infty)$. Write, for $\Im z > 0$,

$$\tilde{U}(z) = \frac{1}{\pi} \int_{-\infty}^{\infty} \frac{x-t}{|z-t|^2} u(t)dt.$$

Then, for each $h > 0$,

$$\int_{-\infty}^{\infty} |\tilde{U}(x+ih)|^p dx \le K_p \int_{-\infty}^{\infty} |u(t)|^p dt.$$

Proof. If $1 < p \le 2$, the result is contained in the above lemma, so assume $2 < p < \infty$, let $\frac{1}{p} + \frac{1}{q} = 1$ - then $1 < q \le 2$. By real variable theory, given any $\varepsilon > 0$ we can find a $v(t) \in L_q(-\infty,\infty)$ of compact support with

$$\int_{-\infty}^{\infty} |v(t)|^q dt = 1$$

such that

$$\|\tilde{U}_h\|_p - \varepsilon \le \left| \int_{-\infty}^{\infty} \tilde{U}_h(x)v(x)dx \right| .$$

But now

$$\int_{-\infty}^{\infty} \tilde{U}_h(x)v(x)dx = \frac{1}{\pi}\int_{-\infty}^{\infty}\int_{-\infty}^{\infty} \frac{(x-t)u(t)v(x)}{(x-t)^2+h^2}\,dt\,dx = \text{by Fubini}$$

$$= -\frac{1}{\pi}\int_{-\infty}^{\infty}\left(\int_{-\infty}^{\infty} \frac{t-x}{(t-x)^2+h^2}\,v(x)dx\right)u(t)dt$$

$$= -\int_{-\infty}^{\infty} u(t)\tilde{V}_h(t)dt,$$

where

$$\tilde{V}(z) = \frac{1}{\pi}\int_{-\infty}^{\infty} \frac{x-s}{|z-s|^2}\,v(s)ds.$$

By the lemma, since $1 < q \leq 2$,

$$\|\tilde{V}_h\|_q \leq \hat{K}_q\|v\|_q = \hat{K}_q,$$

so by Hölder again,

$$\left|\int_{-\infty}^{\infty} u(t)\tilde{V}_h(t)dt\right| \leq \hat{K}_q\|u\|_p.$$

Thus, $\|\tilde{U}_h\|_p - \varepsilon \leq \hat{K}_q\|u\|_p$. Squeeze ε. The theorem is proved.

Q.E.D.

This important theorem has many corollaries.

Corollary. If $u(t) \in L_p(-\infty,\infty)$ and $1 < p < \infty$, then

$$F(z) = \frac{i}{\pi}\int_{-\infty}^{\infty} \frac{u(t)dt}{z-t} \in H_p(\Im z > 0) \text{ and } \Re F(z) \longrightarrow u(t) \text{ as}$$

$z \not\longrightarrow t$ a.e. in t.

Corollary. If $u(t) \in L_p(-\infty,\infty)$ and $1 < p < \infty$, then

$$\tilde{u}(x) = \lim_{\varepsilon \to 0} \frac{1}{\pi}\int_{|t-x| \geq \varepsilon} \frac{u(t)}{x-t}\,dt$$

exists a.e., $\|\tilde{u}\|_p \leq \hat{K}_p\|u\|_p$, and

$$\tilde{U}(z) = \frac{1}{\pi}\int_{-\infty}^{\infty} \frac{x-t}{|z-t|^2}\, u(t)dt = \frac{1}{\pi}\int_{-\infty}^{\infty} \frac{y}{|z-t|^2}\, \tilde{u}(t)dt.$$

Remark. Thus, the H_p-function

$$F(z) = \frac{1}{\pi}\int_{-\infty}^{\infty} \frac{u(t)dt}{z-t}$$

of the previous Corollary has $\Im F(t) = \tilde{u}(t)$ a.e., whilst $\Re F(t) = u(t)$.

Proof of Corollary. Take $\tilde{U}(z)$ - it is harmonic in $\Im z > 0$. By Riesz' Theorem, if $u \in L_p$,

$$\int_{-\infty}^{\infty} |\tilde{U}(x+iy)|^p \leq C, \quad y > 0.$$

Therefore, since $1 < p < \infty$, by a theorem of Section C, there is a function $v(t) \in L_p(-\infty,\infty)$ such that

$$\tilde{U}(z) = \frac{1}{\pi}\int_{-\infty}^{\infty} \frac{yv(t)dt}{|z-t|^2}\ .$$

By Section B, $\tilde{U}(z) \longrightarrow v(t)$ as $z \not\longrightarrow t$ for almost all t. In particular, $\lim_{y\to 0} \tilde{U}(t+iy)$ exists and is finite for almost all $t \in \mathbb{R}$. So, by another result of Section B, for almost all x,

$$\lim_{\varepsilon\to 0} \frac{1}{\pi}\int_{|t-x|\geq\varepsilon} \frac{u(t)dt}{x-t}$$

exists and has the same value as $\lim_{y\to 0} \tilde{U}(x+iy) = v(x)$. So, with the notation for principal values which is now familiar,

$$\frac{1}{\pi}\int_{-\infty}^{\infty} \frac{u(t)dt}{x-t} = v(x) \quad \text{a.e.}$$

This is enough.

Q.E.D.

Definition.

$$\tilde{u}(x) = \frac{1}{\pi} \int_{-\infty}^{\infty} \frac{u(t)dt}{x - t}$$

is called the Hilbert transform of $u \in L_p(-\infty,\infty)$, $1 < p < \infty$.

Corollary. If $1 < p < \infty$ and $u \in L_p(-\infty,\infty)$, $\tilde{\tilde{u}} = -u$.

Proof. Write $F(z) = \frac{i}{\pi} \int_{-\infty}^{\infty} \frac{u(t)dt}{z - t}$, then $F \in H_p$, $\Re F(t) = u(t)$ and $\Im F(t) = \tilde{u}(t)$. Let

$$G(z) = \frac{i}{\pi} \int_{-\infty}^{\infty} \frac{\tilde{u}(t)dt}{z - t} .$$

$$- iF(z) + G(z) = \frac{1}{\pi} \int_{-\infty}^{\infty} \frac{[u(t) + i\tilde{u}(t)]dt}{z - t} = \frac{1}{\pi} \int_{-\infty}^{\infty} \frac{F(t)}{z - t} dt = -2iF(z)$$

by Section 3. So $G(z) = -iF(z)$. Therefore for almost all x,

$$\tilde{\tilde{u}}(x) = \lim_{y \to 0} \Im G(x + iy) = - \lim_{y \to 0} \Re F(x + iy) = -u(x).$$

Q.E.D.

Corollary. If $1 < p < \infty$, there are two constants C_p and D_p with

$$\boxed{C_p \|u\|_p \leq \|\tilde{u}\|_p \leq D_p \|u\|_p,}$$

and the Hilbert transform is an isomorphic mapping of $L_p(-\infty,\infty)$ onto itself.

Proof. By above 2 corollaries.

Remark. In case $p = 2$, we have an isometry,

$$\int_{-\infty}^{\infty} |\tilde{u}(x)|^2 dx = \int_{-\infty}^{\infty} |u(t)|^2 dt.$$

Indeed, $u(t) + i\tilde{u}(t) = F(t)$ with $F \in H_2$, so $F^2 \in H_1$, so $\int_{-\infty}^{\infty} (f(t))^2 dt = 0$. Take real parts.

Corollary. If $\frac{1}{p} + \frac{1}{q} = 1$, $1 < p < \infty$, $u \in L_p$, $v \in L_q$, we have

$$\int_{-\infty}^{\infty} u(t)v(t)dt = \int_{-\infty}^{\infty} \tilde{u}(x)\tilde{v}(x)dx$$

and

$$\int_{-\infty}^{\infty} u(t)\tilde{v}(t)dt = -\int_{-\infty}^{\infty} \tilde{u}(t)v(t)dt,$$

all the integrals being absolutely convergent.

Proof. Put $F(z) = \frac{1}{\pi}\int_{-\infty}^{\infty} \frac{u(t)dt}{z - t}$, $G(z) = \frac{1}{\pi}\int_{-\infty}^{\infty} \frac{v(t)dt}{z - t}$, then $F \in H_p$, $G \in H_q$, $F(t) = u(t) + i\tilde{u}(t)$ a.e., $G(t) = v(t) + i\tilde{v}(t)$ a.e. We have $F(z)G(z) \in H_1$, so by the first lemma of this Section,

$$\int_{-\infty}^{\infty} F(t)G(t)dt = 0.$$

Tkaing real parts gives us the first boxed relation. Taking imaginary parts gives us the second. Absolute convergence of the integrals in question follows by Hölder's inequality.

E. Fourier transforms. The Paley-Wiener Theorem. Let $F(t) \in L_p(-\infty,\infty)$. The Hausdorff-Young theorem says that if $1 \le p \le 2$ (and in general for no larger values of p!), and if we write

$$\hat{F}_N(\lambda) = \int_{-N}^{N} e^{i\lambda t} F(t)dt,$$

then, as $N \to \infty$, the $\hat{F}_N(\lambda)$ tend in $L_q(-\infty,\infty)$ to a function $\hat{F}(\lambda)$, called the Fourier transform of F, where $\frac{1}{q} = 1 - \frac{1}{p}$. The Fourier transform $\hat{F}$ satisfies

$$\|\hat{F}\|_q \le K_p\|F\|_p.$$

If $p = 1$, these statements are obvious; in that case $\hat{F}(\lambda)$ is even continuous and zero at $\pm\infty$. For $p = 2$, they are part of Plancherel's theorem. These two cases ($p = 1$ and $p = 2$) suffice for many applications.

Theorem. Let $F(t) \in L_p(-\infty,\infty)$ with $1 \le p \le 2$ be the boundary-value function of a function $F \in H_p$. Then $\hat{F}(\lambda) = 0$ a.e. for $\lambda \ge 0$.

Proof. First of all, if $p = 1$, i.e., $F \in H_1$, then $e^{i\lambda z}F(z)$ is also in H_1 for $\lambda \ge 0$ ($|e^{i\lambda z}| \le 1$ for $\Im z \ge 0$ then!). So then $\int_{-\infty}^{\infty} e^{i\lambda t}F(t)dt = 0$, i.e., $\hat{F}(\lambda) = 0$, by a lemma of Section D.

If we just have $F \in H_p$ for $1 < p \le 2$, we use $F_{(k)}(z) = \frac{ik}{z+ik} F(z)$ with $k > 0$. Clearly, $F_{(k)} \in H_1$. So $\hat{F}_{(k)}(\lambda) \equiv 0$, $\lambda > 0$. But now $\|F_{(k)} - F\|_p \to 0$ as $k \to \infty$ so, by the Hausdorff-Young theorem (or just plain old Plancherel in case $p = 2$), $\|\hat{F}_{(k)}(\lambda) - \hat{F}(\lambda)\|_q \to 0$ as $k \to \infty$, where $\frac{1}{q} = 1 - \frac{1}{p}$. Since each $\hat{F}_{(k)}(\lambda)$ vanishes identically on $[0,\infty)$, $\hat{F}(\lambda) \equiv 0$ a.e. there.

Q.E.D.

Remark. In case $p = 2$, the converse of the above theorem is true. Namely, if $\Phi(\lambda) \in L_2(-\infty,\infty)$ and $\Phi(\lambda) \equiv 0$ a.e. for $\lambda \geq 0$, there is an $F \in H_2$ with $\hat{F}(\lambda) = \Phi(\lambda)$. It suffices to take

$$F(t) = \frac{1}{2\pi} \underset{N\to\infty}{\text{l.i.m.}} \int_{-N}^{N} e^{-i\lambda t} \Phi(\lambda) d\lambda.$$

(Here, l.i.m = "limit in the mean" denotes a limit in L_2-norm.)

Indeed, if $\Phi(\lambda) \equiv 0$ on $(0,\infty)$ we have

$$F(t) = \frac{1}{2\pi} \underset{N\to\infty}{\text{l.i.m.}} \int_{0}^{N} e^{i\lambda t} \Phi(-\lambda) d\lambda.$$

For each $y > 0$, $e^{-y\lambda} \Phi(-\lambda) \in L_1(-\infty,\infty)$, so

$$\int_0^\infty e^{i\lambda z} \Phi(-\lambda) d\lambda = \int_0^\infty e^{i\lambda x} e^{-y\lambda} \Phi(-\lambda) d\lambda$$

converges absolutely and defines a function $2\pi F_1(z)$ regular in $\Im z > 0$. By Plancherel's theorem,

$$\int_{-\infty}^\infty |F_1(x+iy)|^2 dx = 2\pi \int_0^\infty e^{-2y\lambda} |\Phi(-\lambda)|^2 d\lambda \leq 2\pi \|\Phi\|_2^2,$$

so $F_1(z) \in H_2$.

Finally, $\|e^{-y\lambda}\Phi(-\lambda) - \Phi(-\lambda)\|_2 \to 0$ as $y \to 0$, so by Plancherel again,

$$\int_{-\infty}^\infty |F_1(t+iy) - F(t)|^2 dt \to 0 \quad \text{as} \quad y \to 0.$$

Therefore $F(t)$ is the boundary-value function $F_1(t)$ of a function $F_1 \in H_2$. The Fourier integral inversion formula for L_2 now shows that $\hat{F}(\lambda) = \Phi(\lambda)$.

Theorem (Phragmén-Lindelöf). Let $f(z)$ be analytic in $\Im z > 0$ and continuous in $\Im z \geq 0$. Suppose that

i) $|f(z)| \le \text{const.}\ \exp(\text{const.}\ |z|)$ in $\Im z \ge 0$

ii) $|f(x)| \le M,\quad x \in \mathbb{R}$

iii) $\limsup_{y\to\infty}\{\log|f(iy)|/y\} = a.$

Then $|f(z)| \le Me^{ay}$, $\Im z \ge 0$.

Proof. Pick any $\varepsilon > 0$. In quadrants I and II, $g(z) = e^{i(a+\varepsilon)z}f(z)$ satisfies $\log|g(z)| \le \mathcal{O}(|z|)$ and $|g(z)|$ is bounded on the boundary of each of these quadrants.

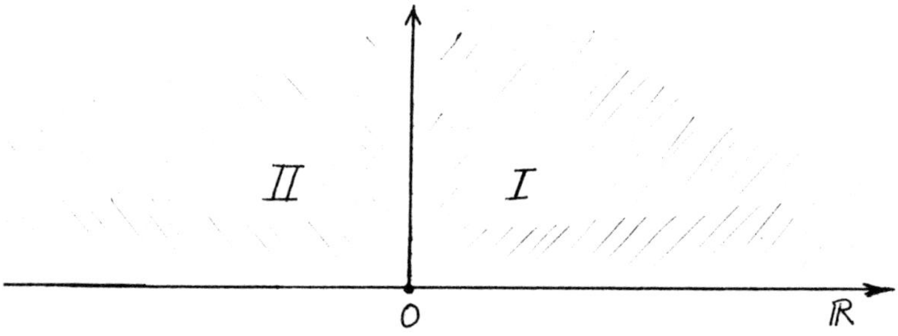

Each quadrant has an opening of $90° < 180°$, so, by the ordinary Phragmén-Lindelöf theorem, $g(z)$ is bounded in quadrants I and II. Therefore $g(z)$ is bounded in $\Im z \ge 0$.

From this, by another elementary Phragmén-Lindelöf theorem,

$$|g(z)| \le \sup_{t\in\mathbb{R}} |g(t)| \le M \quad \text{in} \quad \Im z \ge 0,$$

so $|f(z)| \le Me^{(a+\varepsilon)y}$ there. Squeeze ε.

Definition. An entire function $f(z)$ is said to be of exponential type if there are constants A and B with $|f(z)| \le Ae^{B|z|}$.

Theorem (Paley and Wiener). Let $f(z)$ be entire of exponential type, and suppose that $\int_{-\infty}^{\infty} |f(x)|^2 dx < \infty$.

Let

$$a = \limsup_{y \to -\infty} \frac{\log|f(iy)|}{|y|}$$

$$b = \limsup_{y \to \infty} \frac{\log|f(iy)|}{y} .$$

Then

$$f(z) = \int_{-a}^{b} e^{-iz\lambda}\, \varphi(\lambda)d\lambda,$$

where

$$\int_{-a}^{b} |\varphi(\lambda)|^2 d\lambda < \infty.$$

Proof.

1°. Assume first that we also know f is bounded on $\mathbb{R}$, say $|f(t)| \le M$ for t real. Then, by the above Phragmén-Lindelöf theorem, $|e^{ibz}f(z)| \le M$ for $\Im z \ge 0$, i.e., $e^{ibz}f(z) \in H_\infty(\Im z > 0)$. Therefore, for $\Im z > 0$, by Section C,

$$e^{ibz}f(z) = \frac{1}{\pi}\int_{-\infty}^{\infty} \frac{y}{|z-t|^2}\, e^{ibt}f(t)dt.$$

So, since $f(t)$, and therefore $e^{ibt}f(t)$, belongs to $L_2(-\infty,\infty)$, we have, by Section B,

$$\int_{-\infty}^{\infty} |e^{ib(t+iy)}f(t+iy)|^2 dt \le c, \qquad y > 0,$$

i.e., in fact, $e^{ibz}f(z) \in H_2(\Im z > 0)$.

So, by the first theroem of this section,

$$\hat{f}(\lambda) = \underset{N\to\infty}{\text{l.i.m.}} \int_{-N}^{N} e^{i(\lambda-b)t}[e^{ibt}f(t)]dt$$

is zero a.e. for $\lambda - b \geq 0$, i.e., $\hat{f}(\lambda) \equiv 0$ a.e., $\lambda \geq b$.

Working in $\Im z < 0$ instead of in $\Im z > 0$ we see in the same way that $\hat{f}(\lambda) \equiv 0$ a.e., $\lambda \leq -a$.

The L_2 inversion formula for Fourier transforms now yields, for $x \in \mathbb{R}$,

$$f(x) = \underset{N\to\infty}{\text{l.i.m.}} \frac{1}{2\pi} \int_{-N}^{N} e^{-ix\lambda}\, \hat{f}(\lambda)d\lambda.$$

In the present case, this simply reduces to

$$f(x) = \frac{1}{2\pi} \int_{-a}^{b} e^{-ix\lambda}\, \hat{f}(\lambda)d\lambda.$$

Because $\hat{f}(\lambda) \in L_2$, the integral $\frac{1}{2\pi}\int_{-a}^{b} e^{-iz\lambda}\, \hat{f}(\lambda)d\lambda$ actually converges absolutely for all complex z and represents an entire function of z. This entire function coincides with another, $f(z)$, when z is real. So in fact it is identically equal to $f(z)$. We are done if f is bounded on $\mathbb{R}$.

2°. But $f(x)$ is indeed bounded on $\mathbb{R}$! We are given the inequality

$$f(z) \leq Ae^{B|z|}$$

with two constants A and B. For each $h > 0$, put

$$f_h(z) = \frac{1}{h}\int_0^h f(z+t)dt,$$

then $f_h(z)$ is also entire and satisfies $|f_h(z)| \leq A_h e^{B|z|}$, with the same B as before, irrespective of the value of h.

Since $\|f(x)\|_2 < \infty$, Schwarz' inequality now gives

$$|f_h(x)| \leq \frac{\|f\|_2}{\sqrt{h}},$$

i.e., f_h <u>is bounded on</u> $\mathbb{R}$, and we also have $\|f_h\|_2 \leq \|f\|_2$. Therefore the result already proved in 1° guarantees that each Fourier transform $\hat{f}_h(\lambda)$ <u>vanishes</u> a.e. for $\lambda < -B$ or for $\lambda > B$. So, by the inversion formula already used,

$$f_h(x) = \frac{1}{2\pi}\int_{-B}^{B} e^{-ix\lambda}\,\hat{f}_h(\lambda)d\lambda,$$

whence, by Schwarz,

$$|f_h(x)|^2 \leq \frac{B}{2\pi^2}\int_{-B}^{B} |\hat{f}_h(\lambda)|^2 d\lambda.$$

By Plancherel, the right side equals $\frac{B}{\pi}\|f_h(x)\|_2^2$ which, as we have observed, is $\leq \frac{B}{\pi}\|f\|_2^2$. Thus, for real x,

$$|f_h(x)| \leq \sqrt{\frac{B}{\pi}}\,\|f\|_2$$

for all $h > 0$. Since clearly $f_h(x) \to f(x)$ as $h \to 0$, we get finally $|f(x)| \leq \sqrt{b/\pi}\,\|f\|_2$, and f is bounded on $\mathbb{R}$.

The theorem is completely proved.

<u>Remark</u>. <u>Clearly</u>, <u>if</u> $\varphi \in L_2$ and

$$f(z) = \int_{-a}^{b} e^{-iz\lambda}\,\varphi(\lambda)d\lambda,$$

$f(z)$ <u>is</u> entire of exponential type, $\int_{-\infty}^{\infty} |f(x)|^2 dx < \infty$, and

$$|f(iy)| \leq \text{Const. } e^{by} \quad \text{for} \quad y \to \infty,$$

$$|f(iy)| \leq \text{Const. } e^{a|y|} \quad \text{for} \quad y \to -\infty,$$

(assuming, of course, that $-a < b$!). The Paley-Wiener theorem is a precise converse to this elementary observation.

Remark. If, in the hypothesis of the Paley-Wiener theorem, we suppose $\int_{-\infty}^{\infty} |f(x)|dx < \infty$ instead of $\int_{-\infty}^{\infty} |f(x)|^2 dx < \infty$, but keep the rest of the assumptions the same, we still have

$$f(z) = \frac{1}{2\pi} \int_{-a}^{b} e^{-iz\lambda} \hat{f}(\lambda)d\lambda$$

where $\hat{f}(\lambda)$ is now even continuous.

The proof of this variant is the same as that of the Paley-Wiener theorem.

F. Titchmarsh's Convolution Theorem.

Definition. If $\varphi(\lambda)$, $\psi(\lambda) \in L_1(-\infty,\infty)$, the convolution $\varphi * \psi$ is defined by

$$(\varphi * \psi)(\lambda) = \int_{-\infty}^{\infty} \varphi(\lambda - \tau)\psi(\tau)d\tau.$$

One proves (by use of Fubini's theorem) that the integral on the right converges absolutely for almost all $\lambda \in \mathbb{R}$, and that $\varphi * \psi \in L_1(-\infty,\infty)$. By a change of variable one shows that $\varphi * \psi = \psi * \varphi$. (I am sorry for the heavy use of Greek letters here. The way this chapter has gone, Latin letters serve almost exclusively for Fourier transforms of functions in L_1, especially if these transforms are analytic in some half plane.) As is well known, taking Fourier transforms converts convolution to multiplication, i.e.,

$$\widehat{(\varphi * \psi)}(t) = \hat{\varphi}(t)\hat{\psi}(t).$$

Definition. If $\varphi \in L_1(-\infty,\infty)$, by the supporting interval of φ is meant the minimal closed interval (perhaps infinite!) containing the support of φ, i.e., outside of which $\varphi \equiv 0$ a.e.

Let a be the upper endpoint of the supporting interval of φ, and b that of the supporting interval of ψ. Then direct insepction of the convolution integral shows that $(\varphi * \psi)(\lambda) \equiv 0$ for $\lambda > a + b$. What is remarkable is that if a and $b < \infty$, no cancellation can take place, and a converse to the remark just made holds:

Theorem. If a is the upper endpoint of the supporting interval of φ, b the upper endpoint of the supporting interval of ψ, and if a and b both are finite (sic!), then the upper endpoint of the supporting interval of $\varphi * \psi$ is precisely $a + b$.

Proof. Calling c the upper endpoint of the supporting interval of $\varphi * \psi$, we already know that $c \leq a + b$. We have to prove that $a + b \leq c$.

To this end, use the inverse Fourier transforms

$$f(z) = \hat{\varphi}(-z) = \int_{-\infty}^{a} e^{-iz\lambda}\varphi(\lambda)d\lambda$$

$$g(z) = \hat{\psi}(-z) = \int_{-\infty}^{b} e^{-iz\lambda}\psi(\lambda)d\lambda$$

$$S(z) = \widehat{(\varphi * \psi)}(-z) = \int_{-\infty}^{c} e^{-iz\lambda}(\varphi * \psi)(\lambda)d\lambda.$$

We have $S(z) = f(z)g(z)$.

Because a, b and c are finite, $f(z)$, $g(z)$ and $S(z)$ are analytic for $\Im z > 0$, and because φ, ψ and $\varphi * \psi \in L_1$, f, g and S

are continuous up to $\mathbb{R}$ and bounded on $\mathbb{R}$. In fact, we clearly have, in $\Im z \geq 0$,

$$|f(z)| \leq Ke^{ay}, \quad |g(z)| \leq Le^{by}, \quad |S(z)| \leq Me^{cy}$$

with certain constants K, L and M.

I say that

$$\limsup_{y \to \infty} \frac{\log|g(iy)|}{y} \quad \text{is precisely} \quad b.$$

Call the lim sup in question b', we certainly have $b' \leq b$. Suppose that $b' < b$. For $h > 0$, call $\psi_h(\lambda) = \frac{1}{2h}\int_{-h}^{h} \psi(\lambda + \tau)d\tau$; then $\|\psi_h - \psi\|_1 \to 0$ as $h \to 0$, and, since Fourier transformation takes convolution to multiplication,

$$\hat{\psi}_h(-z) = \frac{\sin hz}{hz}\,\hat{\psi}(-z) = \frac{\sin hz}{hz}\,g(z).$$

But now $\psi_h(\lambda) \in L_2(-\infty,\infty)$, so the Fourier inversion formula for L_2 gives

$$\psi_h(\lambda) = \frac{1}{2\pi}\ \underset{N \to \infty}{\text{l.i.m.}} \int_{-N}^{N} e^{i\lambda t}\,\hat{\psi}_h(-t)dt.$$

Observe, however, that for sufficiently large ℓ,

$$e^{i\ell z}\,\hat{\psi}_h(-z) = e^{i\ell z}\,\frac{\sin hz}{hz}\,g(z)$$

belongs to $H_2(\Im z > 0)$. Indeed,

$$\limsup_{y \to \infty} \frac{\log|g(iy)|}{y} = b'$$

makes $|g(z)| \leq Le^{b'y}$ by the Phragmén-Lindelöf theorem of Section E and we see by direct inspection that $e^{i\ell z}\,\hat{\psi}_h(-z)$ is in H_2 for

$\ell = b' + h$. Therefore, by the first theorem of Section E, $\psi_h(\lambda) \equiv 0$ a.e. for $\lambda \geq b' + h$. Making $h \to 0$, we see that $\psi(\lambda) \equiv 0$ a.e. for $\lambda \geq b'$, and this would contradict our choice of b as the upper limit of the supporting interval of ψ if $b' < b$. So $b' = b$.

In like manner, we prove that $\limsup_{y\to\infty} \frac{\log|f(iy)|}{y}$ is precisely a.

Now, for $k = 1,2,3,\ldots$, let $M(k) = \sup_{x\in\mathbb{R}} |g(x+ki)|$. Given any $\varepsilon > 0$, we must have $M(k+1)/M(k) > e^{b-\varepsilon}$ for some k, otherwise, we would have

$$\limsup_{y\to\infty} \frac{\log|g(iy)|}{y} \leq b - \varepsilon$$

as is easily seen. So, fixing $\varepsilon > 0$, choose a k for which the aforementioned inequality holds, and take $x_0 \in \mathbb{R}$ such that

$$\frac{|g(x_0 + (k+1)i)|}{M(k)} \geq e^{b-\varepsilon}.$$

Without loss of generality, and to simplify the notation, we now assume that $x_0 = 0$.

We work in the half-plane $\Im z \geq k$. By the Phragmén-Lindelöf theorem of Section E, $|g(z)| \leq M(k)e^{b(y-k)}$ for $\Im z \geq k$, so $e^{ibz}g(z)$ is bounded in that half-plane and by Section C we can find a Blaschke product $\mathfrak{B}(z)$ for the upper half-plane with $g(z)/\mathfrak{B}(z-ki)$ free of zeros in $\Im z > k$. We see that there is an $h \in H_\infty(\Im z > 0)$ with $\|h\|_\infty \leq 1$, h free of zeros in $\Im z > 0$, such that

$$g(z) = e^{-ib(z-ki)}M(k)\mathfrak{B}(z-ki)h(z-ki) \quad \text{for } \Im z > k.$$

In the same manner we find a Blaschke product $B(z)$ for the upper half plane and a function $L(z) \in H_\infty(\Im z > 0)$, free of zeros in $\Im z > 0$, with

$$S(z) = e^{-ic(z-ki)}B(z-ki)L(z-ki) \quad \text{for } \Im z > k.$$

The ratio $S(z)/g(z) = f(z)$ is regular for $\Im z > k$. Therefore $B(z)$ must have enough zeros to cancel those of $\mathfrak{B}(z)$! So $B(z)/\mathfrak{B}(z) = B_1(z)$, another Blaschke product, and

$$f(z) = \frac{S(z)}{g(z)} = B_1(z-ki)e^{i(b-c)(z-ki)}\frac{L(z-ki)}{M(k)}\cdot\frac{1}{h(z-ki)} \text{ for } \Im z > k,$$

whence

$$|f(z)| \le \text{const}\cdot\frac{e^{(c-b)y}}{|h(z-ki)|} \quad \text{for } \Im z > k.$$

Now $h(z)$ is free of zeros in $\Im z > 0$ and in modulus ≤ 1 there, so $\log|1/h(z)|$ is positive and harmonic in $\Im z > 0$. Thence, by the Poisson representation given near the beginning of this chapter,

$$\log\left|\frac{1}{h(z)}\right| = \frac{1}{\pi}\int_{-\infty}^{\infty}\frac{y}{|z-t|^2}\,d\mu(t) + \beta y \quad \text{for } \Im z > 0,$$

with a positive measure μ and $\beta \ge 0$. By our choice of k,

$$e^{b-\varepsilon} \le \frac{|g((k+1)i)|}{M(k)} \le e^b|h(i)|$$

(since $x_0 = 0$ and $|\mathfrak{B}(z)| \le 1$), whence $\log|1/h(i)| \le \varepsilon$. That is,

$$\frac{1}{\pi}\int_{-\infty}^{\infty}\frac{d\mu(t)}{t^2+1} + \beta \le \varepsilon\,.$$

But from this we see that for $y \ge 1$

$$\log\left|\frac{1}{h(iy)}\right| = \frac{1}{\pi}\int_{-\infty}^{\infty}\frac{y\,d\mu(t)}{t^2+y^2} + \beta y \le \varepsilon y$$

since $d\mu(t) \ge 0$ and $\beta \ge 0$. Substituted into a previous relation, this makes

$$|f(iy)| \le \text{Const} \cdot e^{(c-b)y} e^{\varepsilon(y-k)}$$

for $y \ge k+1$, i.e.,

$$\limsup_{y\to\infty} \frac{\log|f(iy)|}{y} \le c - b + \varepsilon.$$

However, as we saw above, the lim sup on the left is precisely a. Therefore $a \le c - b + \varepsilon$. Squeezing ε, we get $a + b \le c$, as required.

We are done.

Corollary (Titchmarsh's convolution theorem). Let $\varphi \in L_1$ have finite supporting interval $[a_1, a_2]$ and $\psi \in L_1$ have the finite supporting interval $[b_1, b_2]$. Then the supporting interval of $\varphi * \psi$ is precisely $[a_1 + b_1, a_2 + b_2]$.

Proof. The above theorem shows that the upper endpoint of $\varphi * \psi$'s supporting interval is $a_2 + b_2$. The lower endpoint is similarly seen to be $a_1 + b_1$ - one way of doing this is to first make the change of variable $\lambda \to -\lambda$ and then apply the above theorem again.

Remark. The finiteness of a_1 and b_1 in order to get $a_1 + b_1$ (and of a_2 and b_2 in order to get $a_2 + b_2$) is essential here. This is easily seen by examples.

Remark. Titchmarsh's convolution theorem is a celebrated example of a purely "real variable" result proved by complex variable methods.

Mikusinski and others have given real variable proofs. Most of them are harder than the one given above.

Problem N° 6.

a) Compute $\tilde{k}_\varepsilon(x) = \frac{1}{\pi}\int_{-\infty}^{\infty} \frac{k_\varepsilon(t)dt}{x-t}$ for the function

$$k_\varepsilon(t) = \begin{cases} 0, & |t| < \varepsilon \\ \frac{1}{t}, & |t| \geq \varepsilon . \end{cases}$$

b) If $1 < p < \infty$ and $u(t) \in L_p(-\infty,\infty)$ and $\varepsilon > 0$, put

$$(T_\varepsilon u)(x) = \frac{1}{\pi}\int_{|t-x|\geq\varepsilon} \frac{u(t)dt}{x-t} .$$

Using the function $\tilde{k}_\varepsilon$ found in a), show that there is a C_p independent of ε such that

$$\|T_\varepsilon u\|_p \leq C_p\|u\|_p.$$

VII. Duality for H_p-spaces

A. H_p-spaces and their duals. Sarason's theorem.

1°. We consider mainly the unit circle; similar (and in a sense, more symmetric) results holding for the upper half plane are established by analogous methods, and will be tabulated at the end of this subsection.

By looking just at the boundary values $f(e^{i\theta})$ of $f \in H_p$, we see that H_p can be considered as a $\| \|_p$-closed subspace of $L_p(-\pi,\pi)$. This we do from now on.

We define some more spaces: $\mathcal{C} = \{f$ continuous on $[-\pi,\pi]$; $f(-\pi) = f(\pi)\}$

$\mathcal{A} = \mathcal{C} \cap H_\infty$ — the set of functions in $\mathcal{C}$ which have an analytic extension to $\{|z| < 1\}$, yielding continuous functions on the closed unit disk.

We equip $\mathcal{C}$ and $\mathcal{A}$ with the sup-norm $\| \|_\infty$.

$\mathfrak{M}$ = set of finite complex-valued Radon measures on $\{|\zeta| = 1\}$, equipped with the measure-norm

$$\|\mu\| = \int_{-\pi}^{\pi} |d\mu(e^{i\theta})|.$$

Note that $\mathfrak{M}$ is the dual of $\mathcal{C}$, by a classical theorem of Riesz.

> Notation:
>
> $$H_p(0) = \{f \in H_p;\ \int_{-\pi}^{\pi} f(e^{i\theta})d\theta = 0\} = zH_p.$$

Theorem. If $1 < p < \infty$ and $\frac{1}{p} + \frac{1}{q} = 1$, L_q/H_q has dual $H_p(0)$ and H_p has dual $L_q/H_q(0)$.

<u>Remark</u>. <u>This slight but troublesome asymmetry</u> is one of the results of working in the unit circle. It <u>disappears</u> if we work in the upper half plane.

<u>Proof of Theorem</u>. a) Let Λ be a bounded linear functional on L_q/H_q. Then Λ certainly gives a linear functional on L_q, so for $f \in L_q$, we have

$$\Lambda(f) = \Lambda(f + H_q) = \int_{-\pi}^{\pi} f(e^{i\theta})L(e^{i\theta})d\theta,$$

where L is some <u>function</u> in L_p (Riesz representation theorem). <u>We have</u> $\Lambda(g) = 0$ for $g \in H_p$, <u>in particular</u>,

$$\int_{-\pi}^{\pi} e^{in\theta}L(e^{i\theta})d\theta = 0, \qquad n = 0,1,2,\ldots,$$

so $L(e^{i\theta})$ has a <u>Fourier series</u> of the form $\sum_1^\infty A_n e^{in\theta}$, and, since $L(e^{i\theta}) \in L_p$, $L(e^{i\theta}) \in e^{i\theta}H_p = H_p(0)$. <u>Conversely</u>, any $L(e^{i\theta})$ of this form <u>does give</u> a linear functional Λ on L_q/H_q by the above formula.

b) If $L(e^{i\theta}) \in L_q$, and $f \in H_q(0)$ is <u>arbitrary</u>, the <u>linear form</u>

$$\Lambda g = \int_{-\pi}^{\pi} \{L(e^{i\theta}) + f(e^{i\theta})\}g(e^{i\theta})d\theta ,$$

defined for all $g \in H_p$, <u>does not depend on</u> $f \in H_q(0)$. So $L + H_q(0)$ <u>is</u> a bounded linear functional on H_p.

Conversely, <u>take</u> any linear functional Λ on H_p. By Hahn-Banach, we can first extend Λ to <u>all of</u> L_p, thus getting an $L \in L_q$ with

$$\Lambda g = \int_{-\pi}^{\pi} L(e^{i\theta})g(e^{i\theta})d\theta, \qquad g \in L_p.$$

Restricted back to H_p, Λ is given by the coset $f + H_q(0)$ in the above way.

Q.E.D.

Theorem. The dual of

$$L_1/H_1 \text{ is } H_\infty(0).$$

The dual of

$$H_1 \text{ is } L_\infty/H_\infty(0).$$

Proof. By the same argument used above.

Now, however, comes the

Theorem. The dual of

$$\mathcal{C}/\mathcal{a} \text{ is } H_1(0).$$

Proof. Since the dual of $\mathcal{C}$ is $\mathfrak{m}$, the dual of

$\mathcal{C}/\mathcal{a}$ is $\{\mu \in \mathfrak{m};\ \int_{-\pi}^{\pi} f(e^{i\theta})d\mu(\theta) = 0 \quad \text{for} \quad f \in \mathcal{a}\}$.

In particular, for μ to be in the dual of $\mathcal{C}/\mathcal{a}$, we must have $\int_{-\pi}^{\pi} e^{in\theta}d\mu(\theta) = 0, \quad n = 0,1,2,\ldots$.

Therefore, by the theorem of the brothers Riesz, $d\mu(\theta) = g(e^{i\theta})d\theta$ with a $g \in L_1(-\pi,\pi)$, and we easily check that $g \in H_1(0)$.

Conversely, any $g \in H_1(0)$ clearly defines a functional Λ on $\mathcal{C}/\mathcal{a}$ by putting, for $\varphi \in \mathcal{C}$,

$$\Lambda(\varphi + \mathcal{a}) = \int_{-\pi}^{\pi} g(e^{i\theta})\varphi(e^{i\theta})d\theta.$$

We are done.

Remark. Using an obvious extension of the above notation, write

$$\mathcal{A}(0) = \{e^{i\theta}f(e^{i\theta});\ f \in \mathcal{A}\}.$$

Then, by the same argument used in the above proof, we see that $C/\mathcal{A}(0)$ has the dual H_1. Thus:

$C/\mathcal{A}(0)$ has dual H_1 which has dual $L_\infty/H_\infty(0)$.

In particular, H_1 is a dual, whereas the larger space L_1 is not. This has an important consequence - the unit sphere in H_1 is (sequentially) w^* compact over C for instance, whilst the unit sphere in L_1 has its w^* (sequential) closure equal to $\mathfrak{M}$ (measures!). H_1 also has some other properties resembling those of the (reflexive) spaces L_p, $1 < p < \infty$, rather than those of L_1. There is a paper of D. Newman in Proc. or Bull. A.M.S. - I believe from the early 60's - on that.

Here is a table summarizing the above duality results, together with some others, which are established in exactly the same way:

For Unit Circle

Space	Dual
H_p, $1 \le p < \infty$	$L_q/H_q(0)$, $\frac{1}{q} = 1 - \frac{1}{p}$
$H_p(0)$, $1 \le p < \infty$	L_q/H_q, $\frac{1}{q} = 1 - \frac{1}{p}$
L_p/H_p, $1 \le p < \infty$	$H_q(0)$, $\frac{1}{q} = 1 - \frac{1}{p}$
$L_p/H_p(0)$, $1 \le p < \infty$	H_q, $\frac{1}{q} = 1 - \frac{1}{p}$
$C/\mathcal{A}$	$H_1(0)$
$C/\mathcal{A}(0)$	H_1

Very similar (and more symmetric) results hold in connection with the H_p-spaces for the upper half plane, introduced in Chapter VI. Instead of $\mathcal{C}$ and $\mathcal{A}$, the relevant spaces are here

$$\mathcal{C}_0 = \{F \text{ continuous on } \mathbb{R};\ F(x) \to 0 \text{ as } x \to \pm\infty\}$$

$$\mathcal{A}_0 = \mathcal{C}_0 \cap H_\infty(\Im z > 0).$$

$\mathcal{A}_0$ is the space of functions analytic in $\Im z > 0$, having a continuous extension up to $\mathbb{R}$, and vanishing at ∞ in the closed upper half plane.

Both $\mathcal{C}_0$ and $\mathcal{A}_0$ are equipped with the sup-norm.

Proofs of the following results, presented in tabular form, are very much like the corresponding ones for the unit circle.

For Upper Half Plane

Space	Dual
H_p, $1 \le p < \infty$	L_q/H_q, $\frac{1}{q} = 1 - \frac{1}{p}$
L_p/H_p, $1 \le p < \infty$	H_q, $\frac{1}{q} = 1 - \frac{1}{p}$
$\mathcal{C}_0/\mathcal{A}_0$	H_1

2°. The above duality results give us some theorems about approximation by H_p-functions. We mostly just give results for the unit circle; analogous ones hold for the upper half plane.

Theorem. Let $F \in L_p(-\pi,\pi)$, $1 < p < \infty$, and call $\|F - H_p\|_p = \inf\{\|F - h\|_p;\ h \in H_p\}$. Then:

i) $\|F - H_p\|_p = \sup\{|\int_{-\pi}^{\pi} F(e^{i\theta})g(e^{i\theta})d\theta|;\ g \in H_q(0)\ \&\ \|g\|_q = 1\}$.

ii) There is an $h_0 \in H_p$ with $\|F - H_p\|_p = \|F - h_0\|_p$ (i.e., the minimum is attained).

iii) There is a $g_0 \in H_q(0)$, $\|g_0\|_q = 1$, with

$$\|F - H_p\|_p = \int_{-\pi}^{\pi} F(e^{i\theta})g_0(e^{i\theta})d\theta, \quad \text{(i.e., the sup is attained).}$$

Proof. i) is a metric restatement of the duality between L_p/H_p and $H_q(0)$.

To prove ii), let $h_n \in H_p$ with $\|F - h_n\|_p \xrightarrow[n]{} \|F - H_p\|_p$. Then $\|h_n\|_p$ is bounded, so there is a subsequence $\{h_{n_j}\}$ converging weakly in L_p to a limit h_0; one easily checks that $h_0 \in H_p$. Then $f - h_{n_j} \xrightarrow[j]{} F - h_0$ (weakly). From this, it easily follows that

$$\|F - h_0\|_p \leq \liminf_{j \to \infty} \|F - h_{n_j}\|_p = \|F - H_p\|_p,$$

so $\|F - h_0\|_p = \|F - H_p\|_p$. For the beginner, here is the proof: Take any $\varepsilon > 0$, and take g, $g \in L_q$, $\|g\|_q = 1$ with

$$\|F - h_0\|_p \leq \left| \int_{-\pi}^{\pi} (f(e^{i\theta}) - h_0(e^{i\theta}))g(e^{i\theta})d\theta \right| + \varepsilon.$$

The expression on the right, by w^* convergence of $F - h_{n_j}$ to $F - h_0$, is

$$\lim_{j \to \infty} \left| \int_{-\pi}^{\pi} \{F(e^{i\theta}) - h_{n_j}(e^{i\theta})\}g(e^{i\theta})d\theta \right| + \varepsilon.$$

But

$$\left| \int_{-\pi}^{\pi} \{F(e^{i\theta}) - h_{n_j}(e^{i\theta})\}g(e^{i\theta})d\theta \right| \leq \|F - h_{n_j}\|_p,$$

since $\|g\|_q = 1$.

By going to another subsequence, we get $\|F - h_0\|_p \leq \liminf_{j \to \infty} \|F - h_{n_j}\| + \varepsilon$. Squeeze ε.

Now we must prove iii). By the Hahn-Banach Theorem, there is a linear functional Λ of norm 1 on H_p/H_p such that $\Lambda(F+H_p) = \|F - H_p\|_p$. By a previous theorem, $\Lambda(f+H_p) = \int_{-\pi}^{\pi} f(e^{i\theta})g_0(e^{i\theta})d\theta$ for some $g_0 \in H_q(0)$ of q-norm 1.

We are done.

Theorem. Let $F \in L_1(-\pi,\pi)$. Then there is an $h_0 \in H_1$ with $\|F - H_1\|_1 = \|F - h_0\|_1$, and there is a $g_0 \in H_\infty(0)$ with $\|g_0\|_\infty = 1$ and

$$\{F(e^{i\theta}) - h_0(e^{i\theta})\}g_0(e^{i\theta}) = |F(e^{i\theta}) - h_0(e^{i\theta})|$$

almost everywhere.

Proof. According to a previous remark, H_1 is w^* compact over C, so we can show the existence of an $h_0 \in H_1$ which minimizes $\|F - h_0\|_1$ as in the proof of the previous theorem.

As in the proof of iii) in that theorem, we get a $g_0 \in H_\infty(0)$ with $\|g_0\|_\infty = 1$ and

$$\|F - H_1\|_1 = \|F - h_0\|_1 = \int_{-\pi}^{\pi} F(e^{i\theta})g_0(e^{i\theta})d\theta = \int_{-\pi}^{\pi} \{F(e^{i\theta}) - h_0(e^{i\theta})\}g_0(e^{i\theta})d\theta.$$

Since $|g_0(e^{i\theta})| \le 1$, equality of $\int_{-\pi}^{\pi} |F(e^{i\theta}) - h_0(e^{i\theta})|d\theta$ and $\int_{-\pi}^{\pi} \{F(e^{i\theta}) - h_0(e^{i\theta})\}g_0(e^{i\theta})d\theta$ shows that $g_0(e^{i\theta})$ does the job required of it.

Theorem. Let $F \in C$. Then:

i) $\|F - \mathcal{Q}\|_\infty = \|F - H_\infty\|_\infty = \sup\{|\int_{-\pi}^{\pi} F(e^{i\theta})g(e^{i\theta})d\theta|;\ g \in H_1(0)\ \&\ \|g\|_1 = 1\}$.

ii) There is a $g_0 \in H_1(0)$, $\|g_0\|_1 = 1$, with

$$\|F - \mathcal{Q}\|_\infty = \int_{-\pi}^{\pi} F(e^{i\theta})g_0(e^{i\theta})d\theta.$$

iii) There is an $h_0 \in H_\infty$ with

$$|F(e^{i\theta}) - h_0(e^{i\theta})| \equiv \|F - \mathcal{A}\|_\infty \text{ a.e.}$$

Proof. $H_\infty \supset \mathcal{A}$ so surely $\|F - H_\infty\|_\infty \le \|F - \mathcal{A}\|$. Since $H_1(0)$ is the dual of $C/\mathcal{A}$, we get a $g_0 \in H_1(0)$, $\|g_0\|_1 = 1$, so that ii) holds. But $\int_{-\pi}^{\pi} h(e^{i\theta})g_0(e^{i\theta})d\theta = 0$ for $h \in H_\infty$, so clearly

$$\left|\int_{-\pi}^{\pi} F(e^{i\theta})g_0(e^{i\theta})d\theta\right| \le \|g_0\|_1 \, \|F - H_\infty\|_\infty,$$

proving, by choice of g_0, that $\|F - \mathcal{A}\|_\infty \le \|F - H_\infty\|_\infty$, proving i).

Now H_∞ is a dual (of $L_1/H_1(0)$), so by the argument used above (w^* compactness), there is an $h_0 \in H_\infty$ with $\|F - H_\infty\|_\infty = \|F - h_0\|_\infty$. Then $\|F - h_0\|_\infty =$

$$= \|F - H_\infty\|_\infty = \|F - A_\infty\|_\infty = \int_{-\pi}^{\pi} F(e^{i\theta})g_0(e^{i\theta})d\theta = \int_{-\pi}^{\pi} [F(e^{i\theta}) - h_0(e^{i\theta})]g_0(e^{i\theta})d\theta.$$

That is, since $\int_{-\pi}^{\pi} |g_0(e^{i\theta})|d\theta = 1$,

$$\int_{-\pi}^{\pi} [F(e^{i\theta}) - h_0(e^{i\theta})]g_0(e^{i\theta})d\theta = \|F - h_0\|_\infty \int_{-\pi}^{\pi} |g_0(e^{i\theta})|d\theta \ !$$

So $|F(e^{i\theta}) - h_0(e^{i\theta})| = \|F - h_0\|_\infty$ a.e. on the support of g_0. Since $g_0 \in H_1(0)$, $|g_1(e^{i\theta})| > 0$ a.e. by an old theorem (in Chapter III, Section B). So $|F(e^{i\theta}) - h_0(e^{i\theta})| \equiv \|F - h_0\|_\infty$ a.e.

Q.E.D.

Scholium. If $F \in C$, there is ONLY ONE $h \in H_\infty$ with $\|F - h\|_\infty = \|F - H_\infty\|_\infty$. For, suppose there were two, say, h_1 and h_2. Take the $g_0 \in H_1(0)$ of which the above theorem affirms the existence. Then we get as in the above proof

$$\int_{-\pi}^{\pi} \{F(e^{i\theta}) - h_k(e^{i\theta})\}g_0(e^{i\theta})d\theta = \|F - h_k\|_\infty \int_{-\pi}^{\pi} |g_0(e^{i\theta})|d\theta$$

with $k = 1$ and 2, so, since $|g_0(e^{i\theta})| > 0$ a.e., we must have

$$F(e^{i\theta}) - h_1(e^{i\theta}) = \|F - h_1\|_\infty \frac{|g_0(e^{i\theta})|}{g_0(e^{i\theta})} \quad \text{a.e.},$$

$$F(e^{i\theta}) - h_2(e^{i\theta}) = \|F - h_2\|_\infty \frac{|g_0(e^{i\theta})|}{g_0(e^{i\theta})} \quad \text{a.e.},$$

or, since

$$\|F - h_1\|_\infty = \|F - h_2\|_\infty = \|F - H_\infty\|_\infty,$$

$$h_1(e^{i\theta}) = F(e^{i\theta}) - \|F - H_\infty\|_\infty \frac{|g_0(e^{i\theta})|}{g_0(e^{i\theta})} = h_2(e^{i\theta}) \quad \text{a.e.},$$

as required.

If $F \notin C$, all we have is the

Theorem. Let $F \in L_\infty$. Then there is an $h_0 \in H_\infty$ with

$$\|F - h_0\|_\infty = \|F - H_\infty\|_\infty = \sup \{ | \int_{-\pi}^{\pi} F(e^{i\theta}) g(e^{i\theta}) d\theta | ; \; g \in H_1(0) \;\&\; \|g\|_1 = 1 \}.$$

Proof. Existence of h_0 follows by the w^* compactness of H_∞, and the rest by previous duality results.

Remark. In general, the sup in the above theorem is not attained if $F \in H_\infty$ is not continuous. The work in the above scholium shows that it cannot be attained (by a $g_0 \in H_1(0)$ with $\|g_0\|_1 = 1$) if $F - H_\infty$ contains MORE THAN ONE element of norm equal to $\|F - H_\infty\|_\infty$, i.e., if there is more than one $h \in H_\infty$ which minimizes $\|F - h\|_\infty$. Cases where there are several $h \in H_\infty$ minimizing $\|F - h\|_\infty$ are of practical importance in various questions.

Historical Remark. The above (duality) approach is due to Rogosinski and H. S. Shapiro and, independently, to Havinson. It has been often rediscovered by various people who didn't know about prior work (e.g., by me!). There is a historical note by A. Shields in the A.M.S. Translations, Ser. 2, Vol. 32, 1963.

These results are obviously of great importance in studying approximation problems of various kinds, and continue to have many applications.

If $F \in H_\infty$ is NOT CONTINUOUS, the sup in the above theorem may not be attained even though there is only one $h \in H_\infty$ minimizing $\|F - h\|_\infty$.

Example. Let E_1 and E_2 be two disjoint sets of positive measure on $\{|z| = 1\}$ adding up to $\{|z| = 1\}$, and let

$$F(e^{i\theta}) = \begin{cases} 1 & \text{on } E_1 \\ -1 & \text{on } E_2. \end{cases}$$

Then there is no nonzero $h \in H_\infty$ with $\|F - h\|_\infty \le 1$. But there is no $g_0 \in H_1(0)$ with $\int_{-\pi}^{\pi} F(e^{i\theta}) g_0(e^{i\theta}) d\theta = \|g_0\|_1 = 1$. ($E_1$ and E_2 can even be two complementary arcs, so that F has only two points of discontinuity of very simple form.

Proof. There is no such g_0. Suppose there were. Then we'd have to have

$$g_0(e^{i\theta}) > 0 \quad \text{a.e. on } E_1$$

$$g_0(e^{i\theta}) < 0 \quad \text{a.e. on } E_2,$$

so $g_0(e^{i\theta})$ is real a.e.

So since $g_0 \in H_1$, a generalization of the Schwarz reflection principle given in Vol. II shows that, if we put

$$g_0(z) = \overline{g_0\left(\frac{z}{|z|^2}\right)}$$

for $|z| > 1$, $g_0(z)$ becomes regular everywhere in $\mathbb{C}$ (even on $|z| = 1$). Also, $g_0(0) = 0$ makes $g_0(\infty) = 0$, so $g_0(z)$ is bounded everywhere, hence constant, and the constant must be zero, so $g_0 \equiv 0$. (Another way of seeing this is to observe that $\Im g_0(z)$ must vanish identically, so $g_0(z)$ must be constant and hence $= 0$.)

Suppose now that $h \not\equiv 0$, $h \in H_\infty$ and $|F(e^{i\theta}) - h(e^{i\theta})| \leq 1$ a.e.

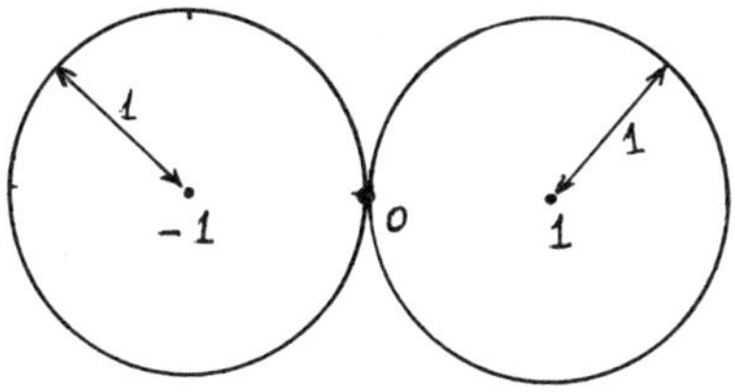

Then, for $e^{i\theta} \in E_1$, $h(e^{i\theta})$ is within the right hand circle, and for $e^{i\theta} \in E_2$, within the left hand one. By a well-known theorem (Chapter III, Section B), $h(e^{i\theta}) = 0$ only on a set of measure zero, so that $h(e^{i\theta})$ takes some values really in the right hand circle (away from 0) and really in the left hand circle.

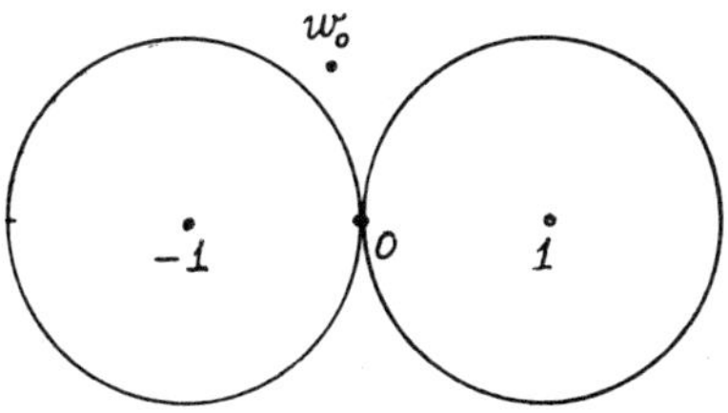

All the values of $h(z)$, for $|z| < 1$, are in these two circles. Indeed, let w_0 be outside them, and suppose $h(z_0) = w_0$. By Runge's theorem, we can get a polynomial $P(w)$ with $|P(w)| \leq 1$ if $|w-1| \leq 1$ or $|w+1| \leq 1$ but $P(w_0) = 2$.

$P(h(z)) \in H_\infty$, and by construction $|P(h(e^{i\theta}))| \leq 1$. Therefore by the Poisson representation for H_∞-functions, $|P(h(z_0))| \leq 1$. But $h(z_0) = w_0$ and $P(w_0) = 2$, yielding a contraduction. So $h(z)$ takes $\{|z| < 1\}$ into the dashed region,

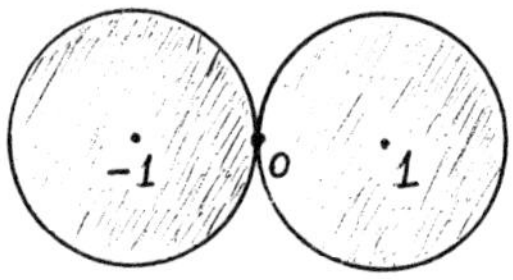

and since it has some boundary values to the right of 0 and some boundary values to the left of 0, it must, in $\{|z| < 1\}$, take values in each of the two circles. Therefore, by connectivity, there must be a z_0, $|z_0| < 1$, with $h(z_0) = 0$. By the principle of conservation of domain, $h(z)$ takes values filling out a neighborhood of 0 for $\{|z| < 1\}$. BUT IT DOESN'T!

So $h \not\equiv 0$ cannot exist, and we are done.

3°. Let us go back to the chain

$$\mathcal{C}/\mathcal{A}(0) \text{ has dual } H_1 \text{ has dual } L_\infty/H_\infty(0).$$

Thus, if B is the Banach space $\mathcal{C}/\mathcal{A}(0)$, B^{**} is $L_\infty/H_\infty(0)$. Now B has a canonical isometric image in B^{**} obtained by identifying linear functionals over B^*. In the present case, the element of

$L_\infty/H_\infty(0)$ corresponding to $F + \mathcal{A}(0)$, $F \in \mathcal{C}$, is the coset $\Phi + H_\infty(0)$, $\Phi \in L_\infty$, determined by

$$\int_{-\pi}^{\pi} F(e^{i\theta})g(e^{i\theta})d\theta = \int_{-\pi}^{\pi} g(e^{i\theta})\Phi(e^{i\theta})d\theta$$

for <u>all</u> $g \in H_1$. We see that this holds iff $\Phi \in F + H_\infty(0)$, which is to say that in the canonical embedding of $\mathcal{C}/\mathcal{A}(0)$ in $L_\infty/H_\infty(0)$, $F + \mathcal{A}(0)$, $F \in \mathcal{C}$, <u>corresponds to</u> $F + H_\infty(0)$. The <u>image</u> of $\mathcal{C}/\mathcal{A}(0)$ in $L_\infty/H_\infty(0)$ under this embedding is <u>thus</u>

$$\mathcal{E} = \{F + H_\infty(0);\ F \in \mathcal{C}\}.$$

<u>In particular</u>, $\mathcal{E}$ <u>is</u> $\| \ \|_\infty$ <u>closed in the quotient space</u> $L_\infty/H_\infty(0)$.

Now the canonical homomorphism $\Theta : L_\infty \longrightarrow L_\infty/H_\infty(0)$ <u>is continuous</u>. <u>Therefore</u> $\Theta^{-1}(\mathcal{E})$ is $\| \ \|_\infty$-closed in L_∞. <u>But</u> $\Theta^{-1}(\mathcal{E})$ is $\mathcal{C} + H_\infty(0)$ which also $= \mathcal{C} + H_\infty$ since $1 \in \mathcal{C}$. Therefore:

<u>Theorem</u> (Sarason). $\mathcal{C} + H_\infty$ is $\| \ \|_\infty$<u>-closed</u>. From this we can prove:

<u>Theorem</u> (Sarason). If F and $G \in \mathcal{C} + H_\infty$, $FG \in \mathcal{C} + H_\infty$, i.e., $\mathcal{C} + H_\infty$ <u>is an algebra</u>.

<u>Proof</u>. It is enough to show that if $F \in \mathcal{C}$ and $G \in H_\infty$, $FG \in \mathcal{C} + H_\infty$. Let the $F_N(e^{i\theta})$ be of the form $\sum_{-N}^{N} A_k(N)e^{ik\theta}$ such that $F_n \longrightarrow F$ <u>uniformly</u>, then $F_NG \longrightarrow FG$ <u>uniformly</u>, so if each $F_NG \in \mathcal{C} + H_\infty$, $FG \in \mathcal{C} + H_\infty$ by the above theorem.

But if

$$G(z) = \sum_0^\infty a_n z^n \in H_\infty,$$

<u>clearly</u> $G_N(z) = \sum_N^\infty a_n z^n \in H_\infty$, and

$$F_N(e^{i\theta})G_N(e^{i\theta}) = \sum_{-N}^{N} A_n(N)e^{in\theta}G_N(e^{i\theta}) \in H_\infty$$

because $e^{in\theta}G_N(e^{i\theta}) \in H_\infty$ for $n = -N, -N+1, -N+2, \ldots$. Finally, $F_N(G - G_N)$, a trigonometric polynomial, is in $\mathcal{C}$. So $F_N G = F_N(G - G_N) + F_N G_N \in \mathcal{C} + H_\infty$, and we're done.

The above proof of these results depends on the fact that H_1 is the dual of $\mathcal{C}/\mathcal{Q}(0)$ and thus, ultimately, on the F. and M. Riesz theorem. As Sarason first showed, norm closure of $\mathcal{C} + H_\infty$ is independent of the F. and M. Riesz theorem. (In fact, Sarason presented this in his 1973 Bull. A.M.S. survey, but the idea itself comes from Zalcman.) In a 1974 lecture at McGill University, Walter Rudin abstracted the Zalcman argument into a general theorem, worthy in its own right:

Theorem. Let B be a Banach space, with E and F norm-closed subspaces of B. Suppose there is a family $\mathfrak{T}$ of linear operators on B with the following properties:

i) $|||T||| \leq M < \infty$ for all $T \in \mathfrak{T}$.

ii) $TB \subseteq E$ for each $T \in \mathfrak{T}$.

iii) $TF \subseteq F$ for each $T \in \mathfrak{T}$.

iv) If $u \in E$ and $\varepsilon > 0$, there is a $T \in \mathfrak{T}$ with $\|Tu - u\| < \varepsilon$.

Then $E + F$ is norm-closed in B.

Proof. Let $x \in$ norm closure of $E + F$. Then we can find $u_n \in E$, $v_n \in F$, with $\|u_n + v_n\| \leq 2^{-n}$ for $n \geq 2$ and $x = \sum_{n=1}^{\infty} (u_n + v_n)$. Write $x_n = u_n + v_n$. Then $x = \sum_{n=1}^{\infty} x_n$, and we have

$$x_n = (u_n - T_n u_n + T_n x_n) + (v_n - T_n v_n)$$

where, for each n, $T_n \in \mathfrak{T}$ is chosen so that $\|u_n - T_n u_n\| < 2^{-n}$. Now $\tilde{u}_n = u_n - T_n u_n + T_n x_n \in E$ and $\|\tilde{u}_n\| \leq 2^{-n} + \|T_n x_n\| \leq 2^{-n}(1+M)$ for $n \geq 2$, since $\|x_n\| \leq 2^{-n}$ for $n \geq 2$. Also, $v_n - T_n v_n = \tilde{v}_n \in F$, and $\|\tilde{v}_n\| \leq \|\tilde{u}_n\| + \|x_n\| \leq (2+M)2^{-n}$ for $n \geq 2$. Therefore $\sum_1^\infty \tilde{u}_n$ converges, to, say, $u \in E$ because E is closed, and $\sum_1^\infty \tilde{v}_n$ converges to $v \in F$ because F is closed. So

$$x = \sum_1^\infty x_n = \sum_1^\infty (\tilde{u}_n + \tilde{v}_n) = u + v \in E + F.$$

Q.E.D.

Corollary. $\mathcal{C} + H_\infty$ is $\| \ \|_\infty$-closed.

Proof. Take $B = L_\infty$, $E = \mathcal{C}$, $F = H_\infty$, and let $\mathfrak{T} = \{T_N, N = 1,2,\ldots\}$ where, if $F \in L_\infty$, if $F(e^{i\theta}) \sim \sum_{-\infty}^\infty A_n e^{in\theta}$,

$$(T_N F)(e^{i\theta}) = \sum_{-N}^{N} \left(1 - \frac{|n|}{N}\right) A_n e^{in\theta}$$

(the Nth Fejer partial sum of the Fourier series of F). Then $|||T_N||| \leq 1$ and for each $F \in \mathcal{C}$, $\|T_N F - F\|_\infty \xrightarrow[N]{} 0$. Clearly $T_N L_\infty \subseteq \mathcal{C}$ and $T_N H_\infty \subseteq H_\infty$. We have the desired result.

Many other similar applications of the general theorem can be given.

B. Elements of constant modulus in cosets of L_∞/H_∞. Marshall's theorem.

1°. Given $F \in L_\infty$, we have seen in the previous section that under certain circumstances (e.g., F continuous), the coset $F + H_\infty$ contains an element of constant modulus equal to $\|F - H_\infty\|_\infty$.

We are interested in seeing the extent to which elements of constant modulus occur in $F + H_\infty$. Around 1920, Nevanlinna proved some deep results about this, using very "hard" methods. In 1976, Garnett found functional-analytic proofs of these results.

Theorem. If $F \in L_\infty$ and $\|F - H_\infty\|_\infty < 1$, $F + H_\infty$ contains an element ω with $|\omega(e^{i\theta})| \equiv 1$ a.e.

Proof (Garnett). The idea is to look for the $\omega \in F + H_\infty$ with $\|\omega\|_\infty \leq 1$ which maximizes $\left|\int_{-\pi}^{\pi} \omega(e^{i\theta})d\theta\right|$ and show that such an ω does the job.

Let

$$a = \sup\{ \left|\int_{-\pi}^{\pi} \omega(e^{i\theta})d\theta\right|; \ \omega \in F + H_\infty, \ \|\omega\|_\infty \leq 1\}.$$

There is indeed a $\omega \in F + H_\infty$, $\|\omega\|_\infty \leq 1$, with $\left|\int_{-\pi}^{\pi} \omega(e^{i\theta})d\theta\right| = a$. For if we take $\omega_n \in F + H_\infty$, $\|\omega_n\|_\infty \leq 1$, with $\left|\int_{-\pi}^{\pi} \omega_n(e^{i\theta})d\theta\right| \underset{n}{\longrightarrow} a$, we can let ω be a w^* (in L_∞) limit of some convergent subsequence of the ω_n, and then $\|\omega\|_\infty \leq 1$, $\left|\int_{-\pi}^{\pi} \omega(e^{i\theta})d\theta\right| = a$, since $1 \in L_1$!

We shall henceforth suppose that $\int_{-\pi}^{\pi} \omega(e^{i\theta})d\theta = a$, which is no restriction, since we can attain this by working with $e^{i\gamma}F$ instead of F, where γ is a real constant.

Now firstly, $\|\omega\|_\infty = 1$. For if $\|\omega\|_\infty = 1 - \varepsilon$ with $\varepsilon > 0$, $\omega + \varepsilon \in \omega + H_\infty = F + H_\infty$ (N.B. $\omega \in F + H_\infty$ because the above ω_n are, and H_∞ is w^* closed!), $\|\omega + \varepsilon\|_\infty \leq 1$, and

$$\int_{-\pi}^{\pi} [\omega(e^{i\theta}) + \varepsilon]d\theta = a + 2\pi\varepsilon > a,$$

a contradiction to the choice of ω.

Secondly, $\|\omega - H_\infty(0)\|_\infty = 1$. For otherwise $\|\omega - H_\infty(0)\|_\infty < 1$, and then there is an $h \in H_\infty(0)$ with $\|\omega - h\|_\infty = 1 - \varepsilon$, $\varepsilon > 0$. Then $-h + \varepsilon \in H_\infty$, so

$$\omega - h + \varepsilon \in F + H_\infty, \ \|\omega - h + \varepsilon\|_\infty \le 1,$$

and because $h \in H_\infty(0)$,

$$\int_{-\pi}^{\pi} [\omega(e^{i\theta}) - h(e^{i\theta}) + \varepsilon]d\theta = \int_{-\pi}^{\pi} \omega(e^{i\theta})d\theta + 2\pi\varepsilon = a + 2\pi\varepsilon > a,$$

again contradicting the choice of $\omega \in F + H_\infty$.

Now, since $\|\omega - H_\infty(0)\|_\infty = 1$, by the above duality theorems, there is a sequence of $f_n \in H_1$, $\|f_n\|_1 = 1$, with

$$\int_{-\pi}^{\pi} \omega(e^{i\theta})f_n(e^{i\theta})d\theta \xrightarrow[n]{} 1.$$

We must have $|f_n(0)| \ge c$ for some $c > 0$. Indeed, if not, then, wlog, $f_n(0) = c_n$ with $c_n \xrightarrow[n]{} 0$. Then $f_n - c_n \in H_1(0)$, $\|f_n - c_n\|_1 \xrightarrow[n]{} 1$, whilst

$$\int_{-\pi}^{\pi} \omega(e^{i\theta})[f_n(e^{i\theta}) - c_n]d\theta \ \text{also} \xrightarrow[n]{} 1.$$

Therefore, by the above duality theorems, $\|\omega - H_\infty\|_\infty \ge 1$, i.e., $\|F - H_\infty\|_\infty \ge 1$ since $\omega \in F + H_\infty$, contradicting our hypothesis that $\|F - H_\infty\|_\infty < 1$. So $|f_n(0)| \ge c > 0$.

We now show that this last inequality implies $|\omega(e^{i\theta})| \equiv 1$ a.e. Assume not. Then, for some $\lambda < 1$ there is a measurable E, $|E| > 0$, with $|\omega(e^{i\theta})| \le \lambda$ on E, wlog, $|E| < 2\pi$ so that also $|\sim E| > 0$. Since in any case $|\omega(e^{i\theta})| \le 1$, we have, because $\|f_n\|_1 = 1$,

$$\left| \int_{-\pi}^{\pi} \omega(e^{i\theta})f_n(e^{i\theta})d\theta \right| \le \lambda \cdot \int_E |f_n(e^{i\theta})|d\theta + \left(1 + \int_E |f_n(e^{i\theta})|d\theta\right),$$

and in order for the left hand side to $\xrightarrow[n]{} 1$ we must have

$$\int_E |f_n(e^{i\theta})|d\theta \xrightarrow[n]{} 0.$$

Therefore

$$\frac{1}{|E|}\int_E \log|f_n(e^{i\theta})|\,d\theta \le \log\left[\frac{1}{|E|}\int_E |f_n(e^{i\theta})|\,d\theta\right] \xrightarrow[n]{} -\infty.$$

At the same time,

$$\frac{1}{|\sim E|}\int_{\sim E}\log|f_n(e^{i\theta})|\,d\theta \le \log\left(\frac{\|f_n\|_1}{|\sim E|}\right) = \log\left(\frac{1}{|\sim E|}\right) < \infty,$$

so finally

$$\int_{-\pi}^{\pi}\log|f_n(e^{i\theta})|\,d\theta \xrightarrow[n]{} -\infty.$$

Therefore the outer factors of the f_n already tend to 0 at the origin, so surely $|f_n(0)| \xrightarrow[n]{} 0$. This contradicts $|f_n(0)| \ge c > 0$ just proven, so we must have $|\omega(e^{i\theta})| \equiv 1$ a.e., Q.E.D.

Corollary. Let $f \in H_\infty$, let $\|f\|_\infty < 1$; and let Ω be any inner function. Then there is another inner function, $\omega \in F + \Omega H_\infty$.

Proof. Apply the above theorem with $F(e^{i\theta}) = (f(e^{i\theta}))/\Omega(e^{i\theta})$. Then $\|F\|_\infty < 1$, so $\|F - H_\infty\|_\infty < 1$, so there is an $h \in H_\infty$ with $|F(e^{i\theta}) + h(e^{i\theta})| \equiv 1$ a.e. Then $f + \Omega h \in H_\infty$ and $|f(e^{i\theta}) + \Omega(e^{i\theta})h(e^{i\theta})| \equiv$ $\equiv 1$ a.e., so $f + \Omega h$ is also an inner function.

Remark. Nevanlinna also proved that if $F \in H_\infty$ and $\|F - H_\infty\|_\infty = 1$, but $F - H_\infty$ contains more than one element of norm 1, then there is an $\omega \in F - H_\infty$ with $|\omega(e^{i\theta})| \equiv 1$ a.e. Garnett has a functional-analytic proof of this fact also, but we do not give it in this course.

2°. Around 1975, D. Marshall verified a long standing conjecture about the uniform approximation of H_∞-functions by linear combinations of Blaschke products. In order to situate his result, let us first establish a much easier and very well known analogous proposition about L_∞.

Theorem. Let $f \in L_\infty(-\pi,\pi)$ and $\|f\|_\infty \le 1$. Then, given any $\varepsilon > 0$ we can find $u_1,\ldots,u_k \in L_\infty$ with $|u_k(\theta)| \equiv 1$ a.e. (so-called unimodular functions) and numbers $\lambda_1,\ldots,\lambda_n \ge 0$, $\sum_k \lambda_k = 1$, with

$$\|f - \sum_k \lambda_k u_k\|_\infty < \varepsilon.$$

In other words:

The norm-closed convex hull of the set of unimodular functions in L_∞ is precisely the unit sphere (ball) of L_∞.

Proof. f can first be uniformly approximated by a measurable function of norm ≤ 1 taking only a finite number of values. This new function can then be uniformly approximated by convex linear combinations of unimodular functions, as elementary duality theory (separation theorems for convex bodies) for finite dimensional spaces shows.

There is another proof, which is more instructive here. Wlog, $\|f\|_\infty \le 1 - \frac{\varepsilon}{2}$, say; otherwise work with $(1 - \frac{\varepsilon}{2})f$ instead of f and observe that $\|f - (1 - \frac{\varepsilon}{2})f\|_\infty \le \frac{\varepsilon}{2}$. Then we may assume $|f(\theta)| \le 1 - \frac{\varepsilon}{2}$ everywhere, so, by CAUCHY'S THEOREM(!),

$$f(\theta) = \frac{1}{2\pi}\int_{-\pi}^{\pi} \frac{e^{it} + f(\theta)}{1 + \overline{f(\theta)}e^{it}}\, dt.$$

Now each of the functions

$$u_t(\theta) = \frac{e^{it} + f(\theta)}{1 + \overline{f(\theta)}e^{it}}$$

is in L_∞, and $|u_t(\theta)| \equiv 1$! Also, since $\|f\| \le 1 - \frac{\varepsilon}{2}$, we have $\|u_t - u_{t'}\|_\infty < \frac{\varepsilon}{2}$ if $|t - t'|$ is less than some positive number δ depending on ε. So for large N,

$$\left\| \frac{1}{2\pi} \int_{-\pi}^{\pi} u_t dt - \frac{1}{N} \sum_{k=1}^{N} u_{2\pi k/N} \right\|_\infty < \frac{\varepsilon}{2}.$$

Therefore $\|f - \frac{1}{N} \sum_{k=1}^{N} u_{2\pi k/N}\|_\infty < \frac{\varepsilon}{2}$, and we're done.

The unimodular functions in H_∞ are just the <u>inner functions</u>, i.e., those having modulus 1 a.e. on the unit circumference. (See Chapter IV, Sections D, E, F and G.) It is very natural to conjecture that, in H_∞, the <u>norm-closed convex hull of the set of inner functions</u> is the <u>unit sphere of</u> H_∞. <u>Marshall</u> proved this.

<u>Lemma</u> (Douglas and Rudin). Let $u \in L_\infty$, $|u(\zeta)| \equiv 1$ a.e. for $|\zeta| = 1$. Then there are <u>inner functions</u> ω and Ω in H_∞ with

$$\left| u(\zeta) - \frac{\omega(\zeta)}{\Omega(\zeta)} \right| < \varepsilon \text{ a.e.}, \qquad |\zeta| = 1.$$

<u>Proof</u> (Marshall). If, for $k = 1,\dots,N$, E_k denotes the subset of $\{|\zeta| = 1\}$ where $\frac{2\pi}{N}(k-1) \le \arg u(\zeta) < \frac{2\pi}{N} k$, then the E_k are disjoint and add up to the unit circumference. Put

$$u_k(\zeta) = \begin{cases} e^{2\pi k i/N}, & \zeta \in E_k \\ 1, & \zeta \notin E_k, \ |\zeta| = 1. \end{cases}$$

Then each u_k is unimodular and <u>takes only two values</u>, and

$$\|u - u_1 \cdot u_2 \cdots u_N\|_\infty \le \frac{2\pi}{N}.$$

So it clearly suffices to establish the result for each factor u_k, i.e., for unimodular functions taking only two values, because the product of inner functions is surely an inner function.

Thus, let the inner function u take only two values, say 1 and γ, $|\gamma| = 1$, $\gamma \neq 1$, say

$$u(\zeta) = 1, \qquad \zeta \in E$$

$$u(\zeta) = \gamma, \qquad \zeta \in \sim E \quad \text{(complement in unit circumference).}$$

Using Poisson's formula, construct $V(z)$ bounded and harmonic in $\{|z| < 1\}$, having boundary values

$$V(\zeta) = 0 \quad \text{a.e.} \quad \text{on} \quad E$$

$$V(\zeta) = -K \quad \text{a.e.} \quad \text{on} \quad \sim E.$$

Then $h = \exp(V + i\tilde{V})$ is analytic and bounded in $|z| < 1$, and there takes values in the open ring $e^{-K} < |h| < 1$. On E, $|h(\zeta)| = 1$ a.e., and on $\sim E$, $|h(\zeta)| = e^{-K}$ a.e.

Let Φ_K be a conformal mapping of the ring $e^{-K} < |w| < 1$ onto the infinite domain (including ∞) obtained from $\mathbb{C} \cup \{\infty\}$ by removing therefrom the two segments $[-\varepsilon, 0]$ and $[\ell, \ell']$, with $\ell = i\,\frac{1-\gamma}{1+\gamma}$ and $\ell' > \ell$. We furthermore take Φ_K so as to map $|w| = 1$ onto $[-\varepsilon, 0]$ and $|w| = e^{-K}$ onto $[\ell, \ell']$.

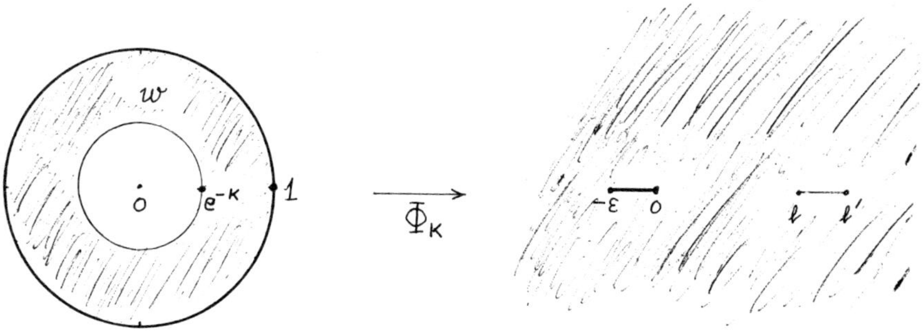

Here, $\varepsilon > 0$ can be chosen as small as we like, but then $\ell' > \ell$ is not free, but depends on the radius e^{-K} of the smaller circle. It is, however, true that $\ell' \to \ell$ as $K \to \infty$, for in that limit Φ_K clearly tends to a conformal mapping of $\{|w| < 1\}$ onto $(\mathbb{C} \cup \{\infty\}) \sim [-\varepsilon, 0]$ which takes 0 to ℓ. Given $\varepsilon > 0$, let us therefore fix K so large that $\ell' < \ell + \varepsilon$.

Having determined K and Φ_K in the manner described, put

$$\Psi(w) = \frac{i - \Phi_K(w)}{i + \Phi_K(w)}$$

for $e^{-K} \le |w| \le 1$. Ψ takes the ring $e^{-K} < |w| < 1$ conformally onto the complement, in $\mathbb{C} \cup \{\infty\}$, of the two arcs $\overparen{\frac{i+\varepsilon}{i-\varepsilon}, 1}$ and $\overparen{\gamma, \gamma'}$ lying on the unit circle.

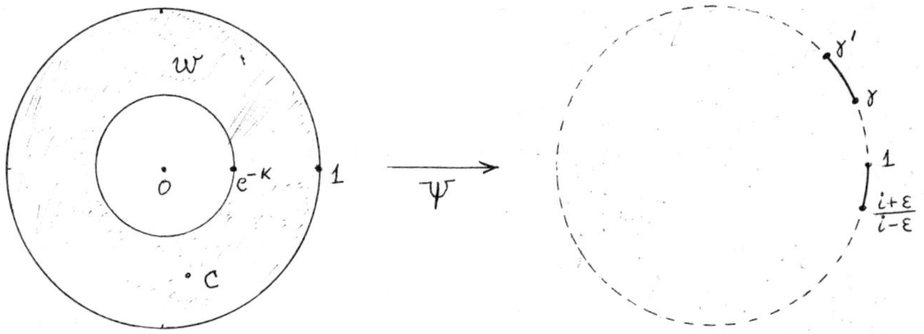

Here, $\gamma' = \frac{i - \ell'}{i + \ell'}$ is close to $\gamma = \frac{i - \ell}{i + \ell}$ by choice of K; in fact, each of the two arcs $\overparen{\frac{i+\varepsilon}{i-\varepsilon}, 1}$ and $\overparen{\gamma, \gamma'}$ has diameter $< 2\varepsilon$.

By construction of the function h we now see that $\Psi(h(\zeta))$ lies on the arc $\overparen{\frac{i+\varepsilon}{i-\varepsilon}, 1}$ for almost all ζ in E, and on the arc $\overparen{\gamma, \gamma'}$ for almost all ζ in $\sim E$. Therefore $|\Psi(h(\zeta)) - u(\zeta)| < 2\varepsilon$ a.e. on $|\zeta| = 1$, and $\Psi(h(\zeta))$ is unimodular, being a.e. in modulus equal to 1 there.

$\Psi(h(z))$ is <u>meromorphic</u> for $|z| < 1$. Indeed, Ψ being <u>conformal</u>, there is precisely <u>one point</u>, say c, $e^{-K} < |c| < 1$, with $\Psi(c) = \infty$, and $\Psi(w)$ has <u>just a simple pole at</u> c. Elsewhere in $e^{-K} < |w| < 1$, $\Psi(w)$ is <u>regular</u>. So $\Psi(h(z))$ is <u>regular</u> at the points z, $|z| < 1$, where $h(z) \neq c$, and has a <u>pole</u> at any z where $h(z) = c$, the <u>order</u> of this <u>pole</u> being the <u>order</u> of the <u>zero</u> that $h(z) - c$ <u>has</u> there.

The function $h(z) - c$ belongs to H_∞; by Chapter IV, Section D we can write

$$h(z) - c = \Omega(z)\mathfrak{G}(z)$$

with $\Omega(z)$ the <u>inner factor</u> of $h(z) - c$ and $\mathfrak{G}(z)$ its <u>outer factor</u>. We have:

$$|\mathfrak{G}(\zeta)| = |h(\zeta) - c| \geq 1 - |c| > 0 \quad \text{a.e. for } \zeta \text{ in } E,$$

$$|\mathfrak{G}(\zeta)| = |h(\zeta) - c| \geq |c| - e^{-K} > 0 \quad \text{a.e. for } \zeta \text{ in } \sim E.$$

Thus, $|\mathfrak{G}(\zeta)|$ is <u>bounded below</u> a.e. on $|\zeta| = 1$, so, by Chapter IV, Section E.2, $1/\mathfrak{G}(z) \in H_\infty$. In other words, $|\Omega(z)/(h(z) - c)|$ is <u>bounded above</u> in $|z| < 1$.

From what we have just seen, it follows easily that $\omega(z) = \Omega(z)\Psi(h(z))$ belongs to H_∞. This function is surely <u>analytic</u> in $|z| < 1$ because any <u>poles</u> of $\Psi(h(z))$ will be <u>cancelled</u> by corresponding <u>zeros</u> of $\Omega(z)$, the inner factor of $h(z) - c$. It is <u>bounded</u> in $|z| < 1$. Indeed, take any small $\delta > 0$; if $e^{-K} < |w| < 1$ and $|w - c| \geq \delta$, $\Psi(w)$ is <u>bounded</u>, hence $\Psi(h(z))$, and therefore $\Omega(z)\Psi(h(z))$, is <u>bounded</u> on the set of z where $|h(z) - c| \geq \delta$. For $|w - c| < \delta$, we have $|\Psi(w)| \leq A/|w - c|$, say, so if $|h(z) - c| < \delta$,

$$|\Omega(z)\Psi(h(z))| \leq A|\Omega(z)/(h(z)-c)|.$$

We have, however, just seen that the expression on the right is bounded in $|z| < 1$.

For almost all ζ, $|\zeta| = 1$, we now have $|\omega(\zeta)| = |\Omega(\zeta)||\Psi(h(\zeta))| = 1$. Therefore $\omega \in H_\infty$ is also an inner function, like Ω. Since $\Psi(h(\zeta)) = \frac{\omega(\zeta)}{\Omega(\zeta)}$, we now have $|u(\zeta) - \frac{\omega(\zeta)}{\Omega(\zeta)}| = |u(\zeta) - \epsilon(h(\zeta))| < 2\varepsilon$ a.e., $|\zeta| = 1$. The lemma is proved with 2ε instead of ε.

Corollary. Let $f \in L_\infty$, $\|f\|_\infty \leq 1$, and let $\varepsilon > 0$. Then we can find inner functions $\omega_1,\dots,\omega_n$; $\Omega_1,\dots,\Omega_n$, and numbers $\lambda_k > 0$, $\Sigma_1^n \lambda_k = 1$, with $|f(\zeta) - \Sigma_1^n \lambda_k\omega_k(\zeta)/\Omega_k(\zeta)| < \varepsilon$ a.e., $|\zeta| = 1$.

Proof. By the lemma and the first theorem of this subsection.

Remark. All the Ω_k can be taken equal - just use a common denominator! For the product of inner functions is an inner function.

Lemma. Let $f \in H_\infty$. Then we can find inner functions $\Omega,\omega,\omega_1,\dots,\omega_n$ and constants $a,a_1,\dots,a_n$ such that

i) $g = \dfrac{a\omega + a_1\omega_1 + \dots + a_n\omega_n}{\Omega}$ is in H_∞

ii) $\|f - g\|_\infty < 2\varepsilon$.

Proof. Wlog, $\|f\|_\infty \leq 1$. Then, by the preceding corollary, we get inner functions $\omega_1,\dots,\omega_n,\Omega$, and numbers $\lambda_k > 0$ with

$$|f(\zeta) - \sum_1^n \lambda_k\omega_k(\zeta)/\Omega(\zeta)| < \varepsilon \text{ a.e. for } |\zeta| = 1.$$

TRICK. Call $F(\zeta) = \sum_1^n \lambda_k\omega_k(\zeta)/\Omega(\zeta)$. Since $f \in H_\infty(!)$, the previous inequality shows that $\|F - H_\infty\|_\infty < \varepsilon$. Therefore by the Nevanlinna theorem in 1° of this Section, there is a $g \in H_\infty$ with $|g(\zeta) - F(\zeta)| \equiv \varepsilon$ a.e. ! Clearly $\|f - g\|_\infty \le \|f - F\|_\infty + \|F - g\|_\infty < 2\varepsilon$. Also, $\Omega(\zeta)g(\zeta) - \Omega(\zeta)F(\zeta) \in H_\infty$, because $\Omega F = \sum_k \lambda_k\omega_k \in H_\infty$. So, since $|\Omega(\zeta)| \equiv 1$ a.e., $|\zeta| = 1$, $\Omega(\zeta)g(\zeta) - \Omega(\zeta)F(\zeta)$ must equal $\varepsilon\omega(\zeta)$ with an inner function ω, being in H_∞ and of constant modulus ε a.e. on $|\zeta| = 1$. So

$$g(\zeta) = \varepsilon \frac{\omega(\zeta)}{\Omega(\zeta)} + \sum_1^n \lambda_k \frac{\omega_k(\zeta)}{\Omega(\zeta)} .$$

We are done.

Theorem (Marshall). Let $f \in H_\infty$ and $\|f\|_\infty \le 1$. Given $\varepsilon > 0$, we can find inner functions, $u_1,\ldots,u_n$ and positive numbers, $\lambda_1,\ldots,\lambda_n$, $\sum_1^n \lambda_k = 1$, with

$$\|f - \sum_k \lambda_k u_k\|_\infty < 4\varepsilon.$$

Proof. Wlog, we may suppose $\|f\|_\infty \le 1 - 2\varepsilon$, otherwise we may work with $(1 - 2\varepsilon)f$ instead of f. By the preceding lemma, we can find a $g \in H_\infty$ of the very special form

$$\sum_k a_k\omega_k/\Omega$$

with constants a_k and inner functions ω_k, Ω, such that $\|f - g\|_\infty < \varepsilon$. In particular, $\|g\|_\infty < 1 - \varepsilon$. Now since, for $|\zeta| = 1$, we have, almost everywhere, $|\omega_k(\zeta)| = 1$, $|\Omega(\zeta)| = 1$, it follows that

$$\overline{g(\zeta)} = \sum_k a_k \frac{\Omega(\zeta)}{\omega_k(\zeta)} \quad \text{a.e.}$$

Let Ω_1 be the product of the ω_k - it is an inner function. Then, although $\overline{g(\zeta)}$ is usually not in H_∞, $\Omega_1(\zeta)\overline{g(\zeta)}$ is!

To functions $g \in H_\infty$ having this property, we can apply a modification of the Cauchy integral argument used to prove the first theorem of the present subsection.

$|g(\zeta)| \leq 1 - \varepsilon$ a.e., so, by Cauchy's theorem,

$$g(\zeta) = \frac{1}{2\pi}\int_{-\pi}^{\pi} \frac{\gamma e^{it} + g(\zeta)}{1 + \overline{g(\zeta)}\gamma e^{it}}\, dt,$$

γ being any number of modulus 1.

Now apply Bernard's trick, and take $\gamma = \Omega_1(\zeta)$!

Then we get

$$g(\zeta) = \frac{1}{2\pi}\int_{-\pi}^{\pi} \frac{\Omega_1(\zeta)e^{it} + g(\zeta)}{1 + \Omega_1(\zeta)\overline{g(\zeta)}e^{it}}\, dt.$$

Each of the functions

$$u_t(\zeta) = \frac{\Omega_1(\zeta)e^{it} + g(\zeta)}{1 + \Omega_1(\zeta)\overline{g(\zeta)}e^{it}}$$

belongs to H_∞, because Ω_1, g, and $\Omega_1\overline{g}$ do, and because $\|\Omega_1\overline{g}\|_\infty \leq 1 - \varepsilon < 1$! Since $|\Omega_1(\zeta)| = 1$ a.e., $|\zeta| = 1$, we have $|u_t(\zeta)| = 1$, a.e., $|\zeta| = 1$, i.e., the functions u_t are inner.

Because $\|g\|_\infty \leq 1 - \varepsilon$, we can apply the argument used to prove the first theorem of this subsection (approximation of the preceding integral by a Riemann sum) to conclude that

$$\|g - \frac{1}{N}\sum_{k=1}^{N} u_{2\pi k/N}\|_\infty < \varepsilon$$

if N is large.

Therefore $\|f - \frac{1}{N}\sum_{k=1}^{N} u_{2\pi k/N}\|_\infty < 2\varepsilon$ for our f with $\|f\|_\infty < 1 - 2\varepsilon$; if we only know $\|f\|_\infty \leq 1$ we get a similar approximation to f to within 4ε instead of 2ε.

Q.E.D.

Recall now Frostman's theorem from Chapter IV, Section G. This says that any inner function can be uniformly approximated by Blaschke products. Combining this fact with the previous result, we immediately have

Marshall's Theorem. Let $f \in H_\infty$, $\|f\|_\infty \leq 1$. Then there are Blaschke products $B_1, \ldots, B_n$ and positive numbers λ_k, $\sum_1^n \lambda_k = 1$, with

$$\|f - \sum_k \lambda_k B_k\|_\infty < \varepsilon,$$

$\varepsilon > 0$ being arbitrary.

Thus, the unit sphere in H_∞ is the norm-closed convex hull of the set of Blaschke products. Truly a most beautiful result.

C. Szegö's Theorem. Let μ be a finite positive measure on $[-\pi, \pi]$, and let $\mathcal{P}(0)$ denote the class of polynomials $P(z)$ with $P(0) = 0$, i.e., polynomials without constant term. If $1 \leq p < \infty$, we are interested in how small we can make

$$\int_{-\pi}^{\pi} |1 - P(e^{i\theta})|^p \, d\mu(\theta) \quad \text{for} \quad p \in \mathcal{P}(0).$$

Theorem. If σ is a positive singular measure,

$$\inf\{ \int_{-\pi}^{\pi} |1 - P(e^{i\theta})|^p \, d\sigma(\theta); \; P \in \mathcal{P}(0)\}$$

is zero.

Proof. Assume the inf is positive. Then, for $\frac{1}{q} + \frac{1}{p} = 1$, there is a $G \in L_q(d\sigma)$ with

$$\int_{-\pi}^{\pi} G(e^{i\theta}) e^{in\theta} d\sigma(\theta) = 0, \qquad n = 1, 2, \ldots,$$

but

$$\int_{-\pi}^{\pi} G(e^{i\theta}) \cdot 1 \, d\sigma(\theta) > 0.$$

By Hölder, $G \in L_1(d\sigma)$, so $ds(\theta) = G(e^{i\theta})d\sigma(\theta)$ is a finite Radon measure on $[-\pi,\pi]$, whose support is contained in that of $d\sigma(\theta)$, i.e., in a set of Lebesgue measure zero. So $ds(\theta)$ is singular. But

$$\int_{-\pi}^{\pi} e^{in\theta} ds(\theta) = 0, \qquad n = 1,2,\ldots,$$

so by the theorem of F. and M. Riesz, $ds(\theta)$ is absolutely continuous!

Therefore $ds(\theta)$ must $= 0$. But on the other hand, $\int_{-\pi}^{\pi} 1 \cdot ds(\theta) > 0$. We have a contradiction. The theorem is proven.

Theorem (Kolmogorov). Let $d\mu(\theta) = w(\theta)d\theta + d\sigma(\theta)$, with $w \in L_1$, $w \geq 0$, $d\sigma \geq 0$ and σ singular. Let $1 \leq p < \infty$.

Then

$$\inf_{P \in \mathcal{P}(0)} \int_{-\pi}^{\pi} |1 - P(e^{i\theta})|^p w(\theta) d\theta = \inf_{P \in \mathcal{P}(0)} \int_{-\pi}^{\pi} |1 - P(e^{i\theta})|^p d\mu(\theta).$$

Remark. Thus, only the absolutely continuous part of μ matters.

Proof of Theorem. Clearly

$$\inf_{P \in \mathcal{P}(0)} \int_{-\pi}^{\pi} |1 - P(e^{i\theta})|^p w(\theta) d\theta \leq \inf_{P \in \mathcal{P}(0)} \int_{-\pi}^{\pi} |1 - P(e^{i\theta})|^p d\mu(\theta),$$

so it suffices to prove the reverse inequality.

<u>Let</u>

$$K = \inf_{P \in \mathcal{P}(0)} \int_{-\pi}^{\pi} |1 - P(e^{i\theta})|^p w(\theta)d\theta.$$

Then there is a $P \in \mathcal{P}(0)$ with

$$\int_{-\pi}^{\pi} |1 - P(e^{i\theta})|^p w(\theta)d\theta < K + \varepsilon.$$

It suffices to find a $Q \in \mathcal{P}(0)$ with

$$\int_{-\pi}^{\pi} |1 - Q(e^{i\theta})|^p (w(\theta)d\theta + d\sigma(\theta)) < K + 4\varepsilon,$$

say.

First of all, by the first theorem of this section, we can find a $P_1 \in \mathcal{P}(0)$ with

$$\int_{-\pi}^{\pi} |1 - P_1(e^{i\theta})|^p d\sigma(\theta) < \varepsilon.$$

Let $P_2(e^{i\theta}) = P_1(e^{i\theta}) - P(e^{i\theta})$, then we have $P_2 \in \mathcal{P}(0)$ and

$$\int_{-\pi}^{\pi} |1 - P(e^{i\theta}) - P_2(e^{i\theta})|^p d\sigma(\theta) < \varepsilon.$$

Take a closed $E \subseteq$ support of σ with

$$\int_{(\sim E)} (1 + |P(e^{i\theta})| + |P_2(e^{i\theta})|)^p d\sigma(\theta) < \varepsilon,$$

this is of course possible. Because $d\sigma$ is <u>singular</u>, $|E| = 0$. Therefore by a construction used in one of the proofs of the F. and M. Riesz theorem given in Chapter II, we can find an $h \in \mathcal{A}$ with $h(e^{i\theta}) \equiv 1$ for $e^{i\theta} \in E$ and $|h(e^{i\theta})| < 1$ if $e^{i\theta} \notin E$. So for $n = 1,2,\ldots,$

$$\int_E |1 - P(e^{i\theta}) - [h(e^{i\theta})]^n P_2(e^{i\theta})|^p d\sigma(\theta)$$

$$= \int_E |1 - P(e^{i\theta}) - P_2(e^{i\theta})|^p d\sigma(\theta) < \varepsilon.$$

Since in any case $|h(e^{i\theta})| \leq 1$, we see, by choice of E, that

$$\int_{-\pi}^{\pi} |1 - P(e^{i\theta}) - (h(e^{i\theta}))^n P_2(e^{i\theta})|^p d\sigma(\theta)$$

$$\leq \varepsilon + \int_{(\sim E)} (1 + |P(e^{i\theta})| + |P_2(e^{i\theta})|)^p d\sigma(\theta) < 2\varepsilon.$$

Since $|h(e^{i\theta})| < 1$ outside E, hence a.e.,

$$1 - P(e^{i\theta}) - (h(e^{i\theta}))^n P_2(e^{i\theta}) \xrightarrow[n]{} 1 - P(e^{i\theta}) \quad \text{a.e.},$$

whence, since $w(\theta) \in L_1$, by Lebesgue's dominated convergence Theorem,

$$\int_{-\pi}^{\pi} |1 - P(e^{i\theta}) - [h(e^{i\theta})]^n P_2(e^{i\theta})|^p w(\theta)d\theta \xrightarrow[n]{}$$

$$\xrightarrow[n]{} \int_{-\pi}^{\pi} |1 - P(e^{i\theta})|^p w(\theta)d\theta < K + \varepsilon.$$

There is thus an n sufficiently large for

$$\int_{-\pi}^{\pi} |1 - P(e^{i\theta}) - [h(e^{i\theta})]^n P_2(e^{i\theta})|^p w(\theta)d\theta \quad \text{to be} \quad < K + 2\varepsilon.$$

Thence, finally,

$$\int_{-\pi}^{\pi} |1 - P(e^{i\theta}) - (h(e^{i\theta}))^n P_2(e^{i\theta})|^p d\mu(\theta) < K + 2\varepsilon + 2\varepsilon = K + 4\varepsilon.$$

Now $h(e^{i\theta}) \in \mathcal{A}$, so $h(e^{i\theta})$ can be uniformly approximated as closely as we want by a polynomial

$$\sum_0^N a_k e^{ik\theta}.$$

We have $(\sum_0^N a_k e^{ik\theta})^n P_2(e^{i\theta}) \in \mathcal{P}(0)$ since $P_2 \in \mathcal{P}(0)$, and if $\sum_0^N a_k e^{ik\theta}$ is <u>close enough</u> to $h(e^{i\theta})$, we will <u>still</u> have

$$\int_{-\pi}^{\pi} |1 - Q(e^{i\theta})|^p d\mu(\theta) < K + 4\varepsilon$$

with $Q \in \mathcal{P}(0)$ given by

$$Q(e^{i\theta}) = P(e^{i\theta}) + (\sum_0^N a_k e^{ik\theta})^n P_2(e^{i\theta}).$$

This does it.

Our study thus reduces to the determination of

$$\inf_{P\in\mathcal{P}(0)} \int_{-\pi}^{\pi} |1 - P(e^{i\theta})|^p w(\theta)d\theta$$

with $w \in L_1(-\pi,\pi)$, $w \geq 0$. This problem is completely solved by the most beautiful and elegant

<u>Theorem of Szegö</u>. If $1 \leq p < \infty$,

$$\boxed{\inf_{P\in\mathcal{P}(0)} \frac{1}{2\pi} \int_{-\pi}^{\pi} |1 - P(e^{i\theta})|^p w(\theta)d\theta = \exp(\frac{1}{2\pi} \int_{-\pi}^{\pi} \log w(\theta)d\theta).}$$

<u>Proof</u>. Suppose first of all that $\int_{-\pi}^{\pi} \log^- w(\theta)d\theta < \infty$, so that $\log w(\theta) \in L_1(-\pi,\pi)$. (<u>Surely</u> $\int_{-\pi}^{\pi} \log^+ w(\theta)d\theta < \infty$ because $w \in L_1$.) It is convenient to work with

$$w_1(\theta) = w(\theta) \cdot \exp(-\frac{1}{2\pi} \int_{-\pi}^{\pi} \log w(t)dt)$$

and

$$K = \exp(\frac{1}{2\pi} \int_{-\pi}^{\pi} \log w(t)dt),$$

so as to have $w(\theta) = Kw_1(\theta)$ with $\int_{-\pi}^{\pi} \log w_1(\theta)d\theta = 0$. In the present case, the desired result will follow if we show that

$$\inf_{P\in\mathcal{P}(0)} \int_{-\pi}^{\pi} |1 - P(e^{i\theta})|^p w_1(\theta)d\theta = 2\pi.$$

Because $\log w_1(\theta) \in L_1(-\pi,\pi)$, we can form the analytic function

$$f(z) = \frac{1}{2\pi}\int_{-\pi}^{\pi} \frac{e^{it}+z}{e^{it}-z} \log w_1(t)dt, \qquad |z| < 1;$$

we have

$$f(0) = \frac{1}{2\pi}\int_{-\pi}^{\pi} \log w_1(t)dt = 0.$$

Since $\log w_1 \in L_1$, by the computation in Chapter IV, Section E.2, $\exp(f(z)/p)$ belongs to H_p. It is outer, and by the material in Chapter I, Section D, $\exp(f(e^{i\theta})/p) = \sqrt[p]{w_1(\theta)}$ a.e.

If $P \in \mathcal{P}(0)$, $G(z) = (1 - P(z))\exp(f(z)/p)$ is also in H_p, and $G(0) = 1$ since $f(0) = 0$. So for $r < 1$, $\frac{1}{2\pi}\int_{-\pi}^{\pi} |G(re^{i\theta})|^p d\theta \geq 1$. Since $G(re^{i\theta}) \longrightarrow G(e^{i\theta})$ in L_p-norm as $r \to 1$ (Chapter IV, Section C), we get $\frac{1}{2\pi}\int_{-\pi}^{\pi} |G(e^{i\theta})|^p d\theta \geq 1$, i.e., $\int_{-\pi}^{\pi} |1 - P(e^{i\theta})|^p w_1(\theta)d\theta \geq 2\pi$ for any $P \in \mathcal{P}(0)$, in view of the relation of $f(e^{i\theta})$ to $w_1(\theta)$. We have thus shown that the inf in question is $\geq 2\pi$.

To prove the reverse inequality, observe that, since $\exp(f(z)/p)$ is outer, there is, by Beurling's theorem (Chapter IV, Section E) a sequence of polynomials $Q_n(z)$ with

$$Q_n(z)\exp(f(z)/p) \xrightarrow[n]{} 1$$

in H_p-norm. Because $\exp(f(0)/p) = 1$, we must have $Q_n(0) \xrightarrow[n]{} 1$, so also

$$\frac{Q_n(z)}{Q_n(0)} \exp \frac{f(z)}{p} \xrightarrow[n]{} 1$$

in H_p-norm. We can write $Q_n(z)/Q_n(0) = 1 - P_n(z)$ with $P_n \in \mathcal{P}(0)$, so we surely have

$$\int_{-\pi}^{\pi} |1 - P_n(e^{i\theta})|^p w_1(\theta)d\theta = \int_{-\pi}^{\pi} \left|1 - P_n(e^{i\theta})\right|^p \left|e^{f(e^{i\theta})}\right| d\theta \xrightarrow[n]{}$$

$$\xrightarrow[n]{} \int_{-\pi}^{\pi} 1^p d\theta = 2\pi,$$

showing that the desired inf is $\leq 2\pi$.

The desired formula is thus proved in the present case.

If now $\int_{-\pi}^{\pi} \log w(\theta)d\theta = -\infty$, let us take

$$w_n(\theta) = \max\{w(\theta);\ \tfrac{1}{n}\};$$

then each $\log w_n \in L_1$. We have $\int_{-\pi}^{\pi} \log w_n(\theta)d\theta \xrightarrow[n]{} -\infty$. For each n, $w_n(\theta) \geq w(\theta)$, so that

$$\inf_{P\in\mathcal{P}(0)} \int_{-\pi}^{\pi} |1 - P(e^{i\theta})|^p w(\theta)d\theta$$

$$\leq \inf_{P\in\mathcal{P}(0)} \int_{-\pi}^{\pi} |1 - P(e^{i\theta})|^p w_n(\theta)d\theta = 2\pi \exp\left(\frac{1}{2\pi}\int_{-\pi}^{\pi} \log w_n(\theta)d\theta\right)$$

by what has already been done. Since this holds for every n, we see, making $n \to \infty$, that

$$\inf_{P\in\mathcal{P}(0)} \int_{-\pi}^{\pi} |1 - P(e^{i\theta})|^p w(\theta)d\theta = 0.$$

But this equals $2\pi \exp\left(\frac{1}{2\pi}\int_{-\pi}^{\pi} \log w(\theta)d\theta\right)$ in the present case.

Szegö's theorem is completely proved.

Q.E.D.

Remark. The connection with Beurling's Theorem (Chapter IV, Section E) is seen to be very close, and the preceding discussion could just as well have been placed at the end of Chapter IV.

D. The Helson-Szegö Theorem. In a 1960 Bologna Annali paper, Helson and Szegö give a characterization of the finite positive measures μ on $[-\pi,\pi]$ having the property that

$$\int_{-\pi}^{\pi} |\tilde{T}(\theta)|^2 d\mu(\theta) \leq \text{const.} \int_{-\pi}^{\pi} |T(\theta)|^2 d\mu(\theta)$$

for all trigonometric polynomials $T(\theta)$.

A trigonometric polynomial $T(\theta)$ is simply a finite sum of the form $\sum_n a_n e^{in\theta}$, and for such a function $T(\theta)$, the harmonic conjugate (Chapter I, Section E) $\tilde{T}(\theta)$ is $-i \sum_n (\text{sgn } n) a_n e^{in\theta}$, where we put $\text{sgn } 0 = 0$.

Definition. A positive measure μ is called a Helson-Szegö measure if

$$\int_{-\pi}^{\pi} |\tilde{T}(\theta)|^2 d\mu(\theta) \leq \text{const.} \int_{-\pi}^{\pi} |T(\theta)|^2 d\mu(\theta)$$

for all trigonometric polynomials T.

Simple direct computation shows that $d\mu(\theta) = d\theta$ is a Helson-Szegö measure.

Theorem. A Helson-Szegö measure is absolutely continuous.

Proof. Suppose E is closed, $|E| = 0$, but $\mu(E) > 0$. Fatou's construction, used in one of the proofs of the theorem of the Brothers Riesz (Chapter II, Section A) gives us an $h \in \mathcal{A}$ (for Definition see Section A.1) with $h(e^{i\theta}) \equiv 1$ on E and $|h(e^{i\theta})| < 1$ elsewhere. We have, surely, $h(0) = a$ with $|a| < 1$. Put $F_n(z) = [h(z)]^n - a^n$.

Then $F_n \in \mathcal{A}$ and $F_n(0) = 0$ so $\Re F_n = -\Im F_n$. If μ is Helson-Szegö, we have therefore

$$\int_{-\pi}^{\pi} |(\Re F_n)(e^{i\theta})|^2 \, d\mu(\theta) \leq C \int_{-\pi}^{\pi} |(\Im F_n)(e^{i\theta})|^2 \, d\mu(\theta)$$

since, because $F_n \in \mathcal{A}$, there is, for each n, a sequence of trigonometric polynomials T_m with $T_m(\theta) \xrightarrow[m]{} (\Im F_n)(e^{i\theta})$ uniformly and $\tilde{T}_m(\theta) \xrightarrow[m]{} -(\Re F_n)(e^{i\theta})$ uniformly. Now, on E,

$$(\Re F_n)(e^{i\theta}) = 1 - \Re a^n \xrightarrow[n]{} 1,$$

so

$$\liminf_{n\to\infty} \int_{-\pi}^{\pi} |(\Re F_n)(e^{i\theta})|^2 \, d\mu(\theta) \geq \mu(E) > 0.$$

However, $|(\Im F_n)(e^{i\theta})| \leq 1 + |a|^n < 2$, whilst, for $e^{i\theta} \in E$,

$$(\Im F_n)(e^{i\theta}) = -\Im a^n \xrightarrow[n]{} 0,$$

and for $e^{i\theta} \notin E$,

$$|(\Im F_n)(e^{i\theta})| \leq |h(e^{i\theta})|^n - \Im a^n \xrightarrow[n]{} 0.$$

So $(\Im F_n)(e^{i\theta}) \xrightarrow[n]{} 0$ boundedly and everywhere, and hence

$$\int_{-\pi}^{\pi} |(\Im F_n)(e^{i\theta})|^2 \, d\mu(\theta) \xrightarrow[n]{} 0.$$

We have reached a contradiction, and are done.

Theorem. A nonzero Helson-Szegö measure is necessarily of the form $d\mu(\theta) = w(\theta)d\theta$ with $w \geq 0$ in $L_1(-\pi,\pi)$ and $\int_{-\pi}^{\pi} \log w(\theta)\,d\theta > -\infty$.

Proof. By the previous theorem $d\mu(\theta) = w(\theta)d\theta$, $w \in L_1$, $w \geq 0$. Suppose $\int_{-\pi}^{\pi} \log w(\theta)d\theta = -\infty$, then we use Szegö's theorem (Section C) to get a contradiction. For then we have a sequence of trigonometric polynomials P_n belonging to our old friend $\rho(0)$ of Section C (the set of finite sums of the form $\sum_{n>0} a_n e^{in\theta}$) such that

$$\int_{-\pi}^{\pi} |1 - P_n(\theta)|^2 w(\theta)d\theta \underset{n}{\longrightarrow} 0 .$$

Put $T_n(\theta) = 1 - P_n(\theta)$. Then $P_n = \approx{T}_n$, so if $w(\theta)d\theta$ is a Helson-Szegö measure,

$$\int_{-\pi}^{\pi} |P_n(\theta)|^2 w(\theta)d\theta = \int_{-\pi}^{\pi} |\approx{T}_n(\theta)|^2 w(\theta)d\theta \leq C \int_{-\pi}^{\pi} |\tilde{T}_n(\theta)|^2 w(\theta)d\theta \leq$$

$$\leq C^2 \int_{-\pi}^{\pi} |T_n(\theta)|^2 w(\theta)d\theta$$

which $\underset{n}{\longrightarrow} 0$. So $\int_{-\pi}^{\pi} |P_n(\theta)|^2 w(\theta)d\theta \underset{n}{\longrightarrow} 0$, and, finally, $\int_{-\pi}^{\pi} 1^2 w(\theta)d\theta = 0$, or $w(\theta) = 0$ a.e. We are done.

Our determination of all Helson-Szegö measures is thus reduced to the examination of those of the form $w(\theta)d\theta$ with positive $w \in L_1$ and $\int_{-\pi}^{\pi} \log w(\theta)d\theta > -\infty$.

Helson and Szegö introduced an auxilliary operation related to ~ .

Definition. If $T(\theta) = \sum_n a_n e^{in\theta}$ is a trigonometric polynomial, call

$$(\Pi T)(\theta) = \sum_{n>0} a_n e^{in\theta}.$$

We use also the following notations:

$$\langle f,g\rangle_w = \int_{-\pi}^{\pi} f(\theta)\overline{g(\theta)}w(\theta)d\theta$$

$$\|f\|_w = \sqrt{\int_{-\pi}^{\pi} |f(\theta)|^2 w(\theta)d\theta} .$$

<u>Lemma</u>. $\|\tilde{T}\|_w \leq K\|T\|_w$ for all trigonometric polynomials T and some K iff $\|\Pi T\|_w \leq C\|T\|_w$ for all such T and some C.

<u>Proof</u>. 1°. If $\|\Pi T\|_w \leq C\|T\|_w$ observe that for $T(\theta) = \sum_n a_n e^{in\theta}$, $\overline{T(\theta)} = \sum_n \overline{a_{-n}}\, e^{in\theta}$, so $(\overline{\Pi\overline{T}}) = \sum_{n<0} a_n e^{in\theta}$, and finally

$$\tilde{T} = -i\Pi T + i(\overline{\Pi\overline{T}}).$$

Clearly $\|\overline{T}\|_w = \|T\|_w$, so

$$\|\tilde{T}\|_w \leq \|\Pi T\|_w + \|\Pi\overline{T}\|_w \leq C\|T\|_w + C\|T\|_w = 2C\|T\|_w.$$

2^0. If $\|\tilde{T}\|_w \leq K\|T\|_w$, observe that $2\Pi T = -\tilde{\tilde{T}} + i\tilde{T}$, so $\|\Pi T\|_w \leq \frac{K^2 + K}{2}\|T\|_w$.

Q.E.D.

<u>Lemma</u>. $w(\theta)d\theta$ is a Helson-Szegö measure iff there is a $\rho < 1$ such that, whenever $P, Q \in \mathcal{P}(0)$,

$$|\Re \int_{-\pi}^{\pi} P(e^{i\theta}) \circ e^{-i\theta}Q(e^{i\theta})w(\theta)d\theta| \leq \rho\|P\|_w\|Q\|_w.$$

<u>Proof</u>. By the above lemma, $w(\theta)d\theta$ is Helson-Szegö iff $\|\Pi T\|_w \leq C\|T\|_w$ for each trigonometric polynomial T. Any such T can be written as $P(\theta) + e^{i\theta}\overline{Q(e^{i\theta})}$ where $P, Q \in \mathcal{P}(0)$ and $P = \Pi T$. Because $|e^{i\theta}| \equiv 1$, we have for such T,

$$\|T\|_w^2 = \|P\|_w^2 + \|Q\|_w^2 + 2\Re \int_{-\pi}^{\pi} P(e^{i\theta})e^{-i\theta}Q(e^{i\theta})w(\theta)d\theta.$$

If, first of all, the inequality in the statement of the lemma holds, this last expression is $\geq \|P\|_w^2 + \|Q\|_w^2 - 2\rho\|P\|_w\|Q\|_w =$ $= (1-\rho^2)\|P\|_w^2 + (\rho\|P\|_w - \|Q\|_w)^2 \geq (1-\rho^2)\|P\|_w^2 = (1-\rho^2)\|\Pi T\|_w^2$, proving $\|\Pi T\|_w^2 \leq \dfrac{1}{1-\rho^2}\|T\|_w^2$.

Conversely, if $\|P\|_w^2 \leq C^2\|T\|_w^2$ (where, wlog, $C > 1!$), we have, if $\|P\|_w = \|Q\|_w = 1$, $\|T\|_w^2 \geq \frac{1}{C^2}$, so that, using the formula for $\|T\|_w^2$ given above,

$$1 + 1 + 2\Re \int_{-\pi}^{\pi} P(e^{i\theta})e^{-i\theta}Q(e^{i\theta})w(\theta)d\theta \geq \frac{1}{C^2},$$

$$\Re \int_{-\pi}^{\pi} P(e^{i\theta})e^{-i\theta}Q(e^{i\theta})w(\theta)d\theta \geq -\left(1 - \frac{1}{2C^2}\right).$$

Repeating the argument with $-P$ instead of P we get, after changing signs,

$$\Re \int_{-\pi}^{\pi} P(e^{i\theta})e^{-i\theta}Q(e^{i\theta})w(\theta)d\theta \leq 1 - \frac{1}{2C^2}.$$

These two inequalities are the same as the one asserted with $\rho = 1 - \frac{1}{2C^2}$. The lemma is proved.

Theorem of Helson and Szegö. A measure $d\mu$ such that

$$\int_{-\pi}^{\pi} |\tilde{T}(\theta)|^2 d\mu(\theta) \leq C \int_{-\pi}^{\pi} |T(\theta)|^2 d\mu(\theta)$$

for all trigonometric polynomials T is necessarily of the form

$$d\mu(\theta) = e^{u(\theta)+\tilde{v}(\theta)}d\theta$$

where u and v are real valued, $|u(\theta)|$ is bounded, and

$$|v(\theta)| \le \frac{\pi}{2} - \varepsilon, \quad \varepsilon > 0.$$

Conversely, if

$$d\mu(\theta) = e^{u(\theta)+\tilde{v}(\theta)}d\theta$$

with u and v as stated, the above inequality holds with some C.

Proof. By the first two theorems of this section, we may restrict ourselves to examination of $d\mu(\theta) = w(\theta)d\theta$ with $w \in L_1$, and $\int_{-\pi}^{\pi} \log w(\theta)d\theta > -\infty$. Take any such w, and form

$$\varphi(z) = \exp\left(\frac{1}{2\pi}\int_{-\pi}^{\pi} \frac{e^{it}+z}{e^{it}-z} \log\sqrt{w(t)}\, dt\right) ;$$

because $w \in L_1$, $\varphi \in H_2$ (and is outer!), and by Chapter I, Section D,

$$|\varphi(e^{i\theta})|^2 = w(\theta) \quad \text{a.e.}$$

Call

$$\frac{w(\theta)}{(\varphi(e^{i\theta}))^2} = e^{iv(\theta)} .$$

By the previous lemma, $w(\theta)d\theta$ is Helson-Szegö iff, for some $\rho < 1$, $P, Q \in \mathcal{P}(0)$ and $\|P\|_w \le 1$, $\|Q\|_w \le 1$ imply

$$\left|\Re \int_{-\pi}^{\pi} \varphi(e^{i\theta})P(e^{i\theta}) \cdot \varphi(e^{i\theta})e^{-i\theta}Q(e^{i\theta})e^{iv(\theta)}d\theta\right| \le \rho.$$

By replacing P with $e^{i\gamma}P$ and varying γ through all real values, the latter condition is seen to be equivalent to

$$\left|\int_{-\pi}^{\pi}\varphi(e^{i\theta})P(e^{i\theta})\cdot\overline{\varphi(e^{i\theta})e^{-i\theta}Q(e^{i\theta})}\cdot e^{iv(\theta)}d\theta\right| \leq \rho\|\varphi P\|_2\|\varphi Q\|_2$$

for all $P, Q \in \mathcal{P}(0)$.

> Now by Beurling's theorem (Chapter IV, Section E) $\varphi\mathcal{P}(0)$ is dense in $H_2(0)$ and $e^{-i\theta}\varphi(e^{i\theta})\cdot\mathcal{P}(0)$ is dense in H_2.

So the previous condition is equivalent to

$$\left|\int_{-\pi}^{\pi} e^{iv(\theta)}F(e^{i\theta})G(e^{i\theta})d\theta\right| \leq \rho\|F\|_2\|G\|_2$$

for all $F \in H_2(0)$ and $G \in H_2$.

By Chapter IV, any $f \in H_1(0)$ can be written as FG with $F \in H_2(0)$, $G \in H_2$, and $\|F\|_2 = \|G\|_2 = \sqrt{\|f\|_1}$.

So finally, $w(\theta)d\theta$ is a Helson-Szegö measure iff, for some $\rho < 1$

$$\left|\int_{-\pi}^{\pi} e^{iv(\theta)}f(\theta)d\theta\right| \leq \rho\|f\|_1, \quad f \in H_1(0).$$

> By Section A.2, this is equivalent to
>
> $$\|e^{iv(\theta)} - H_\infty\|_\infty \leq \rho < 1.$$

Suppose now that $w_0(\theta) = e^{\tilde{v}(\theta)}$ with $|v(\theta)| \leq \frac{\pi}{2} - \varepsilon$, $\varepsilon > 0$. Then, as we see at once, the corresponding outer function $\varphi_0(e^{i\theta})$ satisfies $(\varphi_0(e^{i\theta}))^2 = e^{\tilde{v}(\theta) - iv(\theta)}$, so $(w_0(\theta))/[\varphi_0(e^{i\theta})]^2 = e^{iv(\theta)}$. Since $|v(\theta)| \leq \frac{\pi}{2} - \varepsilon$, we have $|e^{iv(\theta)} - \sin\varepsilon| \leq$ $\leq \cos\varepsilon < 1$, and the constant $\sin\varepsilon$ is in H_∞, so the above boxed relation

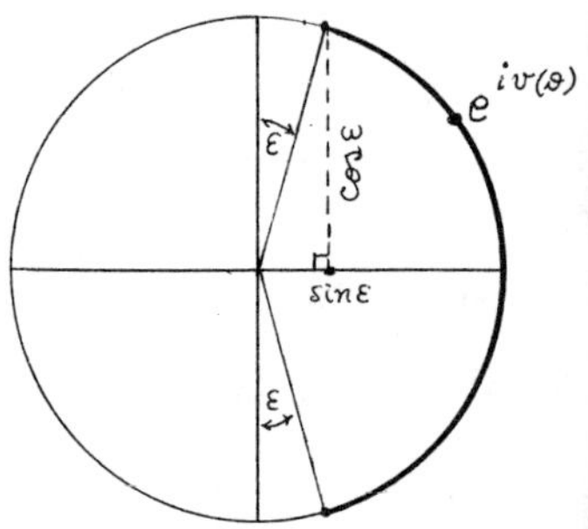

is fulfilled with $\rho = \cos \varepsilon < 1$. Then, $\|\tilde{T}\|_{w_0} \leq K_0 \|T\|_{w_0}$ as we saw above. If now $w(\theta) = e^{u(\theta)} w_0(\theta)$, and $-c \leq u(\theta) \leq c$, we still have

$$\|\tilde{T}\|_w \leq e^{2c} \|T\|_w .$$

So our condition on w is sufficient.

It is necessary. For if the above boxed relation holds, we get an $h \in H_\infty$ with

$$\left| \frac{w(\theta)}{(\varphi(e^{i\theta}))^2} - h(e^{i\theta}) \right| = |e^{iv(\theta)} - h(e^{i\theta})| \leq \rho < 1.$$

Multiplying by $|\varphi(e^{i\theta})|^2 = w(\theta)$, we get, a.e.,

$$|[\varphi(e^{i\theta})]^2 h(e^{i\theta}) - w(\theta)| \leq \rho w(\theta),$$

where $w(\theta) > 0$ a.e. (because $\int_{-\pi}^{\pi} \log w(\theta) d\theta > -\infty$). So the boundary values $[\varphi(e^{i\theta})]^2 h(e^{i\theta})$ of the H_1-function $\varphi^2 h$ belong a.e. to the sector $|\arg Z| \leq \arccos \rho < \frac{\pi}{2}$, and, by Poisson's formula for H_1-functions (Chapter II), the values $[\varphi(z)]^2 h(z)$ also belong to that sector for $|z| < 1$. In particular, $\Re \varphi^2 h \geq 0$ there, so, by a previous exercise, $\varphi^2 h$ is outer. So $\varphi^2 h = e^{i\gamma} e^{u+i\tilde{u}}$ where γ is real. We now have $|\tilde{u}(e^{i\theta}) + \gamma| \leq \arccos \rho$, so by Chapter V we certainly can take the harmonic conjugate

$$\widetilde{\tilde{u} + \gamma} = -u + u(0).$$

Call $-\tilde{u}(e^{i\theta}) - \gamma = v(\theta)$. Then $|v(\theta)| \leq$ arccos ρ and $u(e^{i\theta}) = \tilde{v}(\theta) + c$, c a real constant, so

$$|h(e^{i\theta})||\varphi(e^{i\theta})|^2 = e^{\tilde{v}(\theta)+c}.$$

Finally, $|e^{iv(\theta)} - h(e^{i\theta})| \leq \rho < 1$ makes $\frac{1}{1+\rho} \leq \frac{1}{|h(e^{i\theta})|} \leq \frac{1}{1-\rho}$, so we can call $\frac{e^c}{|h(e^{i\theta})|} = e^{u(\theta)}$ with a bounded (above and below!) real u. Then $w(\theta) = |\varphi(e^{i\theta})|^2 = e^{u(\theta)+\tilde{v}(\theta)}$ with $|v(\theta)| \leq$ arccos $\rho < \frac{\pi}{2}$, Q.E.D.

Remark. This is a most satisfying result. The above line of investigation has been continued by Helson and Sarason, and also by me.

Remark. In 1970 or 71, Hunt, Muckenhoupt and Wheeden determined completely all weights $w(\theta)$ for which

$$\int_{-\pi}^{\pi} |\tilde{T}(\theta)|^p w(\theta)d\theta \leq c \int_{-\pi}^{\pi} |T(\theta)|^p w(\theta)d\theta,$$

where $1 < p < \infty$. Their solution looks entirely different from that of Helson and Szegö, and, for $p = 2$, is as follows:

> $w(\theta)d\theta$ is a Helson-Szegö measure iff, for all intervals I,
>
> $$\left\{\frac{1}{|I|}\int_I w(\theta)d\theta\right\}\left\{\frac{1}{|I|}\int_I \frac{d\theta}{w(\theta)}\right\} \leq c,$$
>
> a finite constant independent of I.

Coifman and Fefferman recently published a simplification of Hunt, Muckenhoupt and Wheeden's work in Studia Mathematica.

The above boxed condition must be equivalent to the fact that $\log w(\theta) - \tilde{v}(\theta) \in L_\infty(-\pi,\pi)$ for some v with $\|v\|_\infty < \frac{\pi}{2}$. Why, is not so evident. The matter is involved with the theory of BMO (the class of functions of bounded mean oscillation), to be treated in Chapter X. At about the time this course was being given, Garnett and Jones found a direct proof of the above mentioned equivalence, which, however, does not quite give the precise upper bound $\frac{\pi}{2}$ for $\|v\|_\infty$. Their work is published in the Annals of Math. for 1976 or 77.

Problem N°7. Let $\omega(\theta) \in L_\infty(-\pi,\pi)$ and $|\omega(\theta)| \equiv 1$ a.e. The problem is to show that there is a nonzero $h \in H_\infty$ such that

(1) $|\omega(\theta) - h(\theta)| \leq 1$ a.e.

if and only if there is a nonzero $f \in H_1$ with

(2) $\omega(\theta) = \dfrac{f(\theta)}{|f(\theta)|}$ a.e.

a) If there is an $f \in H_1$ satisfying (2), show that $h = f/(P + i\tilde{P})$ is in H_∞ and satisfies (1), where $P(\theta) = |f(\theta)|$.

*b) If there is a nonzero $h \in H_\infty$ satisfying (1), show that $he^{\tilde{\psi} - i\psi} \in H_1$, where ψ, $-\frac{\pi}{2} \leq \psi(\theta) \leq \frac{\pi}{2}$ is such that $e^{-i\psi(\theta)}\,\overline{\omega(\theta)}\,h(\theta) > 0$ a.e. Hence get an $f \in H_1$ satisfying (2).

HINT:

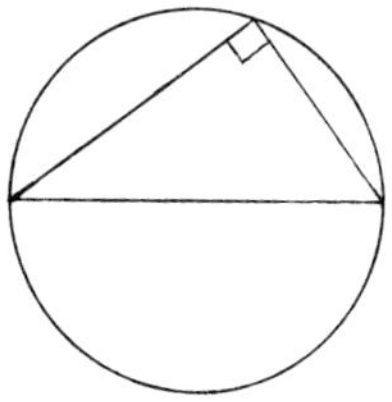

VIII. Application of the Hardy-Littlewood Maximal Function

The present chapter is based on some fairly old work in analysis, but nevertheless marks the beginning of the more recent developments in the theory of H_p-spaces.

A. Use of the distribution function. In this chapter, we only consider functions which are finite almost everywhere.

Definition. If $f(x)$ is complex-valued on $\mathbb{R}$, and finite a.e., we write, for $\lambda > 0$,

$$m_f(\lambda) = |\{x;\ |f(x)| > \lambda\}|.$$

$m_f(\lambda)$ is called the distribution function of f.

The function $m_f(\lambda)$ is clearly decreasing; there is, of course, nothing to prevent its being infinite for some or all values of $\lambda > 0$.

By definition (!) of the Lebesgue integral, we clearly have

$$\boxed{\int_{-\infty}^{\infty} |f(x)|^p dx = \int_{0+}^{\infty} \lambda^p(-dm_f(\mu))}$$

for $p > 0$. We have, moreover, the

Lemma. $\displaystyle\int_{-\infty}^{\infty} |f(x)|^p dx = p\int_0^{\infty} \lambda^{p-1} m_f(\lambda)d\lambda$ for $0 < p < \infty$.

Proof. By integration by parts. There is a slight complication due to the fact that $\mathbb{R}$ has infinite measure.

If, for some $c > 0$, $m_f(\lambda) \geq c$ for all $\lambda > 0$, then $\int_0^{\infty} \lambda^{p-1} m_f(\lambda)d\lambda$ is clearly infinite, but $\int_{-\infty}^{\infty} |f(x)|^p dx$ is clearly $\geq \lambda^p \cdot c$ for all $\lambda > 0$, hence also infinite.

So we need only consider the case where $m_f(\lambda) \to 0$ as $\lambda \to \infty$.

Let $0 < \varepsilon < R < \infty$. Then

$$(*) \qquad -\int_\varepsilon^R \lambda^p dm_f(\lambda) = \varepsilon^p m_f(\varepsilon) - R^p m_f(R) + p\int_\varepsilon^R \lambda^{p-1} m_f(\lambda)d\lambda.$$

i) If $\int_0^\infty \lambda^{p-1} m_f(\lambda)d\lambda < \infty$, then $p\int_0^\varepsilon \lambda^{p-1} m_f(\lambda)d\lambda \to 0$ as $\varepsilon \to 0$, i.e., since $m_f(\lambda)$ <u>decreases</u>,

$$m_f(\varepsilon)\int_0^\varepsilon p\,\lambda^{p-1}d\lambda \to 0$$

as $\varepsilon \to 0$; that is, $\varepsilon^p m_f(\varepsilon) \to 0$ for $\varepsilon \to 0$.

<u>Again</u>, given $\delta > 0$, there is an A so large that for all $R > A$, $p\int_A^R \lambda^{p-1} m_f(\lambda)d\lambda < \delta$. Therefore

$$m_f(R)\int_A^R p\,\lambda^{p-1}d\lambda = (R^p - A^p)m_f(R) < \delta,$$

so, for $R > A$ <u>large enough</u>, $R^p m_f(R) < 2\delta$. This proves $R^p m_f(R) \to 0$ as $R \to \infty$.

Making $\varepsilon \to 0$ and $R \to \infty$ in (*), we thus find

$$p\int_0^\infty \lambda^{p-1} m_f(\lambda)d\lambda = -\int_{0+}^\infty \lambda^p dm_f(\lambda) = \int_{-\infty}^\infty |f(x)|^p dx,$$

<u>in case the extreme left-hand member is finite</u>.

ii) Suppose <u>now</u> that $-\int_{0+}^\infty \lambda^p dm_f(\lambda) < \infty$. As remarked above, we may assume $m_f(\lambda) \to 0$ for $\lambda \to \infty$, and then $R^p m_f(R) = -R^p\int_R^\infty dm_f(\lambda) \le -\int_R^\infty \lambda^p dm_f(\lambda)$, so $R^p m_f(R) \to 0$ as $R \to \infty$. Making $R \to \infty$ in (*) gives, for $\varepsilon > 0$,

$$\varepsilon^p m_f(\varepsilon) + p\int_\varepsilon^\infty \lambda^{p-1} m_f(\lambda)d\lambda = -\int_\varepsilon^\infty \lambda^p dm_f(\lambda),$$

i.e.,

$$p\int_{\varepsilon}^{\infty} \lambda^{p-1} m_f(\lambda)d\lambda \le -\int_{\varepsilon}^{\infty} \lambda^p dm_f(\lambda).$$

Making $\varepsilon \to 0$, we see that $\int_0^{\infty} \lambda^{p-1} m_f(\lambda)d\lambda < \infty$, and we are back in case i). So the lemma holds.

B. The Hardy-Littlewood maximal function

1°. Definition. Let $f(x)$ be measurable on $\mathbb{R}$. The Hardy-Littlewood maximal function, $f^M(x)$, is

$$f^M(x) = \sup_{\xi < x < \xi'} \left\{ \frac{1}{\xi' - \xi} \int_{\xi}^{\xi'} |f(t)|dt \right\}.$$

Evidently, if f is bounded, $|f^M(x)| \le \|f\|_{\infty}$.

Theorem of Hardy and Littlewood.

$$m_{f^M}(\lambda) \le \frac{2}{\lambda} \int_{\{f^M(x) > \lambda\}} |f(x)|dx, \qquad \lambda > 0.$$

Proof. Wlog, $f(x) \ge 0$. Write

$$f_1(x) = \sup_{h>0} \frac{1}{h} \int_x^{x+h} f(t)dt,$$

$$f_2(x) = \sup_{h>0} \frac{1}{h} \int_{x-h}^{x} f(t)dt.$$

For given $\lambda > 0$, call $E_1 = \{x;\ f_1(x) > \lambda\}$, $E_2 = \{x;\ f_2(x) > \lambda\}$, $E = \{x;\ f^M(x) > \lambda\}$; since

$$\frac{1}{h_1+h_2}\int_{x-h_1}^{x+h_2} f(t)dt = \frac{h_1}{h_1+h_2}\left(\frac{1}{h_1}\int_{x-h_1}^{x} f(t)dt\right) + \frac{h_2}{h_1+h_2}\left(\frac{1}{h_2}\int_{x}^{x+h_2} f(t)dt\right),$$

we have $E \subseteq E_1 \cup E_2$, so that $m_{f^M}(\lambda) \leq m_{f_1}(\lambda) + m_{f_2}(\lambda)$, and the theorem will clearly follow as soon as we show that

$$m_{f_1}(\lambda) = \frac{1}{\lambda}\int_{E_1} f(x)dx,$$

$$m_{f_2}(\lambda) = \frac{1}{\lambda}\int_{E_2} f(x)dx.$$

We content ourselves with the proof of the first relation, that of the second being very similar. Write $F(x) = \int_0^x f(t)dt$, using the usual convention if $x < 0$; $F(x)$ is continuous (from $\mathbb{R}$ to $[-\infty,\infty]$), save perhaps at 0, and $F(0) = 0$. $F(x)$ is increasing and E_1 is a countable union of disjoint intervals J_k (one may be of infinite length), obtained from the graph of $F(x)$ vs x by the following construction, due to F. Riesz:

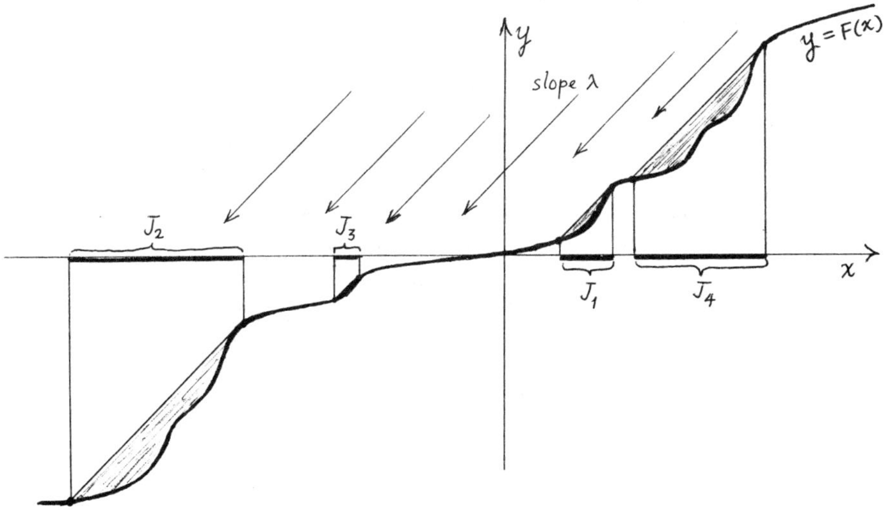

One shines light down along the top of the graph of $F(x)$ vs x, in the direction opposite to that of the lines of slope λ. The J_k are the intervals of the x-axis directly below (or above) the portions of the curve left in shadow.

From the figure, it is manifest that for each J_k,

$$\lambda|J_k| = \text{increase of } F \text{ on } J_k = \int_{J_k} f(t)dt.$$

Hence, adding over the J_k,

$$\lambda|E_1| = \int_{E_1} f(t)dt,$$

i.e.,

$$m_{f_1}(\lambda) = |E_1| = \lambda^{-1}\int_{E_1} f(t)dt,$$

as required.

We are done.

Scholium. The theorem of Hardy and Littlewood can be applied to yield a quick proof of the fact that if $f \in L_1$ and $F(x) = \int_0^x f(t)dt$, then $F'(x)$ exists and equals $f(x)$ almost everywhere (Lebesgue's theorem).

Indeed, if $f \in L_1$ we can construct a sequence of continuous functions $\varphi_n(x)$ with $\|\varphi_n\|_1 \leq 4^{-n}$ for $n \geq 2$, such that the series $\sum_0^\infty \varphi_n(x)$ converges to $f(x)$ both in L_1-norm and a.e. For each $n \geq 2$, let $\mathfrak{S}_n = \{x;\ \varphi_n^M(x) > \frac{1}{2^n}\}$, and put $G_m = \bigcup_{n>m} \mathfrak{S}_n$. If $x \notin G_m$,

$$\sup_{h,h'>0} \frac{1}{h+h'} \int_{x-h'}^{x+h} \left| \sum_{n>m} \varphi_n(t) \right| dt \le \sum_{n>m} \varphi_n^M(x) \le 2^{-m},$$

so, if $\Phi_m(x) = \sum_0^m \varphi_n(x)$, for $x \notin G_m$, we have

$$\Phi_m(x) - 2^{-m} \le \liminf_{h \to 0} \frac{1}{h} \int_x^{x+h} f(t)dt \le \limsup_{h\to 0} \frac{1}{h} \int_x^{x+h} f(t)dt \le \Phi_m(x) + 2^{-m},$$

because $\Phi_m(x) = f(x) - \sum_{n>m} \varphi_n(x)$ is continuous and therefore $\frac{d}{dx} \int_0^x \Phi_m(t)dt$ exists everywhere and equals $\Phi_m(x)$.

Now the Hardy-Littlewood maximal theorem says that

$$|\mathcal{O}_n| \le 2^n \|\varphi_n\|_1 \le 2^{-n} \quad \text{for} \quad n \ge 2,$$

so $|G_m| \le 2^{-m}$. The above calculation shows that if $x \notin G_m$ for all m sufficiently large and $\lim_{m\to\infty} \Phi_m(x)$ exists, then $F'(x)$ also exists and equals that limit. But $\Phi_m(x) \xrightarrow[m]{} f(x)$ a.e., and $G_1 \supset G_2 \supset G_3 \supset \ldots$, with $G = \bigcap_{m>1} G_m$ being of measure zero because $|G_m| \le 2^{-m}$. So $F'(x)$ exists and equals $f(x)$ for almost all $x \notin G$, i.e., almost everywhere.

2°. Norm inequalities for f^M.

Theorem. If $p > 1$ (sic!), $\|f^M\|_p \le C_p \|f\|_p$ with a constant C_p depending only on p.

Proof. One can use the obvious relation $\|f^M\|_\infty \le \|f\|_\infty$ and the Hardy-Littlewood maximal theorem to set up a Marcinkiewicz argument like the one used in Chapter V, Section C.2 in order to prove corresponding inequalities for the harmonic conjugate $\tilde{f}$.

Here, however, a simple trick works. Wlog, $f(x) \geq 0$. We can approximate $f(x)$ from below by bounded functions of compact support. For each φ of the latter, $\varphi^M(x)$ is bounded and $\mathcal{O}(1/|x|)$ at ∞. By making the collection of φ's monotone increasing, we get a sequence of $\varphi^M(x)$ tending to $f^M(x)$ from below, in monotone increasing fashion, so that $\|f^M\|_p$ is the limit of the $\|\varphi^M\|_p$. It thus clearly suffices to show that $\|\varphi^M\|_p \leq C_p\|\varphi\|_p$ for each φ.

Now, however, we are assured that $\|\varphi^M\|_p < \infty$ because $p > 1$, $\varphi^M(x)$ being bounded and $\mathcal{O}(1/|x|)$ at ∞, and we proceed as follows. From Section A,

$$\|\varphi^M\|_p^p = p\int_0^\infty \lambda^{p-1} m_{\varphi^M}(\lambda)d\lambda$$

which, in turn, is

$$\leq 2p\int_0^\infty \left\{ \int_{\{\varphi^M(x)>\lambda\}} \varphi(x)dx \right\} \lambda^{p-2}d\lambda$$

by the Hardy-Littlewood maximal theorem. On changing the order of integration (!), this last is seen to equal

$$2p\int_{-\infty}^\infty \int_0^{\varphi^M(x)} \lambda^{p-2}\varphi(x)d\lambda dx = \frac{2p}{p-1}\int_{-\infty}^\infty \varphi(x)[\varphi^M(x)]^{p-1}dx.$$

Hölder's inequality now shows that the last expression is $\leq \frac{2p}{p-1}\|\varphi\|_p \|\varphi^M\|_p^{p-1}$, so we have

$$\|\varphi^M\|_p^p \leq \frac{2p}{p-1}\|\varphi\|_p \|\varphi^M\|_p^{p-1} .$$

Since $\|\varphi^M\|_p^{p-1} < \infty$, it can be cancelled from both sides (!), yielding $\|\varphi^M\|_p \leq \frac{2p}{p-1}\|\varphi\|_p$, and proving the theorem, with $C_p = \frac{2p}{p-1}$.

Q.E.D.

3°. On $\mathbb{R}$, there is no substitute for the theorem of the preceding subsection in the case $p = 1$. That's because $\mathbb{R}$ has infinite measure.

Take, for instance

$$f(x) = \begin{cases} 1, & |x| \leq 1 \\ 0, & |x| > 1, \end{cases}$$

then, as is easily seen, $f^M(x) \geq \frac{\text{const.}}{|x|}$ for large $|x|$, so $f^M \notin L_1(\mathbb{R})$, despite the fact that f belongs to all L_p-spaces and is of compact support.

The correct L_1 substitute is for $L_1(E)$ with $|E| < \infty$, and it is established by a Marcinkiewicz argument like the one used in proving Zygmund's theorem (Chapter V, Section C.3).

Theorem. If $|E| < \infty$,

$$\int_E f^M(x)dx \leq 2|E| + 4\int_{-\infty}^{\infty} |f(x)|\log^+|2f(x)|dx.$$

Proof. Let $\mu(\lambda) = \{x \in E;\ f^M(x) > \lambda\}$, then

$$\int_E f^M(x)dx \leq |E| + \int_1^{\infty} \lambda(-d\mu(\lambda))$$

which, by integration by parts, is seen to be less than or equal (in fact equal) to

$$|E| + \mu(1) + \int_1^{\infty} \mu(\lambda)d\lambda \leq 2|E| + \int_1^{\infty} \mu(\lambda)d\lambda \leq 2|E| + \int_1^{\infty} m_{f^M}(\lambda)d\lambda.$$

To estimate $m_{f^M}(\lambda)$, we use Marcinkiewicz' trick and write, for given $\lambda > 0$,

$$f(x) = g_\lambda(x) + h_\lambda(x),$$

where

$$g_\lambda(x) = \begin{cases} f(x), & |f(x)| \le \lambda/2 \\ 0\ , & \text{otherwise}\ , \end{cases} \qquad h_\lambda(x) = \begin{cases} 0\ , & |f(x)| \le \lambda/2 \\ f(x), & |f(x)| > \lambda/2. \end{cases}$$

Clearly, $f^M(x) \le g_\lambda^M(x) + h_\lambda^M(x)$, and $\|g_\lambda\|_\infty \le \frac{\lambda}{2}$ so that $\|g_\lambda^M\|_\infty \le \frac{\lambda}{2}$. Therefore $f^M(x) > \lambda$ implies $h_\lambda^M(x) > \frac{\lambda}{2}$, so

$$m_{f^M}(\lambda) \le |\{x;\ h_\lambda^M(x) > \tfrac{\lambda}{2}\}|$$

which, by the Hardy-Littlewood maximal theorem, is

$$\le \frac{2}{(\lambda/2)}\int_{-\infty}^{\infty} |h_\lambda(x)|dx = \frac{4}{\lambda}\int_{\{|f(x)|>\lambda/2\}} |f(x)|dx = \frac{4}{\lambda}\int_{\lambda/2}^{\infty} s(-dm_f(s)).$$

Therefore,

$$\int_E f^M(x)dx \le 2|E| + \int_1^\infty m_{f^M}(\lambda)d\lambda \le 2|E| + \int_1^\infty \left\{\frac{4}{\lambda}\int_{\lambda/2}^{\infty} s(-dm_f(s))\right\} d\lambda =$$

$$= 2|E| + \int_{1/2}^{\infty}\int_1^{2s} \frac{4s}{\lambda}\, d\lambda(-dm_f(s)) = 2|E| + \int_{1/2}^{\infty} (4s \log 2s)(-dm_f(s)) =$$

$$= 2|E| + 4\int_{\{|f(x)|>1/2\}} |f(x)|\log(2|f(x)|)dx =$$

$$= 2|E| + 4\int_{-\infty}^{\infty} |f(x)\log^+(2|f(x)|)dx,$$

Q.E.D.

Since f^M only depends on the <u>modulus</u>, $|f(x)|$, of f, the <u>form</u> of the known partial converse to Zygmund's theorem (Chapter V, Section C.4) should lead us to suspect that the result just proven <u>also</u> has a converse. Despite, however, the length of time that the above theorem has been known (it is in the first edition of Zygmund's book!) its converse was not noticed until 1969, when it was published by E. M. Stein. In fact, the steps in the above proof can practically be reversed!

Theorem. Let $f \in L_1(\mathbb{R})$ be of compact support. Then $\int_E f^M(x)dx$ is finite for every E of finite measure only if

$$\int_{-\infty}^{\infty} |f(x)|\log^+|f(x)|dx < \infty.$$

Proof. If f is of compact support, clearly $f^M(x) \leq \frac{\text{const}}{|x|}$ for large $|x|$, so $E = \{x;\ f^M(x) > 1\}$ is of finite measure. Finiteness of $\int_E f^M(x)dx$ with this E will lead to the desired conclusion.

Let us take the partial maximal function

$$f_1(x) = \sup_{h>0} \frac{1}{h} \int_x^{x+h} |f(t)|dt$$

already used in the proof of the Hardy-Littlewood theorem (1°). Then $f_1(x) \leq f^M(x)$, so

$$\int_{\{f_1(x)>1\}} f_1(x)dx \leq \int_E f^M(x)dx < \infty.$$

Now, however,

$$\int_{\{f_1(x)>1\}} f_1(x)dx = -\int_1^\infty \lambda\, dm_{f_1}(\lambda) = m_{f_1}(\lambda) + \int_1^\infty m_{f_1}(\lambda)d\lambda \geq \int_1^\infty m_{f_1}(\lambda)d\lambda,$$

so the last quantity must be finite.

In proving the Hardy-Littlewood theorem, we actually showed that

$$m_{f_1}(\lambda) = \frac{1}{\lambda}\int_{\{f_1(x)>\lambda\}} |f(x)|dx,$$

so

$$\int_1^\infty m_{f_1}(\lambda)d\lambda = \int_1^\infty \int_{\{f_1(x)>\lambda\}} |f(x)|dx\,\frac{d\lambda}{\lambda} =$$

$$= \int_{\{f_1(x)>1\}} \int_1^{f_1(x)} |f(x)|\frac{d\lambda}{\lambda}\,dx = \int_{\{f_1(x)>1\}} |f(x)|\log f_1(x)dx\,.$$

But $f_1(x) \geq |f(x)|$ a.e., so that the last integral is

$$\geq \int_{\{f(x)>1\}} |f(x)|\log f_1(x)dx \geq \int_{\{f(x)>1\}} |f(x)|\log|f(x)|dx = \int_{-\infty}^\infty |f(x)|\log^+|f(x)|dx.$$

In short,

$$\int_{\{f_1(x)>1\}} f_1(x)dx \geq \int_{-\infty}^\infty |f(x)|\log^+|f(x)|dx,$$

and if the quantity on the left is finite, so is the one on the right. We are done.

C. Application to functions analytic or harmonic in the upper half plane, or in the unit circle.

1°. Lemma. Let $\int_{-\infty}^\infty \frac{|f(t)|dt}{1+t^2} < \infty$, and put for $\Im z > 0$,

$$V(z) = \frac{1}{\pi}\int_{-\infty}^\infty \frac{y}{(x-t)^2+y^2}\,f(t)dt.$$

Then

$$|V(x+iy)| \leq \left(\frac{|x|}{y} + 2\right)f^M(0).$$

<u>Proof</u>. By a calculation which is essentially the same as one already carried out in Chapter I, Section D.3, We may assume $f^M(0) < \infty$, since otherwise there is nothing to prove.

Integration by parts yields:

$$|V(x+iy)| \leq \frac{1}{\pi}\int_{-\infty}^{\infty} \frac{y}{(x-t)^2+y^2}\,|f(t)|\,dt =$$

$$= \frac{1}{\pi}\frac{y}{(x-t)^2+y^2}\int_0^t |f(s)|\,ds\Bigg]_{-\infty}^{\infty} + \frac{1}{\pi}\int_{-\infty}^{\infty}\left\{\frac{2y(t-x)}{[(x-t)^2+y^2]^2}\int_0^t |f(s)|\,ds\right\}dt.$$

Because $f^M(0) < \infty$, the <u>integrated</u> term is <u>zero</u>. The <u>second</u> term can be rewritten as

$$\frac{1}{\pi}\int_{-\infty}^{\infty} \frac{2y(t-x)t}{[(x-t)^2+y^2]^2}\cdot\left(\frac{1}{t}\int_0^t |f(s)|\,ds\right)dt,$$

with $t^{-1}\int_0^t |f(s)|\,ds \leq f^M(0)$ <u>by definition</u>. We have

$$\frac{2y(t-x)t}{[(x-t)^2+y^2]^2} = \frac{2y(t-x)^2}{[(t-x)^2+y^2]^2} + \frac{2y(t-x)\cdot x}{[(t-x)^2+y^2]^2},$$

and this is in absolute value

$$\leq \frac{2y+|x|}{(t-x)^2+y^2}\ .$$

Since $\frac{1}{\pi}\int_{-\infty}^{\infty} \frac{y}{(t-x)^2+y^2}\,dt = 1$, substitution into the previous expression yields

$$|V(x+iy)| \leq \left(2 + \frac{|x|}{y}\right)f^M(0).$$

Q.E.D.

2°. Definition. If $F(z)$ is defined for $\Im z > 0$, put, for $x \in \mathbb{R}$,

$$F^*(x) = \sup\{|F(\xi + i\eta); 0 \le |\xi - x| < \eta\}.$$

Thus, $F^*(x)$ is the sup of $|F(\zeta)|$ in the 90° sector S_x, symmetric about the vertical, with vertex at x:

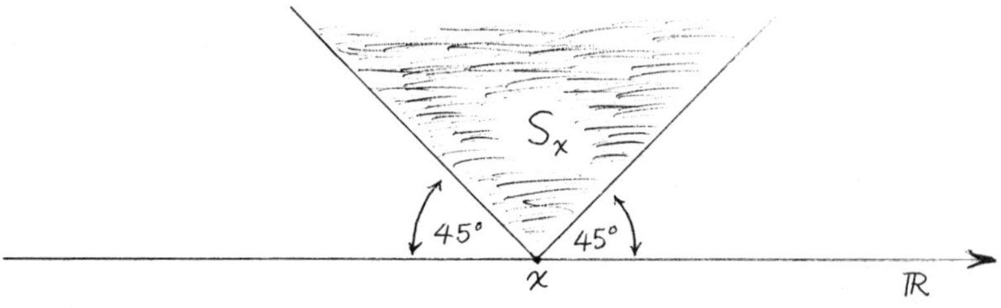

Any other angle less than 180° could have been chosen instead of the 90°, which we take here merely for convenience in writing.

$F^*(x)$ is frequently referred to as the nontangential maximal function of $F(z)$.

Theorem. Let $p > 1$ (sic!), let $V(z)$ be harmonic in $\Im z > 0$, and suppose that

$$\|V\|_p^p = \sup_{h>0} \int_{-\infty}^{\infty} |V(x + ih)|^p dx$$

is finite.

Then

$$\int_{-\infty}^{\infty} [V^*(x)]^p dx \le K_p \|V\|_p^p,$$

with a constant K_p depending only on p.

Proof. By Chapter VI, Section C there is an $f \in L_p(\mathbb{R})$, with $\|f\|_p$ equal to the norm $\|V\|_p$ defined above, such that

$$V(z) = \frac{1}{\pi}\int_{-\infty}^{\infty} \frac{y}{|z-t|^2} f(t)dt.$$

By the lemma of 1°, we have, making a translation along the x-axis:

$$V^*(x) \leq 3f^M(x).$$

But $\|f^M\|_p \leq C_p\|f\|_p$ by Section B.2. So $\|V^*\|_p \leq 3C_p\|f\|_p = 3C_p\|V\|_p$.

Q.E.D.

Theorem. Let $p > 0$ and $F(z) \in H_p(\Im z > 0)$. Then

$$\int_{-\infty}^{\infty} [F^*(x)]^p dx \leq K_p\|F\|_p^{(p)}$$

with a K_p depending only on p, where (as in Chapter IV),

$$(p) = \begin{cases} 1, & 0 < p < 1 \\ p, & p \geq 1. \end{cases}$$

Proof. By Chapter VI, Section C we can write $F(z) = B(z)G(z)$, where $B(z)$ is a Blaschke product for the upper half plane (so that, in particular, $F^*(x) \leq G^*(x)$!), $G(z)$ has no zeros in $\Im z > 0$, and $G \in H_p(\Im z > 0)$ with $\|G\|_p = \|F\|_p$.

Therefore $V(z) = [G(z)]^{2/p}$ can be defined so as to be analytic, hence harmonic (complex valued) in $\Im z > 0$, and $\|V\|_2^2 = \|G\|_p^{(p)} = \|F\|_p^{(p)}$. We have $(F^*(x))^p = (G^*(x))^p = (V^*(x))^2$. Our desired result now follows from the previous theorem.

Corollary. For $F(z) \in H_1(\Im z > 0)$, we have, with $U(z) = \Re F(z)$ and $V(z) = \Im F(z)$,

$$\int_{-\infty}^{\infty} U^*(x)dx \leq C\|F\|_1, \quad \int_{-\infty}^{\infty} V^*(x)dx \leq C\|F\|_1.$$

3°. Definition. If $f(z)$ is defined in $\{|z| < 1\}$ and θ is real we put $f^*(\theta) = \sup\{|f(z)|;\ z \in S_\theta\}$, where S_θ is the dashed region shown here:

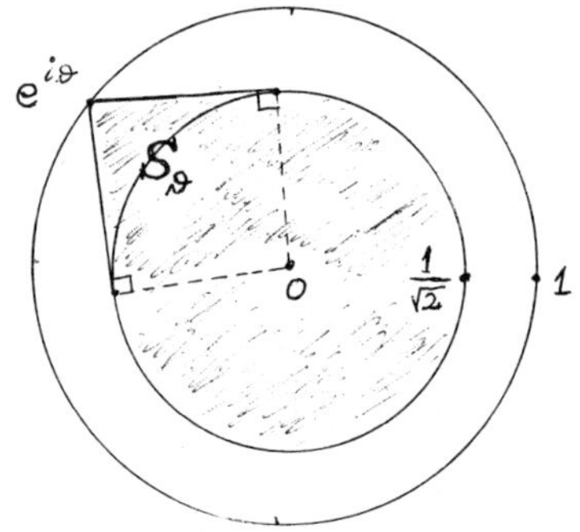

By arguments similar to those used in 2°, we can establish the following results:

Theorem. Let $p > 1$ (sic!), let $v(z)$ be harmonic in $\{|z| < 1\}$ and suppose that

$$\|v\|_p^p = \sup_{r<1} \int_{-\pi}^{\pi} |v(re^{i\theta})|^p d\theta$$

is finite. Then, with a k_p depending only on p,

$$\int_{-\pi}^{\pi} [v^*(\theta)]^p d\theta \leq k_p\|v\|_p^p .$$

<u>Theorem</u>. Let $f(z) \in H_p(|z| < 1)$ for some $p > 0$. Then, with a constant K_p depending only on p,

$$\int_{-\pi}^{\pi} [f^*(\theta)]^p \, d\theta \le K_p \|f\|_p^{(p)} ,$$

where $(p) = \sup(1,p)$.

<u>Corollary</u>. For $f \in H_1$ and $u = \Re f$,

$$\int_{-\pi}^{\pi} u^*(\theta) d\theta \le C\|f\|_1 .$$

4°. The maximal Hilbert transform.

<u>Lemma</u>.

$$\left| \int_{|t|>y} \frac{f(t)}{t} \, dt - \int_{-\infty}^{\infty} \frac{tf(t)}{t^2 + y^2} \, dt \right| \le (1 + 2\pi) f^M(0).$$

<u>Proof</u>. The expression on the <u>left</u> equals

$$\left| \int_{|t|>y} \frac{y^2}{t(y^2 + t^2)} f(t) dt - \int_{-y}^{y} \frac{tf(t)dt}{y^2 + t^2} \right| \le$$

$$\le \frac{1}{2y} \int_{-y}^{y} |f(t)| dt + \int_{-\infty}^{\infty} \frac{y}{t^2 + y^2} |f(t)| dt \le$$

$$\le f^M(0) + \sup_{y>0} \int_{-\infty}^{\infty} \frac{y}{t^2 + y^2} |f(t)| dt \le (1 + 2\pi) f^M(0)$$

by the lemma of 1°.

Q.E.D.

<u>Theorem</u>. Let $1 < p < \infty$ (sic!), and $f \in L_p(\mathbb{R})$. The so-called <u>maximal Hilbert transform</u>

$$\check{f}(x) = \sup_{\varepsilon>0} \left| \frac{1}{\pi} \int_{|t-x|>\varepsilon} \frac{f(t)}{x-t}\, dt \right|$$

satisfies $\|\check{f}\|_p \leq K_p \|f\|_p$.

Proof. For $\Im z > 0$, put

$$V(z) = \frac{1}{\pi} \int_{-\infty}^{\infty} \frac{x-t}{|z-t|^2} f(t)dt,$$

then $V(z)$ is harmonic for $\Im z > 0$, and by Chapter VI, Section D, for each $h > 0$,

$$\int_{-\infty}^{\infty} |V(x+ih)|^p dx \leq C_p \|f\|_p^p,$$

with a C_p depending only on p (Riesz' theorem). Hence, by 2° above, $\|V^*\|_p \leq K_p \|f\|_p$ with some constant K_p. Making a translation, we see by the lemma just proven that

$$\left| \frac{1}{\pi} \int_{|t-x|>y} \frac{f(t)}{x-t}\, dt - V(x+iy) \right| \leq \left(\frac{1}{\pi} + 2\right) f^M(x)$$

for each $y > 0$.

Therefore,

$$\check{f}(x) \leq \left(\frac{1}{\pi} + 2\right) f^M(x) + V^*(x).$$

But we know that $\|f^M\|_p \leq \tilde{C}_p \|f\|_p$ here, according to Section B.2. Since $\|V^*\|_p \leq K_p \|f\|_p$, we're done.

Remark. An analogous result holds for the unit circle.

D. Maximal function characterization of $\Re H_1$. If $F(z) \in H_1(\Im z > 0)$ and $U(z) = \Re F(z)$, a corollary of Section C.2 says that $U^*(x) \in L_1(\mathbb{R})$. In 1971, Burkholder, Gundy and Silverstein discovered (with the help of probability theory) that the CONVERSE result holds!

1°. Theorem. If $U(z)$ is real valued and harmonic in $\Im z > 0$ and $\int_{-\infty}^{\infty} U^*(x)dx < \infty$, then $U = \Re F$ for an $F \in H_1(\Im z > 0)$.

Proof. Based on two ideas in Fefferman and Stein's 1972 Acta paper, and worked up especially for these lectures.

For $h > 0$ and $\Im z \geq 0$ write $U_h(x) = U(z + ih)$. Note that since $U^* \in L_1$, $\|U_h\|_1$ (notation of Section C.2) is surely finite for each $h > 0$. This being the case, U has a harmonic conjugate V, given, for any half plane $\Im z > h > 0$, by

$$V(z) = \frac{1}{\pi}\int_{-\infty}^{\infty} \frac{x - t}{(x - t)^2 + (y - h)^2} U_h(t)dt.$$

The claim is that $U + iV \in H_1$.

In order to prove this, we will show that

$$\int_{-\infty}^{\infty} |V_h(x)|dx \leq 4 \int_{-\infty}^{\infty} U^*(x)dx$$

for each $h > 0$. In fact, we will show that

$$\int_{-\infty}^{\infty} |V_h(x)|dx \leq 4 \int_{-\infty}^{\infty} U_h^*(x)dx$$

which is stronger, since $U_h^*(x) \leq U^*(x)$ as the following diagram makes clear:

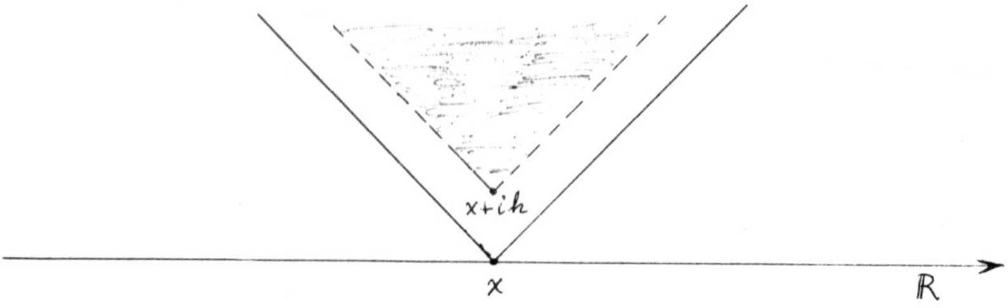

Since $\int_{-\infty}^{\infty} U^*(x)dx < \infty$, $\int_{-\infty}^{\infty} |U(x+ih)|dx$ is bounded for $h > 0$, and therefore, by Chapter VI, Section C, $|U_h(z)| \le \frac{\text{const.}}{h}$ for $\Im z > 0$ which, since $\|U\|_1 < \infty$, implies that in fact

$$\int_{-\infty}^{\infty} [U_h(x+iy)]^2 dx \le C_h < \infty \quad \text{for} \quad y \ge 0.$$

(C_h may be enormous if $h > 0$ is small, but that is of no concern to us here!) By Chapter VI, Section D we now see that $\int_{-\infty}^{\infty} [V_h(x+iy)]^2 dx \le C_h$ for $y \ge 0$, so that in fact $U_h + iV_h \in H_2(\Im z > 0)$ for each $h > 0$. The function $U_h(z) + iV_h(z)$ is continuous in $\Im z \ge 0$ and tends to zero when $z \to \infty$ there, since (Chapter VI)

$$U_h(z) + iV_h(z) = \frac{i}{\pi}\int_{-\infty}^{\infty} \frac{1}{z - t + ih/2}\, U_{h/2}(t)dt$$

for $\Im z \ge 0$.

Let, for $\lambda > 0$

$$m(\lambda) = |\{x;\ U_h^*(x) > \lambda\}|,$$

$$\mu(\lambda) = |\{x;\ |V_h(x)| > \lambda\}|.$$

We proceed to estimate $\mu(\lambda)$ in terms of $m(\lambda)$. Call $\mathcal{O}_\lambda = \{x;\ U_h^*(x) > \lambda\}$, and $E_\lambda = \mathbb{R} \sim \mathcal{O}_\lambda$. We have $|\mathcal{O}_\lambda| = m(\lambda)$, and clearly

$$\mu(\lambda) \le |\{x \in E_\lambda;\ V_h(x) > \lambda\}| + |\mathcal{O}_\lambda| = m(\lambda) + |\{x \in E_\lambda;\ |V_h(x)| > \lambda\}|.$$

We proceed to estimate the second term on the right.

$\mathcal{O}_\lambda$ is a bounded open set, hence a disjoint union of finite open intervals J_k:

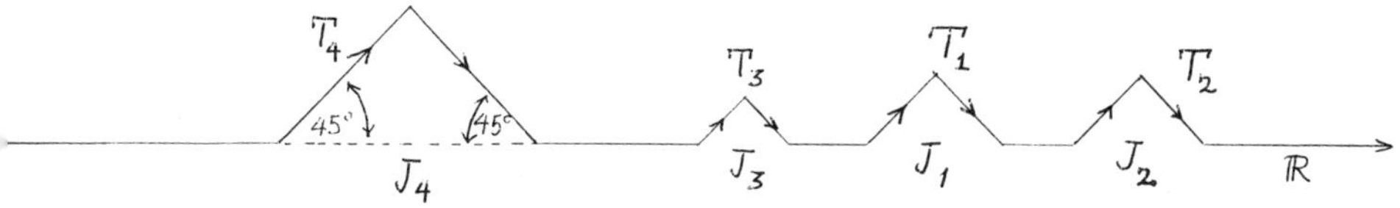

Above J_k, let T_k be the $45°$ roof constructed as shown, let $T = \bigcup_k T_k$, and let Γ be the curve consisting of the T_k and $E_\lambda = \mathbb{R} \sim \mathcal{O}_\lambda$, so oriented that $x = \Re z$ increases as z moves along Γ in the positive sense:

Since $U_h + iV_h \in H_2$ and $\mathcal{O}_\lambda$, hence T, is bounded, by Cauchy's theorem and the lemma at the beginning of Chapter VI, Section D:

$$\int_\Gamma [U_h(z) + iV_h(z)]^2 dz = 0.$$

On taking real parts, this becomes

$$\int_{E_\lambda} (U_h^2 - V_h^2)dx + \int_T (U_h^2 - V_h^2)dx - 2\int_T U_h V_h dy = 0.$$

Now on each piece T_k of T, $dy = \pm dx$, so

$$\left| 2 \int_T U_h V_h dy \right| \le \int_T (U_h^2 + V_h^2) dx$$

which, substituted in the previous, yields, on transposition:

$$\int_{E_\lambda} V_h^2 dx \le \int_{E_\lambda} U_h^2 dx + 2 \int_T U_h^2 dx.$$

> <u>But on each segment of</u> T, $|U_h(z)| \le \lambda$,
> <u>since any such segment has a foot in</u>
> $E_\lambda = \{x; U_h^*(x) \le \lambda\}$. (This is the <u>first</u>
> idea of Fefferman and Stein.)

Therefore

$$\int_{E_\lambda} V_h^2 dx \le \int_{E_\lambda} (U_h^*)^2 dx + 2\lambda^2 \int_T dx = \int_{E_\lambda} (U_h^*)^2 dx + 2\lambda^2 m(\lambda),$$

since

$$\int_T dx = |\mathcal{O}_\lambda| = m(\lambda).$$

We have now

$$\int_{E_\lambda} (U_h^*)^2 dx = \int_{\{U_h^*(x) \le \lambda\}} [U_h^*(x)]^2 dx = \int_0^\lambda s^2(-dm(s)),$$

hence (and this is the <u>second</u> idea of Fefferman and Stein),

$$|\{x \in E_\lambda ; |V_h(x)| > \lambda\}| \le \frac{1}{\lambda^2} \int_{E_\lambda} (V_h)^2 dx \le \frac{1}{\lambda^2} \int_0^\lambda s^2(-dm(s)) + 2m(\lambda).$$

By the above inequality for $\mu(\lambda)$, this yields

$$\boxed{\mu(\lambda) \le \frac{1}{\lambda^2} \int_0^\lambda s^2(-dm(s)) + 3m(\lambda).}$$

The lemma of Section A now gives

$$\int_{-\infty}^{\infty} |V_h(x)|\,dx = \int_0^{\infty} \mu(\lambda)\,d\lambda \leq \int_0^{\infty}\int_0^{\lambda} \frac{s^2}{\lambda^2}\,(-dm(s))\,d\lambda + 3\int_0^{\infty} m(\lambda)\,d\lambda =$$

$$= 3\int_{-\infty}^{\infty} U_h^*(x)\,dx + \int_0^{\infty}\int_s^{\infty} \frac{s^2}{\lambda^2}\,d\lambda\,(-dm(s)) = 3\int_{-\infty}^{\infty} U_h^*(x)\,dx + \int_0^{\infty} s(-dm(s)) =$$

$$= 4\int_{-\infty}^{\infty} U_h^*(x)\,dx,$$

as required.

The theorem is completely proved.

In like manner we can establish:

Theorem. Let $u(z)$ be harmonic in $\{|z| < 1\}$. Then $u + i\tilde{u} \in H_1$ provided that $\int_{-\pi}^{\pi} u^*(\theta)\,d\theta < \infty$.

(For definition of $u^*(\theta)$, see Section C.3.)

2°. The theorem of Burkholder, Gundy and Silverstein can be sharpened. In order to do this, we use a remarkable result as a substitute for (lacking) subharmonicity of $|u(z)|^p$ (u harmonic) when $0 < p < 1$.

Lemma (Fefferman and Stein, 1972). Let $u(z)$ be harmonic in a region containing $\{|z| \leq R\}$ in its interior, and let $0 < p < 1$. There is a constant C_p depending only on p such that

$$|u(0)|^p \leq \frac{C_p}{R^2}\int_0^R\int_0^{2\pi} |u(re^{i\theta})|^p\, r\,dr\,d\theta.$$

Proof. By homogeneity considerations, we reduce first the general situation to the case where $R = 1$ and

$$\int_0^1\int_0^{2\pi} |u(re^{i\theta})|^p\, r\,dr\,d\theta = 1.$$

Write, for $0 < r < 1$, $M(r) = \sup_\theta |u(re^{i\theta})|$, and $I(r) = \int_0^{2\pi} |u(re^{i\theta})|^p d\theta$. According to our preliminary reductions, we have $\int_0^1 I(r)r\,dr = 1$, so surely $\int_{1/2}^1 I(r)\,\frac{dr}{r} \leq 4$, and, by the inequality between geometric and arithmetic means,

$$\frac{1}{\log 2} \int_{1/2}^1 \log I(r)\,\frac{dr}{r} \leq \log\left\{\frac{\int_{1/2}^1 I(r)\frac{dr}{r}}{\log 2}\right\} \leq \log\left(\frac{4}{\log 2}\right),$$

whence

$$\boxed{\int_{1/2}^1 \log I(r)\,\frac{dr}{r} \leq K,}$$

a pure number whose exact value need not concern us here.

Let $\alpha > 1$. By Poisson's formula (Chapter I!), we easily see that

$$M(r^\alpha) \leq \frac{2}{\pi(1 - r^\alpha/r)} \int_0^{2\pi} |u(re^{i\theta})|\,d\theta \leq \frac{2}{\pi(1 - r^{\alpha-1})}\,(M(r))^{1-p}\,I(r).$$

Taking logarithms, we get

$$\log M(r^\alpha) \leq (1-p)\log M(r) + \log I(r) + \log\left\{\frac{2}{\pi(1 - r^{\alpha-1})}\right\}.$$

<u>Now</u> (TRICK!) <u>multiply both sides by</u> $\frac{dr}{r} = \frac{1}{\alpha}\frac{d(r^\alpha)}{r^\alpha}$ <u>and integrate from</u> $1/2$ <u>to</u> 1. Putting $r^\alpha = \rho$ on the <u>right</u>, and using the boxed inequality proved above, we obtain

$$\frac{1}{\alpha}\int_{(1/2)^\alpha}^1 \log M(\rho)\frac{d\rho}{\rho} \leq (1-p)\int_{1/2}^1 \log M(r)\frac{dr}{r} + K + \int_{1/2}^1 \log\left\{\frac{2}{\pi(1 - r^{\alpha-1})}\right\}\frac{dr}{r}.$$

Here,

$$K + \int_{1/2}^{1} \log\left\{\frac{2}{\pi(1 - r^{\alpha-1})}\right\} \frac{dr}{r} = C_\alpha$$

is finite and depends only on α. Now choose α so close to 1 that $\frac{1}{\alpha} > 1 - p$ - we can do this since $0 < p < 1$. Then we find

$$\frac{1}{\alpha} \int_{(1/2)^\alpha}^{1/2} \log M(r) \frac{dr}{r} + [\tfrac{1}{\alpha} - (1 - p))] \int_{1/2}^{1} \log M(r) \frac{dr}{r} \leq C_\alpha .$$

From this we see that there is a number ℓ_α, depending only on α, with $\log M(r) \leq \ell_\alpha$ for some r, $(\frac{1}{2})^\alpha \leq r \leq 1$. Then, by the principle of maximum, $|u(0)| \leq \exp \ell_\alpha$. Taking $C_p = \exp p\,\ell_\alpha$ after having chosen α as above, we see that the lemma holds as stated.

Q.E.D.

Definition. If $U(z)$ is defined for $\Im z > 0$ and $x \in \mathbb{R}$, put

$$U^+(x) = \sup_{y>0} |U(x + iy)|.$$

$U^+(x)$ is the so-called "radial" maximal function of U. Clearly $U^+ \leq U^*$.

Theorem (Fefferman and Stein, 1972). If $U(z)$ is harmonic and real in $\Im z > 0$ and $U^+(x) \in L_1(\mathbb{R})$, then $U = \Re F$ for an $F \in H_1(\Im z > 0)$.

Proof. We show that $U^+ \in L_1$ implies that $U^* \in L_1$, which is enough by the theorem of Section 1.

Let, for $\Im z > 0$, $W(z) = |U(z)|^{1/2}$. The picture shows that if $x \in \mathbb{R}$ and $0 \le |\xi - x| < \eta$, the disk $\{Z;\ |Z - (\xi + i\eta)| < \eta\}$ lies in $\Im z > 0$, so, by the **lemma**, with $p = 1/2$,

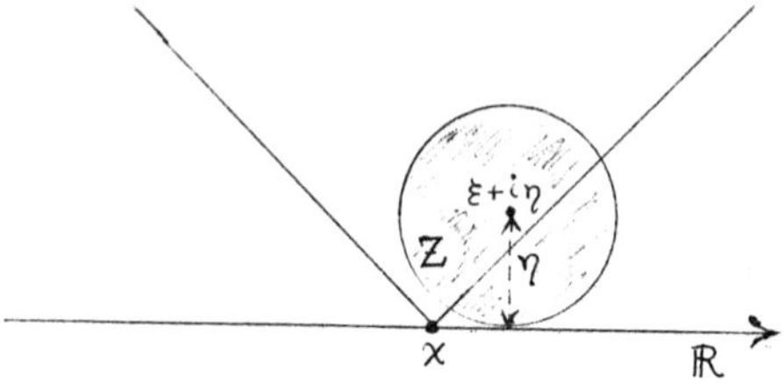

$$W(\xi + i\eta) \le \frac{C}{\eta^2} \iint_{\{|Z-(\xi+i\eta)|<\eta\}} W(Z)\, dX\, dY,$$

where we write $Z = X + iY$.

Therefore, a fortiori,

$$W(\xi + i\eta) \le \frac{C}{\eta^2} \int_0^{2\eta}\int_{\xi-\eta}^{\xi+\eta} W(Z)dX\,dY \le \frac{2C}{\eta}\int_{\xi-\eta}^{\xi+\eta} W^+(X)dX, \quad \text{since } W(Z) \le W^+(X).$$

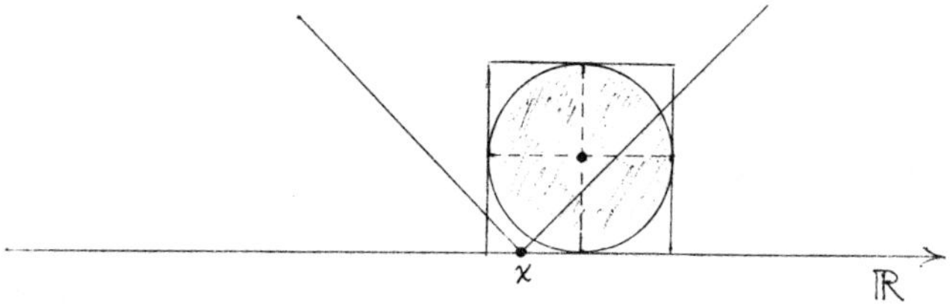

Thence, if $|\xi - x| < \eta$,

$$W(\xi + i\eta) \le \frac{2C}{\eta} \int_{x-2\eta}^{x+2\eta} W^+(X)dX \le 8C(W^+)^M(x),$$

by definition of the Hardy-Littlewood maximal function (for W^+)! Taking the sup over $\xi + i\eta$ with $0 \le |\xi - x| < \eta$, we see finally that $W^*(x) \le 8C(W^+)^M(x)$. Now, however, $W(z) = \sqrt{|U(z)|}$, so

$W^+(x) = \sqrt{U^+(x)}$ is in L_2 by hypothesis. From Section B.2 we get $\|(W^+)^M\|_2 \le K\|W^+\|_2 = K\sqrt{\|U^+\|_1}$, so, by the previous relation, $\sqrt{\|U^*\|_1} = \|W^*\|_2 \le 8CK\sqrt{\|U^+\|_1}$, proving $\|U^*\|_1 < \infty$ if $\|U^+\|_1 < \infty$.

We are done.

In the same way one can prove:

Theorem. Let $u(z)$ be harmonic in $|z| < 1$ and let $u^+(\theta) = \sup_{0 \le r < 1} |u(re^{i\theta})|$ belong to $L_1(-\pi,\pi)$. Then $u + i\tilde{u} \in H_1(|z| < 1)$.

3°. Remark. Note the different positions of the absolute value signs in the definitions of the two maximal functions

$$u^+(\theta) = \sup_{0 \le r < 1} \left| \frac{1}{2\pi} \int_{-\pi}^{\pi} \frac{1 - r^2}{1 + r^2 - 2r\cos t} u(\theta - t)dt \right| ,$$

$$u^M(\theta) = \sup_{h,k > 0} \frac{1}{h + k} \int_{-h}^{k} |u(\theta - t)|dt.$$

(Here we write $u(s)$ instead of $u(e^{is})$.) Obviously, the first of these functions depends in a much more sensitive manner on the properties of u (possible oscillatory behaviour) than the second. It is easy to give examples of functions $u \in L_1(-\pi,\pi)$ for which $u^+(\theta) \in L_1(-\pi,\pi)$ but $u^M(\theta) \notin L_1(-\pi,\pi)$. Indeed, by the second theorem of Section B.3 together with a Corollary in Section C.3, any $f \in H_1(|z| < 1)$ with $u(\theta) = \Re f(e^{i\theta})$ of variable sign, and such that $u \log^+|u| \notin L_1(-\pi,\pi)$, provides us with such a u.

One is tempted to define a Hardy-Littlewood maximal function which is more sensitively related to u by putting, for instance,

$$\tilde{u}(\theta) = \sup_{h,k>0} \left| \frac{1}{h+k} \int_{-h}^{k} u(\theta - t)dt \right| .$$

In their 1972 Acta paper, Fefferman and Stein show that this definition is not useful. Other approximate identities besides the Poisson kernel, used in defining u^+, can be used, but a certain smoothness seems to be required of them.

E. Carleson measures

Definition. A positive measure μ (not necessarily finite) defined in $\Im z > 0$ is called a Carleson measure iff there is a constant K such that, for all $F \in H_1(\Im z > 0)$,

$$(*) \qquad \iint_{\Im z > 0} |F(z)|d\mu(z) \le K\|F\|_1.$$

Carleson measures are important in the next two chapters. Their geometric characterization is given by the

Theorem (Carleson). μ is a Carleson measure iff, for all $x \in \mathbb{R}$ and all $h > 0$,

$$(\ast\ast) \qquad \boxed{\mu((x,x+h) \times (0,h)) \le Ch}$$

with a constant C independent of x and h.

Proof.

Only if) Let $x_0 \in \mathbb{R}$ and $h > 0$ be given. The test function

$$f(z) = \frac{h}{(z - x_0 + ih)^2}$$

belongs to $H_1(\Im z > 0)$ and $\|f\|_1 = \pi$. Also, $|f(z)| \geq \frac{1}{2h}$ for $z \in S_h = (x_0, x_0 + h) \times (0,h)$, so if (*) holds,

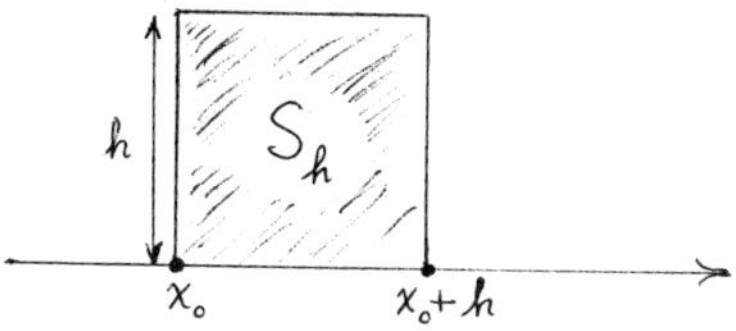

$$\frac{1}{2h}\ \mu(S_h) \leq \iint_{\Im z > 0} |f(z)|\, d\mu(z) \leq \pi K,$$

i.e., $\mu(S_h) \leq 2\pi K \cdot h$, proving ($\ast\!\ast$) with $C = 2\pi K$.

<u>If</u>) It is enough to prove (*) for all $F \in H_1$ of the special form $F(z) = f_h(z) = f(z + ih)$, where $f \in H_1(\Im z > 0)$ and $h > 0$. Take any such F; by the Poisson representation for f in the upper half plane (Chapter VI) it is <u>clear</u> that $F(z)$ is <u>continuous</u> for $\Im z \geq 0$, and tends to zero as $z \to \infty$ in the upper half plane. We must establish (*) with a K independent of F.

To this end, put, for $\lambda > 0$,

$$M(\lambda) = \mu(\{z;\ \Im z > 0\ \&\ |F(z)| > \lambda\}).$$

Using an idea of Hörmander, we prove that $M(\lambda) \leq Cm_{F^*}(\lambda)$ with the C from ($\ast\!\ast$). Here, $F^*(x)$ is the nontangential maximal function of F (Section C.2) and $m_{F^*}(\lambda)$ is the distribution function of F^* (Section A).

Let $E_\lambda = \{x \in \mathbb{R};\ F^*(x) \leq \lambda\}$ and $\mathcal{O}_\lambda = \mathbb{R} \sim E_\lambda = \{x \in \mathbb{R};\ F(x) > \lambda\}$ so that $m_{F^*}(\lambda) = |\mathcal{O}_\lambda|$. If $x_0 \in E_\lambda$, <u>every</u> $z = x + iy$ in the sector $\bar{S}_{x_0} = \{z;\ |x - x_0| \leq y\ \&\ y > 0\}$ must satisfy $|F(z)| \leq \lambda$, therefore, $\Omega_\lambda = \{z;\ \Im z > 0,\ |F(z)| > \lambda\}$ must be contained in Ω'_λ, the complement, in $\Im z > 0$, of $\bigcup_{x_0 \in E_\lambda} \bar{S}_{x_0}$.

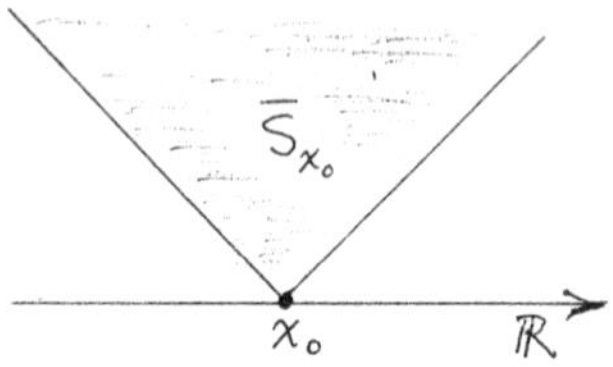

In the present case, since $F(z)$ is continuous in $\Im z \geq 0$ and 0 at ∞ there, $\mathcal{O}_\lambda = \mathbb{R} \sim E_\lambda$ is a bounded open set on $\mathbb{R}$, hence a disjoint union of finite open intervals J_k. Let, for each k, Δ_k be the open $45°$ isoceles triangle lying in $\Im z > 0$ with base J_k.

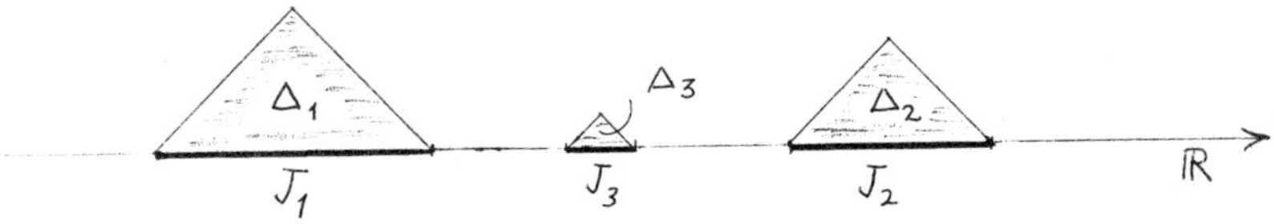

What we have here is the analogue for the upper half plane of Privalov's construction, described in Section D.1 of Chapter III. Arguing as in that place, we see that the set Ω'_λ introduced above is equal to the union of the Δ_k. Therefore $\Omega_\lambda \subseteq \bigcup_k \Delta_k$, and $M(\lambda) = \mu(\Omega_\lambda) \leq \sum_k \mu(\Delta_k)$. But for each k, by $(\ast\ast)$, $\mu(\Delta_k) \leq C|J_k|$,

so

$$M(\lambda) \leq C\sum_k |J_k| = C|\mathcal{O}_\lambda| = Cm_{F^*}(\lambda).$$

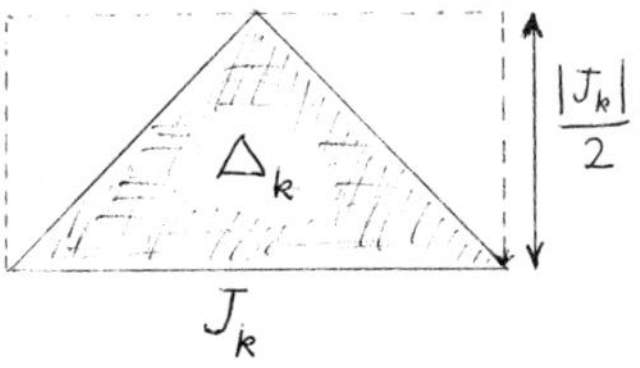

Now we just bring in the material of Section A. The lemma of that section is not particular to Lebesgue measure on the line, but holds (with the same proof) for distribution functions defined by means of quite general σ-finite measures. Therefore

$$\iint_{\Im z>0} |F(z)|\,d\mu(z) = \int_0^\infty M(\lambda)\,d\lambda.$$

By the above work, the integral on the right is

$$\leq C\int_0^\infty m_{F^*}(\lambda)\,d\lambda = C\int_{-\infty}^\infty F^*(x)\,dx,$$

again by the lemma of Section A.

But here $(F \in H_1!)$,

$$\int_{-\infty}^\infty F^*(x)\,dx \leq C_1\|F_1\|$$

by the second theorem of Section C.2. So finally,

$$\iint_{\Im z>0} |F(z)|\,d\mu(z) \leq CC_1\|F\|_1$$

for $F \in H_1$ of the special form given, hence for all $F \in H_1$.

We are done.

In like manner one can prove an analogue of the above result for the unit circle:

<u>Theorem</u>. Let μ be a positive Radon measure on $\{|z| < 1\}$.

Then

$$\iint_{|z|<1} |f(z)|\,d\mu(z) \leq k\|f\|_1$$

for all $f \in H_1(|z| < 1)$ <u>iff</u>

$$\mu(B_h) \leq Ch$$

for every curvelinear box B_h of the form shown:

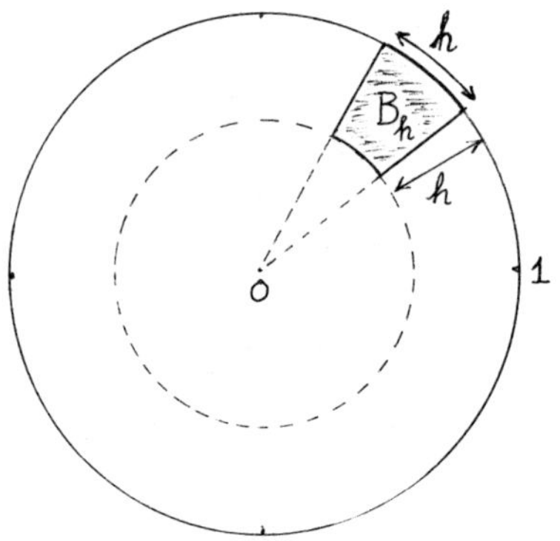

Problem No. 8.

Let $0 = x_0 < x_1 < x_2 < \ldots < x_n = 1$ be any partition whatsoever of $[0,1]$ into nonoverlapping intervals $J_k = [x_{k-1}, x_k]$; put $|J_k| = 2\Delta_k$, $\bar{x}_k = \frac{1}{2}(x_{k-1} + x_k)$.

If

$$\Delta(x) = \sum_{k=1}^{n} \frac{\Delta_k^2}{(x - \bar{x}_k)^2 + \Delta_k^2},$$

show that, for $\lambda > 0$,

$$|\{x \in [0,1];\ \Delta(x) > \lambda\}| \leq 2e^{-C\lambda}$$

with a numerical constant C independent of λ or of the particular choice of the x_k. (I think $C = \frac{1}{6\pi e}$ works here.)

Hint: If $q = 1,2,3,\ldots$, $\frac{1}{q} + \frac{1}{p} = 1$, and $f \in L_p(0,1)$, estimate $\int_0^1 \Delta(x) f(x)\,dx$ in terms of $\|f^*\|_p$, and thence obtain an estimate for $\int_0^1 [\Delta(x)]^q dx$. Note that

$$e^{\rho\Delta(x)} = \sum_0^{\infty} [\rho\Delta(x)]^q / q!\ .$$

IX. Interpolation

Definition. If $\Im z_n > 0$, $n = 1,2,\ldots$, $\{z_n\}$ is called an interpolating sequence for the upper half plane iff, given any bounded sequence $\{c_n\}$, there is an $F \in H_\infty(\Im z > 0)$ such that

$$F(z_n) = c_n, \qquad n = 1,2,3,\ldots .$$

It is possible to determine all interpolating sequences in terms of a simple geometric condition.

A. Necessary conditions

1°. Lemma. $\{z_n\}$ is an interpolating sequence iff there is a constant K, such that for any N, if $|c_n| \leq 1$ for $n = 1,2,\ldots,N$, there is an $F \in H_\infty$ with $\|F\|_\infty \leq K$ such that $F(z_n) = c_n$, $n = 1,2,\ldots,N$ (K being independent of N).

Proof. If) Given such a K, let $\{c_n\}$, $n = 1,2,\ldots$ be an arbitrary sequence with $|c_n| \leq 1$, and, for each N, let $F_N \in H_\infty$ with $\|F_N\|_\infty \leq K$ be such that

$$F_N(z_n) = c_n \quad \text{for} \quad n = 1,2,\ldots,N.$$

Since $\|F_N\|_\infty \leq K$, a subsequence of $\{F_N\}$ converges u.c.c. in $\Im z > 0$ to a function $F \in H_\infty$ with $\|F\|_\infty \leq K$. Clearly $F(z_n) = c_n$ for all n.

Only if) If $\{z_n\}$ is an interpolating sequence, let, for each L, S_L be the subset of $\ell_\infty(\mathbb{N})$ consisting of all the sequences $\{F(z_n)\}$ formed from the $F \in H_\infty$ with $\|F\|_\infty \leq L$. (As usual, we denote by $\mathbb{N}$ the set of positive integers.)

By assumption $\bigcup_{L=1}^{\infty} S_L = L_\infty(\mathbb{N})$. Also, each S_L is closed in $\ell_\infty(\mathbb{N})$, for if $F_k \in H_\infty$ and $\|F_k\|_\infty \le L$, and if $\{c_n\} \in \ell_\infty(\mathbb{N})$ is such that $\|\{F_k(z_n) - c_n\}\| = \sup_{n\in\mathbb{N}} |F_k(z_n) - c_n|$ goes to zero with k, then, extracting a subsequence of $\{F_k\}$ converging u.c.c. in $\Im z > 0$, the limit, F, of that subsequence belongs to H_∞ and has $\|F\|_\infty \le L$, with $F(z_n) = c_n$, $n \in \mathbb{N}$. So $\{c_n\} \in S_L$, and S_L is closed in $\ell_\infty(\mathbb{N})$.

The Baire category theorem now says that some S_L contains a sphere, say of radius $\rho > 0$, about one of its points, say the sequence $\{F_0(z_n)\}$, $F_0 \in H_\infty$, $\|F_0\|_\infty \le L$. That is, if $|c_n - F_0(z_n)| \le \rho$ for all n, then there is an $F \in H_\infty$, $\|F\|_\infty \le L$, with $F(z_n) = c_n$, $n \in \mathbb{N}$. Therefore, if $|d_n| \le \rho$ for all n, we can find a $G \in H_\infty$, $\|G\|_\infty \le 2L$, with $G(z_n) = d_n$, $n \in \mathbb{N}$. (Just put $c_n = d_n + F_0(z_n)$ and then put $G = F - F_0$ with F from the preceding statement.)

It follows that if $|a_n| \le 1$, $n \in \mathbb{N}$, we can find an $H \in H_\infty$ with $\|H\|_\infty \le 2L/\rho$ and $H(z_n) = a_n$ for all n. The lemma holds with $K = 2L/\rho$.

2°. Lemma. $\{z_n\}$ is an interpolating sequence for $\Im z > 0$ only if there is a $\delta > 0$ such that, for each n,

$$(*) \qquad \prod_{k\neq n} \left|\frac{z_n - z_k}{z_n - \bar{z}_k}\right| \ge \delta > 0.$$

Proof. If $\{z_n\}$ is an interpolating sequence there is, by the previous lemma, a K such that, for each n we can find an $F \in H_\infty$ with $\|F\|_\infty \le K$ and

$$F(z_k) = \begin{cases} 1, & k = n \\ 0, & k \neq n. \end{cases}$$

According to Chapter VI, Section C, we can form a Blaschke product $B(z)$ from the zeros z_k, $k \neq n$ of $F(z)$, and we'll have $F(z) = B(z)G(z)$ with a function $G \in H_\infty$ and $\|G\|_\infty = \|F\|_\infty \leq K$. We have

$$|B(z_n)| = \prod_{k \neq n} \left| \frac{z_n - z_k}{z_n - \bar{z}_k} \right| ,$$

so $1 = |F(z_n)| = |B(z_n)|\ |G(z_n)| \leq K|B(z_n)|$ immediately yields (*) with $\delta = 1/K$.

B. Carleson's theorem

The necessary condition (*) for $\{z_n\}$ to be an interpolating sequence is also sufficient. This remarkable result is due to L. Carleson, and goes back to 1958 or thereabouts. Later, Shapiro and Shields found an easier proof of Carleson's theorem by making successive reductions with the help of the duality theory presented in Chapter VII, Sections A.1 and A.2. This proof is given in the books by Duren and Hoffman.

Here, we give a proof which is more like Carleson's original one. It is simplified by applying directly the theorem on Carleson measures given in Chapter VIII, Section E. Thanks to an idea of Hörmander, it was possible to establish that result rather easily.

1° Lemma. If, for every n,

$$\prod_{k \neq n} \left| \frac{z_n - z_k}{z_n - \bar{z}_k} \right| \geq \delta > 0,$$

we have, for each n,

$$\sum_{k\neq n} \frac{\Im z_n \, \Im z_k}{|z_n - \bar{z}_k|^2} \leq \frac{1}{2} \log \frac{1}{\delta}.$$

<u>Proof</u>. Write $z_k = x_k + iy_k$, $z_n = x_n + iy_n$; we have

$$0 < \left| \frac{z_n - z_k}{z_n - \bar{z}_k} \right| < 1,$$

and

$$1 - \left| \frac{z_n - z_k}{z_n - \bar{z}_k} \right|^2 = \frac{|z_n - \bar{z}_k|^2 - |z_n - z_k|^2}{|z_n - \bar{z}_k|^2} =$$

$$= \frac{(x_n - x_k)^2 + (y_n + y_k)^2 - (x_n - x_k)^2 - (y_n - y_k)^2}{|z_n - \bar{z}_k|^2}$$

$$= 4y_n y_k |z_n - \bar{z}_k|^{-2}.$$

Writing $p_k = |(z_n - z_k)/(z_n - \bar{z}_k)|$, we have

$$2 \log \frac{1}{p_k} = \log \frac{1}{1 - (1-p_k^2)} = \sum_{m=1}^{\infty} \frac{(1-p_k^2)^m}{m} \geq 1 - p_k^2 = 4\Im z_n \Im z_k |z_n - \bar{z}_k|^{-2},$$

by the calculation just made.

Since $\sum_k 2 \log (1/p_k) \leq 2 \log (1/\delta)$, the lemma follows.

2° <u>Lemma</u>. If, for all n,

$$(*) \qquad \prod_{k\neq n} \left| \frac{z_n - z_k}{z_n - \bar{z}_k} \right| \geq \delta > 0,$$

the measure μ on $\Im z > 0$ given by

$$d\mu(z) = \sum_n \Im z_n d\delta_{z_n}(z),$$

(<u>in other words, assigning mass</u> $\Im z_n$ <u>to each point</u> z_n) <u>is a Carleson measure</u>.

Proof. We will show that for any $x_0 \in \mathbb{R}$ and any $h > 0$,

$$\mu([x_0, x_0 + h] \times (0,h)) \le (1 + 5 \log \tfrac{1}{\delta})h;$$

this makes μ a Carleson measure by the theorem of Chapter VIII, Section E.

Pick any $x_0 \in \mathbb{R}$ and any $h > 0$, and denote the square $[x_0, x_0 + h] \times (0,h)$ by S:

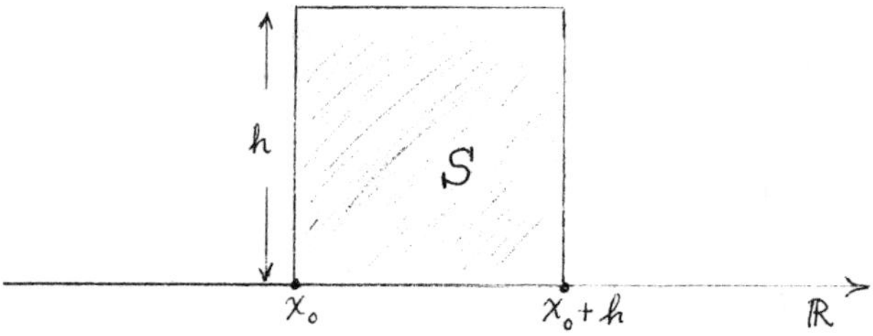

We have to show that

$$\sum_{z_n \in S} \Im z_n \le (1 + 5 \log \tfrac{1}{\delta})h;$$

this will evidently follow if, for every N we have

$$\sum_{\substack{z_n \in S \\ n \le N}} \Im z_n \le (1 + 5 \log \tfrac{1}{\delta})h.$$

Fix a large integer N. If there are no z_k, $1 \le k \le N$, with $z_k \in S$, the desired inequality is surely true.

Suppose now that there are some $z_k \in S$, $1 \le k \le N$. If, for one of them, say z_n, we have $\Im z_n \ge h/2$, then the desired inequality is true. Indeed, take the result of the preceding lemma:

$$\sum_{k\neq n} \Im z_k \cdot \frac{\Im z_n}{|z_n - \bar{z}_k|^2} \leq \frac{1}{2} \log \frac{1}{\delta}.$$

Observe that if, for $k \neq n$, $z_k \in S$, we certainly have

$$\frac{\Im z_n}{|z_n - \bar{z}_k|^2} \geq \frac{1}{10h},$$

because $|z_n - \bar{z}_k|^2 = (x_n - x_k)^2 + (y_n + y_k)^2 \leq h^2 + 4h^2 = 5h^2$ whilst $\Im z_n \geq h/2$. Therefore, by the previous relation,

$$\frac{1}{10h} \sum_{\substack{z_k \in S \\ k\neq n}} \Im z_k \leq \frac{1}{2} \log \frac{1}{\delta},$$

and, since $\Im z_n \leq h$,

$$\sum_{z_k \in S} \Im z_k \leq h + 5h \log \frac{1}{\delta},$$

implying the desired inequality (for every N).

It may be, however, that there are $z_k \in S$ for $1 \leq k \leq N$, but that none of them has $\Im z_k \geq h/2$. In that case, we decompose the square S into an upper rectangle R and two lower quarter squares, $S_{1,1}, S_{1,2}$, in the manner shown:

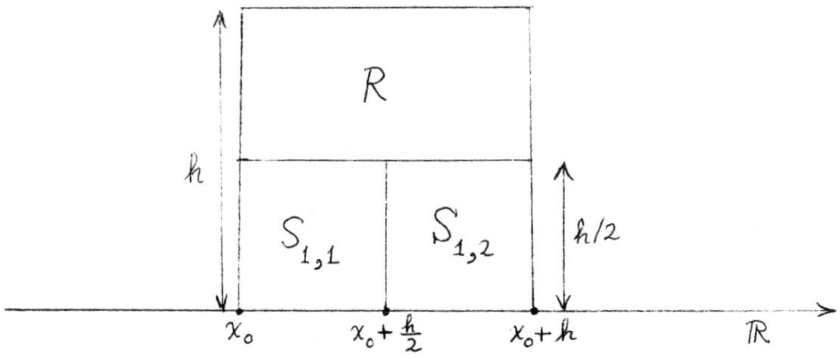

Because there is no z_k, $1 \le k \le n$, with $z_k \in S$ and $\Im z_k \ge h/2$, there is no such z_k in R, and therefore

$$\sum_{\substack{z_k \in S \\ 1 \le k \le N}} \Im z_k = \Sigma(S_{1,1}) + \Sigma(S_{1,2}),$$

where we write, during the rest of this discussion,

$$\Sigma\,(A) = \sum_{\substack{z_k \in A \\ 1 \le k \le N}} \Im z_k \,.$$

If, now, there is a $z_n \in S_{1,1}$, $1 \le k \le N$, with $\Im z_n \ge h/4$, the above discussion shows that $\Sigma(S_{1,1}) \le (1 + 5 \log \frac{1}{\delta})\frac{h}{2}$. And if $S_{1,1}$ has no z_k in it at all for $1 \le k \le N$, $\Sigma(S_{1,1}) = 0$.

Similarly, $\Sigma(S_{1,2}) \le (1 + 5 \log \frac{1}{\delta})\frac{h}{2}$ if $S_{1,2}$ has in it a z_n, $1 \le n \le N$, with $\Im z_n \ge h/4$. If $S_{1,2}$ has no z_k at all in it for $1 \le k \le N$, then $\Sigma(S_{1,2}) = 0$.

If, now, there are z_k, $1 \le k \le N$ in $S_{1,1}$ but all have $\Im z_k < h/4$, we look at the two lower quarter squares $S_{1,1,1}, S_{1,1,2}$ in $S_{1,1}$, each of side $h/4$, and see whether either has in it a z_n, $1 \le n \le N$, with $\Im z_n \ge h/8$, or else has no z_k in it at all for $1 \le k \le N$.

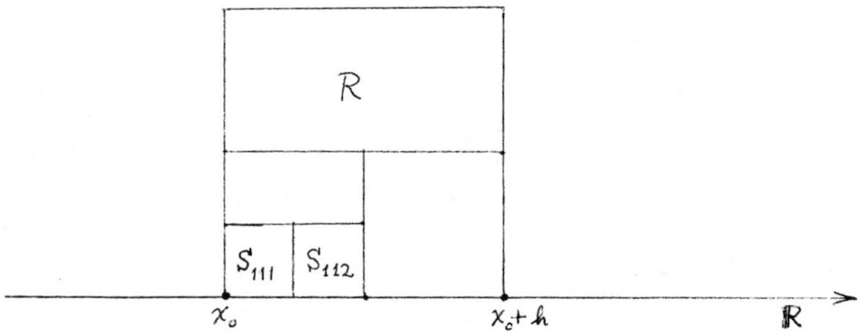

We do the same construction in $S_{1,2}$ if there are z_k, $1 \le k \le N$ in it, but none of them has $\Im z_k \ge h/4$, and we keep on going in this fashion, stopping whenever we first get to a lower quarter-square which either has no z_k, $1 \le k \le n$ in it at all, or else has one with $\Im z_k \ge 1/2$ of that quarter square's side. This process cannot go on indefinitely, because we are looking at only a finite number (N) of z_k.

Here is an example of how the process could work out. In the figure, each shaded lower quarter square has in it a z_k, $1 \le k \le N$, with $\Im z_k \ge 1/2$ of that quarter square's side. The unshaded squares and rectangles have no z_k in them for $1 \le k \le N$.

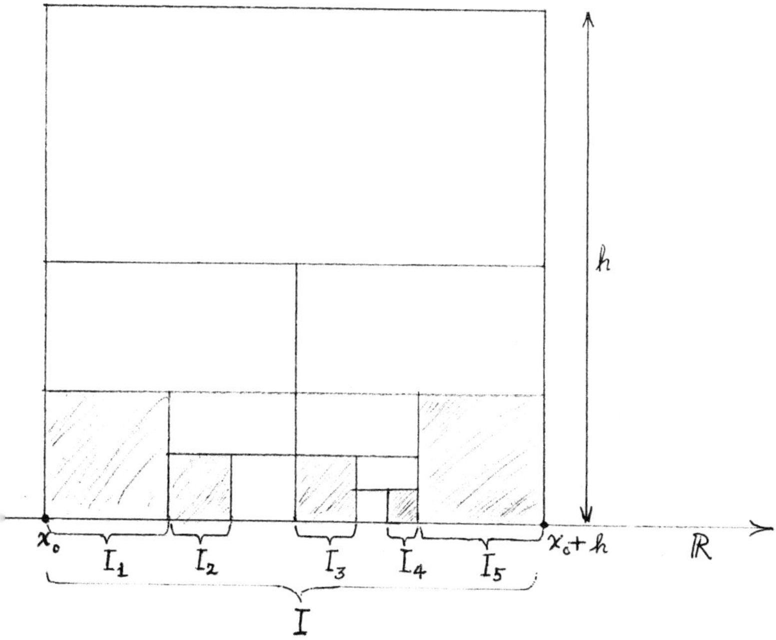

The construction just described leads to a finite number of <u>non-overlapping intervals</u>, say $I_1, I_2, \ldots, I_p$ inside the interval $I = [x_0, x_0 + h]$, each having length equal to that of I divided by some power of 2. Each I_k is the <u>base</u> of a <u>square</u>, say $S^{(k)}$, lying in S, and <u>all the</u> z_k, $1 \le k \le N$, belonging to S lie in the <u>union</u> of the $S^{(k)}$. Each $S^{(k)}$ has in it a z_k, $1 \le k \le N$, with $\Im z_k \ge \frac{1}{2}|I_k|$.

By the discussion at the beginning of this proof, we now have

$$\Sigma(S^{(k)}) \le (1 + 5 \log \tfrac{1}{\delta})\, |I_k|, \qquad k = 1,2,\ldots,p.$$

Therefore

$$\Sigma(S) \le \sum_{k=1}^{p} \Sigma(S^{(k)}) \le (1 + 5 \log \tfrac{1}{\delta}) \sum_{k=1}^{p} |I_k| \le (1 + 5 \log \tfrac{1}{\delta})|I|,$$

that is,

$$\sum_{\substack{z_k \in S \\ 1 \le k \le N}} \Im z_k \le (1 + 5 \log \tfrac{1}{\delta})h,$$

our desired inequality. The lemma is completely proved.

3°. <u>Carleson's Theorem.</u> A sequence $\{z_n\}$ in the upper half plane with

$$(*) \qquad \prod_{k \ne n} \left| \frac{z_n - z_k}{z_n - \bar{z}_k} \right| \ge \delta > 0$$

for every n is an <u>interpolating sequence</u>.

<u>Proof</u>. By Section A.1 it is enough to show that there is a fixed constant K such that, for any positive integer N and any numbers c_n, $n = 1,2,\ldots,N$ with $|c_n| \le 1$, there is an $F \in H_\infty$ with $\|F\|_\infty \le K$ and

$$F(z_n) = c_n, \qquad n = 1,2,\ldots,N.$$

Having fixed N, let us take the finite Blaschke product

$$B(z) = \prod_{k=1}^{N} \left(\frac{z - z_k}{z - \bar{z}_k} \right) .$$

For $1 \leq n \leq N$,

$$B'(z_n) = \frac{\beta_n}{2i \Im z_n} ,$$

where

$$\beta_n = \prod_{\substack{1 \leq k \leq N \\ k \neq n}} \left(\frac{z_n - z_k}{z_n - \bar{z}_k} \right) ,$$

so that surely $|\beta_n| \geq \delta$ for $1 \leq n \leq N$ if (*) holds.

Given the numbers c_n, $|c_n| \leq 1$, for $1 \leq n \leq N$, here is <u>one</u> $F \in H_\infty$ with $F(z_n) = c_n$ for $1 \leq n \leq N$:

$$F(z) = F_0(z) = B(z) \sum_{n=1}^{N} \frac{c_n}{B'(z_n)(z-z_n)} = 2iB(z) \sum_{n=1}^{N} \frac{\Im z_n}{\beta_n} \cdot \frac{c_n}{z-z_n} .$$

<u>Any other</u> $F \in H_\infty$ <u>with</u> $F(z_n) = c_n$ <u>for</u> $1 \leq n \leq N$ <u>is of the form</u> $F_0 + BG$ <u>where</u> $G \in H_\infty$, <u>and conversely</u>. We therefore proceed to <u>see if</u> $\|F_0 - BH_\infty\|_\infty$ <u>has a bound independent of</u> N; <u>if it has one</u>, say K, <u>then there will BE an</u> $F \in H_\infty$ <u>with</u> $\|F\|_\infty \leq K$ and $F(z_n) = c_n$, $n = 1,2,\ldots,N$, <u>so we will be done</u>, K being <u>independent</u> of N.

The idea now (due to D. J. Newman) is to use the duality theory of Chapter VII, Section A.2 to compute $\|F_0 - BH_\infty\|_\infty$. Since $|B(x)| \equiv 1$, $x \in \mathbb{R}$ (That's just why we <u>used</u> B!), we have

$$\|F_0 - BH_\infty\|_\infty = \inf\{\|(F_0/B) - G\|_\infty;\ G \in H_\infty\}.$$

By the analogue of a theorem in Chapter VII, Section A.2 for the upper half plane (see table at end of Section A.1, Chapter VII), the last inf is equal to

$$\sup\left\{ \left| \int_{-\infty}^{\infty} \frac{F_0(x)}{B(x)} f(x)dx \right| ; \; f \in H_1, \; \|f\|_1 \leq 1 \right\} .$$

For each $f \in H_1$, we have

$$\int_{-\infty}^{\infty} \frac{F_0(x)}{B(x)} f(x)dx = 2i \sum_{n=1}^{N} \int_{-\infty}^{\infty} \frac{\Im z_n}{\beta_n} \frac{c_n}{x-z_n} f(x)dx = -4\pi \sum_{n=1}^{N} \frac{c_n}{\beta_n} \Im z_n f(z_n).$$

Since $|\beta_n| \geq \delta$ and $|c_n| \leq 1$, the last expression is in absolute value $\leq 4\pi\delta^{-1}\sum_{n=1}^{N} \Im z_n |f(z_n)|$. But according to the lemma of 2° above, the measure assigning mass $\Im z_n$ to each point z_n is Carleson. Therefore, for $f \in H_1$,

$$4\pi\delta^{-1} \sum_{n=1}^{\infty} \Im z_n |f(z_n)| \leq 4\pi\delta^{-1}C\|f\|_1,$$

so, substituting in the previous relation, we see that $\|F_0 - BH_\infty\|_\infty \leq 4\pi\delta^{-1}C$, a constant independent of N.

The proof is complete.

Remark. The proofs of the lemma in 2° and of the theorem on Carleson measures in Chapter VIII, Section E show that we can take the constant C figuring at the end of the above demonstration equal to $\tilde{C}(1 + 5 \log \frac{1}{\delta})$, where $\tilde{C}$ is a purely numerical constant, independent of δ, hence of the sequence $\{z_n\}$. This means that whenever $|c_n| \leq 1$ for all n we can find an $F \in H_\infty$ with $\|F\|_\infty \leq K$ and $F(z_n) = c_n$, $n = 1,2,\ldots$, where the constant K depends on the δ figuring in (*) like a numerical multiple of $\frac{1}{\delta} \log \frac{1}{\delta}$.

Carleson's theorem can be carried over directly to the unit circle by conformal transformation. We just state the result:

Theorem. Let $|z_n| < 1$, $n = 1,2,\ldots$. A necessary and sufficient condition that there exist, for any bounded sequence $\{c_n\}$, an $f \in H_\infty(|z| < 1)$ with $f(z_n) = c_n$, $n = 1,2,\ldots,$ is that

$$\prod_{k \neq n} \left| \frac{z_n - z_k}{1 - \bar{z}_k z_n} \right| \geq \text{some } \delta > 0$$

for all n with δ independent of n.

It was indeed in this form that Carleson first published his result.

C. Weighted interpolation by functions in other H_p-spaces

Theorem (Shapiro and Shields). Let $1 \leq p \leq \infty$ and $\Im z_n > 0$. The sequences $\{(\Im z_n)^{1/p} F(z_n)\}$ fill out $\ell_p(\mathbb{N})$ as F ranges over $H_p(\Im z > 0)$ iff $\{z_n\}$ is an interpolating sequence.

Proof. By arguing precisely as in the proof of the lemma in Section A.1, we see that the sequences $\{(\Im z_n)^{1/p} F(z_n)\}$ fill out $\ell_p(\mathbb{N})$ as F ranges over H_p iff there is a $K < \infty$ such that, for any positive integer N and any c_n, $1 \leq n \leq N$, with $\sum_n |c_n|^p \leq 1$, we can find an $F \in H_p$ with $\|F\|_p \leq K$ and

$$(\Im z_n)^{1/p} F(z_n) = c_n, \qquad n = 1,2,\ldots,N.$$

I claim first of all that if this property holds, then $\{z_n\}$ is an interpolating sequence.

Indeed, fix any n and put

$$c_k = \begin{cases} 0, & k \neq n, \\ 1, & k = n. \end{cases}$$

For arbitrarily large N we can find an $F \in H_p$, $\|F\|_p \leq K$, with

$$(\Im z_k)^{1/p} F(z_k) = c_k, \qquad 1 \leq k \leq N.$$

Clearly, then, $F = BG$ where $G \in H_p$, $\|G\|_p \leq K$, and $B(z)$ is the partial Blaschke product

$$B(z) = \prod_{\substack{k \neq n \\ 1 \leq k \leq N}} \left(\frac{z - z_k}{z - \bar{z}_k} \right) .$$

Since $1 = |F(z_n)| = |B(z_n)|(\Im z_n)^{1/p}|G(z_n)|$, we must have

$$\prod_{\substack{k \neq n \\ 1 \leq k \leq N}} \left| \frac{z_n - z_k}{z_n - \bar{z}_k} \right| \geq \frac{1}{(\Im z_n)^{1/p}|G(z_n)|} .$$

Because $G \in H_p$ we can use Poisson's formula (Chapter VI) and we find, with $\frac{1}{q} = 1 - \frac{1}{p}$,

$$(\Im z_n)^{1/p}|G(z_n)| = \left| \frac{1}{\pi} \int_{-\infty}^{\infty} \frac{(\Im z_n)^{1+1/p} G(t)\,dt}{|z_n - t|^2} \right| \leq$$

$$\leq \frac{1}{\pi} \|G\|_p \left\{ \int_{-\infty}^{\infty} \frac{y_n^{q+q/p}}{[(x_n - t)^2 + y_n^2]^q} \, dt \right\}^{1/q} .$$

Now $\|G\|_p \leq K$, and the integral in $\{ \}$ equals

$$\int_{-\infty}^{\infty} \frac{y_n^{2q-1}}{(y_n^2 + \tau^2)^q} \, d\tau = \int_{-\infty}^{\infty} \frac{ds}{(s^2 + 1)^q} ,$$

a <u>finite quantity</u>, say C_p^q, <u>depending only on</u> p. Hence $(\Im z_n)^{1/p}|G(z_n)| \leq \pi^{-1}KC_p$ for $\|G\|_p \leq K$, and from the previous paragraph we get

$$\prod_{\substack{k\neq n\\1\leq k\leq N}} \left|\frac{z_n - z_k}{z_n - \bar{z}_k}\right| \geq \frac{\pi}{KC_p},$$

independently of N. Making now $N \to \infty$, we see that condition (*) of Carleson's theorem (Section B.3) <u>is</u> fulfilled, so $\{z_n\}$ <u>is</u> an <u>interpolating sequence</u>.

<u>Conversely, suppose</u> $\{z_n\}$ <u>is an interpolating sequence</u>; <u>say</u>

$$\prod_{k\neq n} \left|\frac{z_n - z_k}{z_n - \bar{z}_k}\right| \geq \delta > 0$$

for all n. Let N be any finite positive integer - <u>fix</u> it - and let numbers c_n, $n = 1,2,\ldots,N$ be given with $\sum_1^N |c_n|^p \leq 1$. Write, as in Section B.3,

$$B(z) = \prod_{n=1}^{N} \left(\frac{z - z_n}{z - \bar{z}_n}\right),$$

$$\beta_n = \prod_{\substack{1\leq k\leq N\\k\neq n}} \left(\frac{z_n - z_k}{z_n - \bar{z}_k}\right).$$

Put $\gamma_n = (\Im z_n)^{-1/p}c_n$, $n = 1,\ldots,N$. Then, as in the proof of Carleson's theorem,

$$F_0(z) = 2iB(z)\sum_{n=1}^{N} \frac{\gamma_n}{\beta_n}\,\frac{\Im z_n}{z - z_n}$$

has $F_0(z_n) = \gamma_n$, i.e., $(\Im z_n)^{1/p}F_0(z_n) = c_n$, and we will be finished if we can show that there is a $K < \infty$ independent of N and the particular numbers c_n, $n = 1,2,\ldots,N$, such that $\|F_0 - BH_p\|_p \leq K$.

Since $|B(x)| \equiv 1$, $x \in \mathbb{R}$, we have, by the duality theory of Chapter VII, Section A,

$$\|F_0 - BH_p\|_p = \|F_0/B - H_p\|_p = \sup\left\{\left|\int_{-\infty}^{\infty} \frac{F_0(t)}{B(t)} f(t)dt\right|;\ f \in H_q,\ \|f\|_q \leq 1\right\}.$$

As in Section B.3, for $f \in H_q$, $\int_{-\infty}^{\infty}(F_0(t)f(t)/B(t))dt$ works out to

$$-4\pi \sum_{n=1}^{N} \Im z_n\, \gamma_n\, f(z_n)/\beta_n,$$

which, in the present case, is seen by Hölder's inequality to be in absolute value

$$\leq 4\pi \left[\sum_{n=1}^{N} \Im z_n |\gamma_n|^p/|\beta_n|^p\right]^{1/p} \left[\sum_{n=1}^{N} \Im z_n |f(z_n)|^q\right]^{1/q}.$$

Here, $\Im z_n |\gamma_n|^p = |c_n|^p$ and $|\beta_n|^p \geq \delta^p$, so, since $\sum_n |c_n|^p \leq 1$, the last expression is seen to be

$$\leq 4\pi\delta^{-1}\left[\sum_{n=1}^{\infty} \Im z_n |f(z_n)|^q\right]^{1/q}.$$

For $f \in H_q$ and $\|f\|_q \leq 1$ we can certainly write $f(z) = b(z)g(z)$ where $b(z)$ is a Blaschke product (hence in modulus ≤ 1 in $\Im z > 0$) and $g \in H_q$, $\|g\|_q = \|f\|_q$, is free of zeros in the upper half plane. So $g^q \in H_1$ with $\|g^q\|_1 \leq 1$, therefore, by the lemma of Section B.2,

$$\sum_{n=1}^{\infty} \Im z_n |f(z_n)|^q \leq \sum_{n=1}^{\infty} \Im z_n |g(z_n)|^q \leq C\|g^q\|_1 \leq C,$$

the measure assigning mass $\Im z_n$ to each point z_n being Carleson.

We see finally that

$$\|F_0 - BH_p\|_p \le 4\pi\delta^{-1} C^{1/q},$$

a number independent of N and of the particular numbers c_n chosen, $1 \le n \le N$, with $\sum_n |c_n|^p \le 1$. The theorem is completely proved.

D. Relations between some conditions on sequences $\{z_n\}$.

In proving the lemma of Section B.2 we really showed that if

$$(*)\qquad \sum_k \frac{\Im z_n \, \Im z_k}{|z_n - \bar{z}_k|^2} \le K,$$

a constant independent of n, for all n, then the measure assigning mass $\Im z_n$ to each point z_n is Carleson.

The converse is true.

Lemma. If the measure assigning mass $\Im z_n$ to each point z_n is Carleson, then there is a constant K with

$$(*)\qquad \sum_k \frac{\Im z_n \, \Im z_k}{|z_n - \bar{z}_k|^2} \le K$$

for every n.

Proof (Garnett). The test functions $F_n(z) = \Im z_n/(z - \bar{z}_n)^2$ belong to H_1 and $\|F_n\|_1 = \pi$ for every n. So by the Carleson measure property,

$$\sum_k \Im z_k |F_n(z_k)| = \sum_k \frac{\Im z_k \, \Im z_n}{|z_n - \bar{z}_k|^2} \le K,$$

a constant independent of n.

Q.E.D.

<u>Lemma</u> (Garnett). If there is an $\eta > 0$ such that $|(z_n - z_m)/(z_n - \bar{z}_m)| \geq \eta$ for $n \neq m$ (in other words, if the <u>hyperbolic distance</u> between different points of the sequence $\{z_n\}$ is <u>bounded below</u>), and if

$$(\ast\!\ast)\qquad \sum_k \frac{\Im z_n\, \Im z_k}{|z_n - \bar{z}_k|^2} \leq K$$

for all n, <u>then</u>

$$(\ast)\qquad \prod_{k \neq n} \left| \frac{z_n - z_k}{z_n - \bar{z}_k} \right| \geq \delta > 0$$

for all n with a suitable δ, and $\{z_n\}$ is an interpolating sequence.

<u>Proof</u>. Take any n and k with $k \neq n$, and put for the moment $r = |(z_n - z_k)/(z_n - \bar{z}_k)|$; then $0 \leq r < 1$, so if also $r > 0$,

$$\log \frac{1}{r^2} = \log \frac{1}{1 - (1 - r^2)} = \sum_{\ell=1}^{\infty} \frac{(1 - r^2)^\ell}{\ell} = (1 - r^2) \sum_{m=0}^{\infty} \frac{(1 - r^2)^m}{m + 1} .$$

If $r \geq \eta$, the last expression is

$$\leq (1 - r^2) \sum_{m=0}^{\infty} (1 - \eta^2)^m = \frac{1 - r^2}{\eta^2} .$$

By the computation in Section B.1,

$$1 - r^2 = \frac{4 \Im z_n\, \Im z_k}{|z_n - \bar{z}_k|^2} ,$$

so we get by the previous calculation

$$\left| \frac{z_n - z_k}{z_n - \bar{z}_k} \right|^2 \geq \exp\left(- \frac{1 - r^2}{\eta^2} \right) = \exp\left(- \frac{4 \Im z_n\, \Im z_k}{\eta^2 |z_n - \bar{z}_k|^2} \right) .$$

Using $(\overset{*}{*})$, we now find

$$\prod_{k\neq n} \left|\frac{z_n - z_k}{z_n - \bar{z}_k}\right|^2 \geq e^{-4K/\eta^2} ,$$

proving $(*)$ with $\delta = e^{-4K/\eta^2}$. We're done.

The above result can now be combined with the lemma of Section B.1 to yield the

Theorem. For sequences $\{z_n\}$ in the upper half plane, the condition that

$$(*) \qquad \prod_{k\neq n} \left|\frac{z_n - z_k}{z_n - \bar{z}_k}\right| \geq \text{some } \delta > 0$$

for all n is equivalent to

$$(\overset{*}{*}) \qquad \sum_k \frac{\Im z_n \, \Im z_k}{|z_n - \bar{z}_k|^2} \leq K$$

independent of n for all n plus the uniform hyperbolic separation condition

$$\left|\frac{z_n - z_m}{z_n - \bar{z}_m}\right| \geq \eta > 0, \qquad n \neq m.$$

It is good to summarize the relations between various properties of sequences $\{z_n\}$ in $\Im z > 0$ by means of the following table:

$$\sum_k \frac{\Im z_k \, \Im z_n}{|z_n - \bar{z}_k|^2} \leq K \quad \text{independent of } n$$

EQUIVALENT TO

$$\sum_{z_k \in S_h} \Im z_k \leq Ch \quad \text{for any square}$$

$$S_h = [x_0, x_0 + h] \times (0,h)$$

EQUIVALENT TO

$$\sum_k \Im z_k |F(z_k)| \leq \tilde{C} \|F\|_1 \quad \text{for}$$

$F \in H_1$ (Carleson measure property)

\+

$$\left| \frac{z_n - z_m}{z_n - \bar{z}_m} \right| \geq \eta > 0$$

if $n \neq m$.

TOGETHER EQUIVALENT TO

$$\prod_{k \neq n} \left| \frac{z_n - z_k}{z_n - \bar{z}_k} \right| \geq \delta > 0 \quad \text{independent of } n$$

EQUIVALENT TO

$\{z_n\}$ is an interpolating sequence for

$\Im z > 0$.

E. Interpolation by bounded harmonic functions. Garnett's theorem.

In establishing the following result, we have to consider the average of an expression $(\pm a_1 \pm a_2 \pm a_3 \pm \cdots \pm a_N)^2$ for all possible choices of the + and - signs (there are 2^N such choices).

This average is simply

$$a_1^2 + a_2^2 + a_3^2 + \cdots + a_N^2,$$

because the cross terms $2(\pm a_k)(\pm a_\ell)$ obtained in the squaring operation average out to zero.

Garnett's Theorem (ca. 1970). Let $\Im z_n > 0$. Suppose for any bounded sequence $\{c_n\}$ there is a $U(z)$ bounded and harmonic (not necessarily analytic!) in $\Im z > 0$ with $U(z_n) = c_n$. Then $\{z_n\}$ is an interpolating sequence for $\Im z > 0$.

Proof (Varopoulos, ca. 1972 or 1973). If the sequences $\{U(z_n)\}$ fill out $\ell_\infty(\mathbb{N})$ as U ranges over the bounded harmonic functions (on the upper half plane), the argument used to prove the lemma of Section A.1 shows that there is a K such that, for any sequence $\{c_n\}$ with $|c_n| \leq 1$, there is a harmonic function $U(z)$ with $|U(z)| \leq K$, $\Im z > 0$, and $U(z_n) = c_n$. If $\{c_n\}$ is real, this holds for a real valued U, for if $U(z)$ is (complex valued) harmonic, so is $\Re U(z)$.

Given any n, let

$$c_k = \begin{cases} 0, & k \neq n \\ 1, & k = n, \end{cases}$$

and take a real valued harmonic U with $|U(z)| \leq K$ in $\Im z > 0$ and $U(z_k) = c_k$. Then $U(z) + K$ is harmonic and ≥ 0 in $\Im z > 0$, $U(z_n) + K = K + 1$, and $U(z_k) + K = K$ for $k \neq n$. Harnack's theorem now gives us

$$\frac{K+1}{K} = \frac{U(z_n) + K}{U(z_k) + K} \leq \frac{1 + |(z_n - z_k)/(z_n - \bar{z}_k)|}{1 - |(z_n - z_k)/(z_n - \bar{z}_k)|},$$

from which we see that

$$(\dagger) \qquad \left| \frac{z_n - z_k}{z_n - \bar{z}_k} \right| \geq \frac{1}{2K+1}, \qquad k \neq n.$$

The equivalences established in Section D show that <u>if we can establish</u>

$$\sum_{z_k \in S_h} \Im z_k \leq Ch$$

for <u>all</u> S_h of the form $[x_0, x_0 + h] \times (0,h)$, <u>we will be done</u>, thanks to $(\dagger)$.

Observe first of all that

$$(\S) \qquad \sum_n |a_n| \leq \frac{K}{\pi} \int_{-\infty}^{\infty} \left| \sum_n \frac{a_n \Im z_n}{|z_n - t|^2} \right| dt$$

for any sequence $\{a_n\}$ in $\ell_1(\mathbb{N})$. <u>Indeed</u>, we can find numbers c_n, $|c_n| = 1$, with $\sum_n |a_n| = \sum_n a_n c_n$, and then we can get $U(z)$ harmonic in $\Im z > 0$, $|U(z)| \leq K$ there, with $U(z_n) = c_n$. By Chapter VI,

$$U(z_n) = \frac{1}{\pi} \int_{-\infty}^{\infty} \frac{\Im z_n}{|z_n - t|^2} U(t) dt$$

with $|U(t)| \leq K$ a.e., so

$$\sum_n |a_n| = \sum_n a_n c_n = \sum_n a_n U_n(z) = \frac{1}{\pi} \int_{-\infty}^{\infty} \left(\sum_n \frac{a_n \Im z_n}{|z_n - t|^2} \right) U(t) dt,$$

which is in absolute value

$$\leq \frac{K}{\pi} \int_{-\infty}^{\infty} \left| \sum_n \frac{a_n \Im z_n}{|z_n - t|^2} \right| dt.$$

Writing $z_n = x_n + iy_n$, let S_h be any square of side h whose base, I, lies on $\mathbb{R}$.

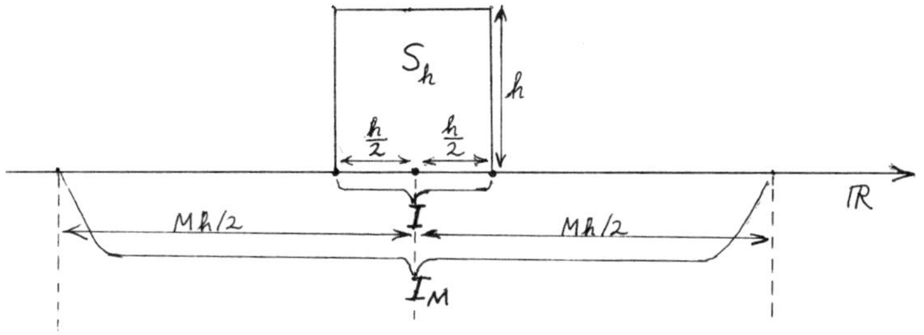

We are to show that

$$\sum_{z_n \in S_h} y_n \leq Ch$$

with a constant C independent of h or the position of I on $\mathbb{R}$.

Let the numerical constant M be so large that

$$\frac{K}{\pi} \int_{(M-1)/2}^{\infty} \frac{ds}{s^2 + 1} \leq \frac{1}{4},$$

fix such an M once and for all. Whatever I, the base of S, may be, let I^M be an interval on $\mathbb{R}$ having the same midpoint as I, but M times the length of I.

If $z_n \in S_h$, our choice of M gives

$$\frac{K}{\pi}\int_{\sim I^M} \frac{\Im z_n}{|z_n - t|^2}\,dt \le \frac{1}{4} + \frac{1}{4} = \frac{1}{2},$$

as is easily seen by making the change of variable $(t - x_n)/y_n = s$.

> <u>In order to estimate</u> $\sum_{z_n \in S_h} y_n$, <u>Varopoulos' idea is to use</u> (§) <u>with</u> $a_n = \pm y_n$ <u>if</u> $z_n \in S_h$ <u>and</u> $a_n = 0$ <u>otherwise, then to average over all possible choices of plus and minus signs</u>!

For technical reasons, we also <u>limit</u> n <u>to the range</u> $1,2,\ldots,N$ where N is some arbitrary large integer. Our estimates will <u>not depend</u> on N, so at the end we can make $N \to \infty$. <u>Thus, for the time being, in all sums</u>, n <u>is restricted to the range</u> $1 \le n \le N$, <u>but this restriction is not explicitly stated</u>. From (§), we have

$$\sum_{z_n \in S_h} y_n = \sum_{z_n \in S_h} |\pm y_n| \le \frac{K}{\pi}\int_{-\infty}^{\infty} \left| \sum_{z_n \in S_h} \frac{\pm y_n^2}{|z_n - t|^2} \right| dt.$$

As we have just seen, our choice of M makes the last integral

$$\le \frac{K}{\pi}\int_{I^M} \left| \sum_{z_n \in S_h} \frac{\pm y_n^2}{|z_n - t|^2} \right| dt + \sum_{z_n \in S_h} \frac{y_n K}{\pi} \int_{\sim I^M} \frac{y_n}{|z_n - t|^2}\,dt \le$$

$$\le \frac{K}{\pi}\int_{I^M} \left| \sum_{z_n \in S_h} \frac{\pm y_n^2}{|z_n - t|^2} \right| dt + \frac{1}{2}\sum_{z_n \in S_h} y_n .$$

Subtracting $\frac{1}{2}\Sigma_{z_n \in S_h} y_n$ and multiplying by 2, we find

$$\sum_{z_n \in S_h} y_n \le \frac{2K}{\pi}\int_{I^M} \left| \sum_{z_n \in S_h} \pm \frac{y_n^2}{|z_n - t|^2} \right| dt.$$

LET $\mathcal{E}$ DENOTE THE OPERATION OF AVERAGING OVER ALL POSSIBLE CHOICES OF PLUS AND MINUS SIGNS.

Then

$$\sum_{y_n \in S_h} y_n = \mathcal{E} \sum_{y_n \in S_h} |\pm y_n| \leq \frac{2K}{\pi} \int_{I^M} \mathcal{E} \left| \sum_{z_n \in S_h} \pm \frac{y_n^2}{|z_n - t|^2} \right| dt.$$

By Schwarz, this last is

$$\leq \frac{2K}{\pi} \left[\int_{I^M} dt \right]^{1/2} \left[\int_{I^M} \left\{ \mathcal{E} \left| \sum_{z_n \in S_h} \pm \frac{y_n^2}{|z_n - t|^2} \right| \right\}^2 dt \right]^{1/2} .$$

<u>The square of the average is less than or equal to the average of the square</u>, so the preceding expression is

$$\leq \frac{2K}{\pi} \sqrt{Mh} \left[\int_{I^M} \mathcal{E} \left(\sum_{z_n \in S_h} \pm \frac{y_n^2}{|z_n - t|^2} \right)^2 dt \right]^{1/2} =$$

$$= \frac{2K}{\pi} \sqrt{Mh} \left[\int_{I^M} \sum_{z_n \in S_h} \frac{y_n^4}{|z_n - t|^4} dt \right]^{1/2} ,$$

using the observation made at the beginning of this Section.

But

$$\int_{I^M} \frac{y_n^4}{|z_n - t|^4} dt \leq \int_{-\infty}^{\infty} \frac{y_n^4 \, dt}{|z_n - t|^4} = y_n \int_{-\infty}^{\infty} \frac{ds}{(s^2 + 1)^2} = c y_n,$$

say, where c is a <u>numerical constant</u> whose <u>value</u> need not concern us here.

Substituting into the previous expression and going back to the chain of inequalities from which it came, we see that

$$\sum_{z_n \in S_h} y_n \leq \frac{2K}{\pi} \sqrt{Mh \cdot c \sum_{z_n \in S_h} y_n} .$$

As stipulated above, the sum here is really for $1 \leq n \leq N$, hence it is certainly finite, so, after squaring and cancelling,

$$\sum_{\substack{z_n \in S_h \\ 1 \leq n \leq N}} y_n \leq \frac{4K^2 Mc}{\pi^2} h.$$

The coefficient on the right does not depend on N, so, now making $N \to \infty$, we have

$$\sum_{z_n \in S_h} y_n \leq \frac{4K^2 Mc}{\pi^2} h.$$

Since (†) also holds, the theorem is now completely proved.

Problem N° 9

In this problem, $\mathfrak{C}_0$ denotes the subspace of H_∞ consisting of functions continuous in $\Im z \geq 0$ and tending to zero for $z \to \infty$ in the closed upper half plane. c_0 denotes the subspace of ℓ_∞ consisting of sequences tending to zero.

Let $\Im z_n > 0$ and $z_n \xrightarrow[n]{} \infty$.

a) If, for each $\{\gamma_n\} \in c_0$ there is a $\Phi \in \mathfrak{C}_0$ with $\Phi(z_n) = \gamma_n$, then $\{z_n\}$ is an interpolating sequence.

*b) If $\{z_n\}$ is an interpolating sequence, then, given $\{\gamma_n\} \in c_0$ there is a $\Phi \in \mathfrak{C}_0$ with $\Phi(z_n) = \gamma_n$.

(Hint: First show how to get a $\varphi \in \mathfrak{C}_0$ with $\|\varphi\|_\gamma \leq K \sup_k |\gamma_k|$ such that $|\varphi(z_n) - \gamma_n| \leq \frac{1}{2} \sup_k |\gamma_k|$. Pay attention to the continuity properties of the Blaschke product having the z_k as its zeros.)

X. Functions of Bounded Mean Oscillation

During this whole chapter we do the work for the circle $\{|z| < 1\}$. Analogous results (with analogous proofs) hold for the half-plane $\Im z > 0$.

In Chapter VII we saw that $H_1(0) = \{zf(z);\ f \in H_1\}$ has the dual L_∞/H_∞. Towards the end of the 1960's, C. Fefferman saw that the dual of $\Re H_1(0)$ could be represented as an actual space of functions, rather than as a quotient space. The functions in this space are characterized by a simple geometric property, namely, that of having bounded mean oscillation. This property was discovered some years ago by Nirenberg and John, in connection with rather unrelated work on differential equations - it was somewhat of a surprise to see that it is relevant to the study of H_p-spaces.

A. Dual of $\Re H_1(0)$.

1. If $\psi(\theta) \in L_\infty$ is periodic of period 2π, we know from Chapter I, Section E that its harmonic conjugate

$$\tilde{\psi}(\theta) = \frac{1}{2\pi}\int_{-\pi}^{\pi} \frac{\psi(\theta - t)}{\tan(t/2)}\, dt$$

exists a.e. and belongs to $L_2(-\pi,\pi)$. In fact, we saw in Chapter V that $\tilde{\psi} \in L_p(-\pi,\pi)$ for all $p < \infty$ (and more!).

Lemma. Let $\psi \in L_\infty$ be periodic of period 2π. Then for each $f \in H_1(0)$,

$$\lim_{r\to 1} \int_{-\pi}^{\pi} \tilde{\psi}(\theta)\, \Re f(re^{i\theta})\, d\theta$$

exists and equals $\displaystyle -\int_{-\pi}^{\pi} \psi(\theta)\, \Im f(e^{i\theta})\, d\theta.$

Proof. We may suppose ψ real - the general case follows from this one by superposition.

Since $\psi + i\tilde{\psi} \in H_2$, for instance (by Chapter I, Section E!), if $f \in H_1(0)$, for each $r < 1$, the function

$$[\psi(\theta) + i\tilde{\psi}(\theta)]f(re^{i\theta})$$

certainly belongs to $H_1(0)$, so by Cauchy's theorem,

$$\int_{-\pi}^{\pi} [\psi(\theta) + i\tilde{\psi}(\theta)]f(re^{i\theta})d\theta = 0.$$

Taking imaginary parts, we find

$$\int_{-\pi}^{\pi} \tilde{\psi}(\theta)\,\Re f(re^{i\theta})d\theta = -\int_{-\pi}^{\pi} \psi(\theta)\,\Im f(re^{i\theta})d\theta.$$

Since $f \in H_1$, we have

$$\int_{-\pi}^{\pi} |f(re^{i\theta}) - f(e^{i\theta})|d\theta \longrightarrow 0 \quad \text{as} \quad r \longrightarrow 1$$

by Chapter II, Section B. (Once again we are using the theorem of the Brothers Riesz!) So, since $\psi \in L_\infty$,

$$\int_{-\pi}^{\pi} \psi(\theta)\,\Im f(re^{i\theta})d\theta \longrightarrow \int_{-\pi}^{\pi} \psi(\theta)\,\Im f(e^{i\theta})d\theta$$

as $r \to 1$. The lemma is proved.

2°. Theorem. Every real valued linear functional L on $H_1(0)$ can be written in the form

$$Lf = \lim_{r \to 1} \int_{-\pi}^{\pi} [\varphi(\theta) + \tilde{\psi}(\theta)] \cdot \Re f(re^{i\theta})d\theta, \qquad f \in H_1(0),$$

with real valued φ and ψ belonging to L_∞ and of period 2π.

Conversely, if φ and ψ are real, of period 2π, and belong to L_∞, the limit on the right side of the above formula exists for every $f \in H_1(0)$ and defines a real linear functional on $H_1(0)$.

Proof. The converse statement follows directly from the above lemma and the theorem of Chapter II, Section B used in its proof.

To get the direct statement, let L be a real linear functional on $H_1(0)$. Then, for $f \in H_1(0)$,

$$Lf = \Re \Lambda f$$

where Λ is some complex-valued linear functional on $H_1(0)$. (Bonenblust-Sobczyk theorem). By Chapter VII, Λ is identified with an element of L_∞/H_∞ so there is certainly a 2π-periodic function in L_∞ - call it $\varphi + i\psi$, with φ and ψ real - such that

$$\Lambda f = \int_{-\pi}^{\pi} [\varphi(\theta) + i\psi(\theta)] f(e^{i\theta}) d\theta, \qquad f \in H_1(0).$$

Taking real parts, we get

$$Lf = \int_{-\pi}^{\pi} \{\varphi(\theta) \Re f(e^{i\theta}) - \psi(\theta) \Im f(e^{i\theta})\} d\theta.$$

According to the lemma and the theorem of Chapter II, Section B already used, this equals

$$\lim_{r \to 1} \int_{-\pi}^{\pi} \{\varphi(\theta) + \tilde{\psi}(\theta)\} \Re f(re^{i\theta}) d\theta.$$

We are done.

Scholium. There is only one $f \in H_1(0)$ having a given real part $\Re f$. Therefore we may look on $\Re H_1(0)$ as a real Banach space with the norm

$$\|\Re f\| = \|f\|_1, \qquad f \in H_1(0).$$

A linear functional on $\Re H_1(0)$ corresponds to a real linear functional on $H_1(0)$ and conversely.

By the theorem, the (real) dual of $\Re H_1(0)$ is thus identified with the set of sums $\Re L_\infty + \widetilde{\Re L_\infty}$. When does such a sum $\varphi + \tilde{\psi}$ correspond to the zero functional on $\Re H_1(0)$? In other words, when is

$$\lim_{r \to 1} \int_{-\pi}^{\pi} \{\varphi(\theta) + \tilde{\psi}(\theta)\} \Re f(re^{i\theta}) d\theta = 0$$

for all $f \in H_1(0)$? Taking, first $f(z) = z^n$ and then $f(z) = iz^n$ with $n = 1,2,3,\ldots$ we see that

$$\int_{-\pi}^{\pi} \{\varphi(\theta) + \tilde{\psi}(\theta)\} \left\{ \begin{matrix} \sin n\theta \\ \cos n\theta \end{matrix} \right\} d\theta = 0, \qquad n = 1,2,3,\ldots$$

so $\varphi(\theta) + \tilde{\psi}(\theta) = \text{const.}$ The converse statement is also clear.

Thus:

> The dual of $\Re H_1(0)$ is $(\Re L_\infty + \widetilde{\Re L_\infty})$ modulo the constant functions.

B. Introduction of BMO.

A function which can be written as $\varphi(\theta) + \tilde{\psi}(\theta)$ with φ and ψ real, 2π-periodic, and bounded can of course be thus written in more than one way. That's why intrinsic characterizations of such sums are important. Fefferman found such a characterization.

1°. Notation. If $G(\theta)$ is locally integrable and I is any interval, we henceforth write

$$G_I = \frac{1}{|I|}\int_I G(\theta)d\theta \ .$$

G_I is the average value of G on I.

Definition. A locally integrable 2π-periodic function $G(\theta)$ is said to be of bounded mean oscillation, in symbols, $G \in \mathrm{BMO}$, provided that

$$\frac{1}{|I|}\int_I |G(\theta) - G_I|d\theta \leq \text{ some finite constant}$$

for all intervals $I \subseteq \mathbb{R}$.

We write, for $G \in \mathrm{BMO}$,

$$\|G\|_* = \sup_I \int_I |G(\theta) - G_I|d\theta,$$

the sup being taken over all intervals I.

Comment. Although it's horrible to represent a space of functions by three letters (BMO), this designation has now become standard usage!

Remark 1. If C is a constant, $\|C\|_* = 0$ and $\|G - C\|_* = \|G\|_*$ for $G \in \mathrm{BMO}$. This is evident from the definition.

Remark 2. For 2π-periodic locally integrable functions G (the only ones considered systematically in this Chapter), in order to check that $\|G\|_* < \infty$, it is enough to verify that

$$\frac{1}{|I|}\int_I |G(\theta) - G_I|\,d\theta$$

is bounded above for all intervals I of length $\le$ an arbitrary given $\ell > 0$.

The proof of this fact, important in applications, is left as an exercise.

Remark 3. In studying the dual of $\Re H_1(\Im z > 0)$, one deals with non-periodic functions of bounded mean oscillation. Their definition is formally the same as that of the periodic ones given above. Of course, in their case, Remark 2 does not apply.

2°. Lemma. Let I be any interval. Then if C is any constant,

$$\frac{1}{|I|}\int_I |G(t) - G_I|\,dt \le \frac{2}{|I|}\int_I |G(t) - C|\,dt.$$

Proof. $G(t) - G_I = [G(t) - C] - [G_I - C]$. But $|G_I - C| \le \frac{1}{|I|}\int_I |G(t) - C|\,dt$, so

$$\frac{1}{|I|}\int_I |G(t) - G_I|\,dt \le \frac{1}{|I|}\int_I |G(t) - C|\,dt + |G_I - C| \le \frac{2}{|I|}\int_I |G(t) - C|\,dt.$$

Q.E.D.

Our first connection between the result of Section A and BMO is provided by the

Theorem. Let $G = \varphi + \tilde{\psi}$, φ, ψ periodic of period 2π and bounded. Then $G \in \mathrm{BMO}$.

Proof. It is obvious that $\|\varphi\|_* \leq 2\|\varphi\|_\infty$, so it is enough to verify that $\|\tilde{\psi}\|_* < \infty$. By Remark 2 of 1° it is enough to show that

$$\frac{1}{|J|}\int_J |\tilde{\psi}(t) - \tilde{\psi}_J|\,dt \leq \text{some } C < \infty$$

for any interval J of length $\leq \frac{2\pi}{3}$.

We do this by the rudimentary application of a typical BMO technique. Given J of length $\leq \frac{2\pi}{3}$, let J' be an interval having the same midpoint as J, but three times its length:

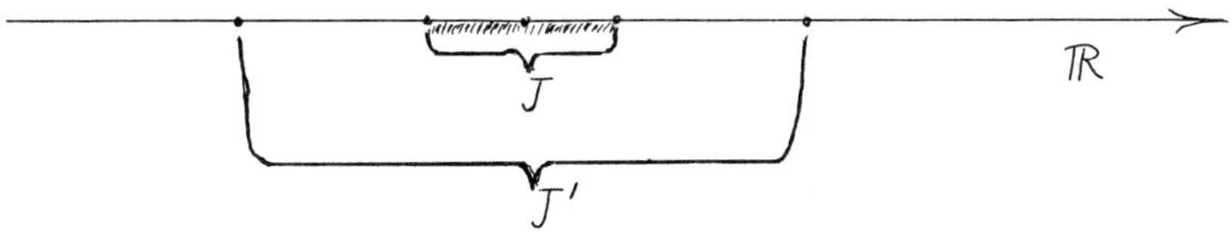

We then take

$$\psi_1(t) = \begin{cases} \psi(t), & t \in \bigcup_{n=-\infty}^{\infty} (J' + 2\pi n) \\ 0, & \text{elsewhere} \end{cases}$$

$$\psi_2(t) = \psi(t) - \psi_1(t).$$

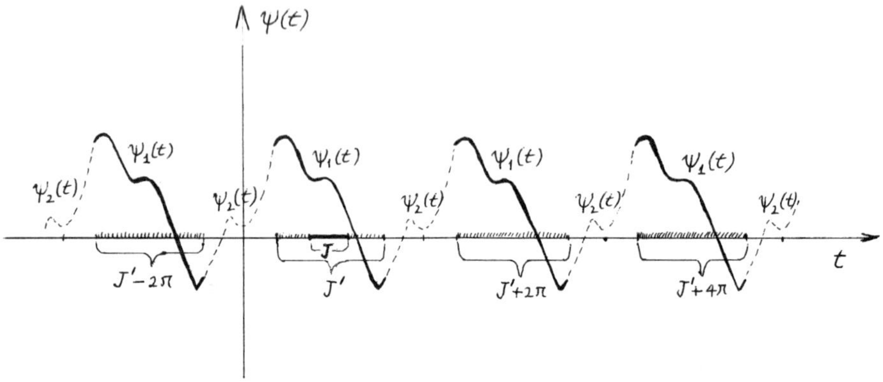

Since $\tilde{\psi} = \tilde{\psi}_1 + \tilde{\psi}_2$, $\tilde{\psi}_J = (\tilde{\psi}_1)_J + (\tilde{\psi}_2)_J$,

$$\frac{1}{|J|}\int_J |\tilde{\psi}(t) - \tilde{\psi}_J|dt \le \frac{1}{|J|}\int_J |\tilde{\psi}_1(t) - (\tilde{\psi}_1)_J|dt + \frac{1}{|J|}\int_J |\tilde{\psi}_2(t) - (\tilde{\psi}_2)_J|dt.$$

By the lemma,

$$\frac{1}{|J|}\int_J |\tilde{\psi}_1(t) - (\tilde{\psi}_1)_J|dt \le \frac{2}{|J|}\int_J |\tilde{\psi}_1(t)|dt,$$

and we use Hilbert's inequality to estimate the right hand side. We have:

$$\int_J |\tilde{\psi}_1(t)|dt \le \left[|J| \int_{-\pi}^{\pi} |\tilde{\psi}_1(t)|^2 dt \right]^{1/2}$$

which by Hilbert's inequality (part of theorem in Chapter I, Section E.4) is $\le \left[|J| \int_0^{2\pi} |\psi_1(t)|^2 dt \right]^{1/2}$. Since, on $[0,2\pi]$, $\psi_1(t)$ vanishes outside a set of measure $|J'| = 3|J|$, the last expression is $\le (|J| \cdot 3|J| \|\psi\|_\infty^2)^{1/2} = \sqrt{3}\, |J| \|\psi\|_\infty$, and therefore

$$\frac{1}{|J|}\int_J |\tilde{\psi}_1(t) - (\tilde{\psi}_1)_J|\,dt \le 2\sqrt{3}\,\|\tilde{\psi}\|_\infty .$$

We turn now to $\frac{1}{|J|}\int_J |\tilde{\psi}_2(t) - (\tilde{\psi}_2)_J|\,dt$ which, by the lemma, is

$$\le \frac{2}{|J|}\int_J |\tilde{\psi}_2(t) - c|\,dt,$$

where, here, we use a constant c equal to $\tilde{\psi}_2(m)$, where m is the midpoint of J. Without loss of generality, take J to be $(-\alpha,\alpha)$, $0 < \alpha \le \frac{\pi}{3}$, then $m = 0$, so $c = \tilde{\psi}_2(0)$ and, on $[-\pi,\pi]$, $\psi_2(t)$ vanishes on $J' = [-3\alpha,3\alpha]$. Thus, we are using

$$c = \psi_2(0) = -\frac{1}{2\pi}\int_{3\alpha\le|t|\le\pi} \frac{\psi(t)}{\tan(t/2)}\,dt.$$

We thus have to estimate

$$\frac{1}{2\alpha}\int_{-\alpha}^{\alpha}\left|\frac{1}{2\pi}\int_{3\alpha\le|t|\le\pi}\left\{\frac{1}{\tan((\theta-t)/2)} + \frac{1}{\tan(t/2)}\right\}\psi(t)\,dt\right|d\theta$$

which we simply majorize by

$$\frac{\|\psi\|_\infty}{4\pi\alpha}\int_{-\alpha}^{\alpha}\int_{2\alpha\le|t|\le\pi}\left|\frac{1}{\tan((\theta-t)/2)} + \frac{1}{\tan(t/2)}\right|dt\,d\theta .$$

This computation would be easier if we were working with the harmonic conjugate for the upper half plane. Be that as it may, we have

$$\frac{1}{\tan((\theta-t)/2)} + \frac{1}{\tan(t/2)} = \frac{1+\tan\frac{\theta}{2}\tan\frac{t}{2}}{\tan\frac{\theta}{2}-\tan\frac{t}{2}} + \frac{1}{\tan\frac{t}{2}}$$

$$= \frac{\tan\frac{\theta}{2}\sec^2\frac{t}{2}}{\tan\frac{t}{2}\,[\tan\frac{\theta}{2}-\tan\frac{t}{2}]} ,$$

whence, for $-\alpha \le \theta \le \alpha$,

$$\int_{2\alpha\le|t|\le\pi} \left|\frac{1}{\tan((\theta-t)/2)} + \frac{1}{\tan(t/2)}\right| dt =$$

$$= 2\left|\tan\frac{\theta}{2}\right| \int_{2\alpha\le|t|\le\pi} \frac{d\tan\frac{t}{2}}{\tan^2\frac{t}{2} - \tan\frac{t}{2}\tan\frac{\theta}{2}} \le$$

$$\le 8\left|\tan\frac{\theta}{2}\right| \int_{\tan\alpha}^{\infty} \frac{d\tau}{\tau^2} = \frac{8\left|\tan\frac{\theta}{2}\right|}{\tan\alpha},$$

so finally,

$$\frac{1}{|J|}\int_J |\tilde{\psi}_2(\theta) - (\tilde{\psi}_2)_J| d\theta \le \frac{2}{|J|}\int_J |\tilde{\psi}_2(\theta) - c| d\theta =$$

$$= \frac{1}{\alpha}\int_{-\alpha}^{\alpha} |\tilde{\psi}_2(\theta) - \tilde{\psi}_2(0)| d\theta \le \frac{4\|\psi\|_\infty}{\pi\alpha\tan\alpha}\int_{-\alpha}^{\alpha} \left|\tan\frac{\theta}{2}\right| d\theta \le \frac{4}{\pi}\|\psi\|_\infty .$$

Putting this together with the preceding estimate for $\frac{1}{|J|}\int_J |\tilde{\psi}_1(\theta) - (\tilde{\psi}_1)_J| d\theta$, we obtain

$$\frac{1}{|J|}\int_J |\tilde{\psi}(t) - \psi_J| dt \le (2\sqrt{3} + \tfrac{4}{\pi})\|\psi\|_\infty ,$$

valid whenever $|J| \le \frac{2\pi}{3}$.

We are done.

Remark. The above calculation, in conjunction with the one behind Remark 2 of 1° (that one was left to the reader!) shows that

$$\|\varphi + \tilde{\psi}\|_* \le K(\|\varphi\|_\infty + \|\psi\|_\infty)$$

for 2π-periodic functions φ and ψ, with a strictly numerical constant K.

C. Garsia's norm.

Garsia and others have observed that there is another norm for functions in BMO which is easier to work with.

If $\varphi(\theta)$ is locally integrable and 2π-periodic, put, for $|z| < 1$,

$$U_\varphi(z) = \frac{1}{2\pi}\int_{-\pi}^{\pi} \frac{1 - |z|^2}{|e^{it} - z|^2}\,\varphi(t)dt;$$

U_φ is a harmonic extension of $\varphi(t)$ to $\{|z| < 1\}$.

1°. The norm $\mathfrak{n}(\varphi)$.

Lemma. For $|z| < 1$,

$$U_{\varphi^2}(z) - [U_\varphi(z)]^2 = \frac{1}{8\pi^2}\int_{-\pi}^{\pi}\int_{-\pi}^{\pi} [\varphi(t) - \varphi(s)]^2 \cdot \frac{1 - |z|^2}{|z - e^{it}|^2} \cdot \frac{1 - |z|^2}{|z - e^{is}|^2}\,dt\,ds.$$

Proof. Multiply out in the integrand.

Definition. If φ is real valued and periodic of period 2π,

$$\mathfrak{n}(\varphi) = \sup_{|z|<1} \{U_{\varphi^2}(z) - [U_\varphi(z)]^2\}^{1/2}.$$

We call $\mathfrak{n}(\varphi)$ the Garsia norm of φ. It is indeed a norm.

Lemma. If a is a real constant, $\mathfrak{n}(a\varphi) = |a|\mathfrak{n}(\varphi)$, And $\mathfrak{n}(\varphi + \psi) \leq \mathfrak{n}(\varphi) + \mathfrak{n}(\psi)$.

Proof. The first relation is obvious. The second follows from the previous lemma and the triangle inequality for the Hilbert space norm. For, by that lemma, $\{U_{\varphi^2}(z) - [U_\varphi(z)]^2\}^{1/2}$ can, for each fixed z, be looked upon as a Hilbert space norm of $\varphi(s) - \varphi(t)$ on $[-\pi,\pi]\times[-\pi,\pi]$, using a certain positive measure on that square.

2°. Two simple inequalities for $\eta(\varphi)$.

It turns out that $\eta(\varphi)$ is equivalent to the BMO norm $\|\varphi\|_*$ introduced in Section B.

For the moment, let us just observe how much easier $\eta(\varphi)$ is to work with than $\|\varphi\|_*$.

Theorem. If φ and ψ are real, 2π-periodic, and bounded,

$$\eta(\varphi + \tilde{\psi}) \le \sqrt{2}\,(\|\varphi\|_\infty + \|\psi\|_\infty).$$

Proof. We see easily (from the first lemma of 1°, for instance), that

$$\eta(\varphi) \le \sqrt{2}\,\|\varphi\|_\infty .$$

Observe now the elementary relation $(\psi + i\tilde{\psi})^2 = \tilde{\psi}^2 - \psi^2 + 2i\psi\tilde{\psi}$, which makes $[U_\psi(z)]^2 - [U_{\tilde{\psi}}(z)]^2$ harmonic in $|z| < 1$. Since $\psi + i\tilde{\psi} \in H_p$ for every $p < \infty$ (Chapter V), the harmonic function $[U_\psi(z)]^2 - [U_{\tilde{\psi}}(z)]^2$ can be recovered from its boundary values $\psi^2 - \tilde{\psi}^2$ by Poisson's formula, i.e.,

$$[U_\psi(z)]^2 - [U_{\tilde{\psi}}(z)]^2 = U_{\psi^2 - \tilde{\psi}^2}(z) = U_{\psi^2}(z) - U_{\tilde{\psi}^2}(z),$$

and we thus have the identity

$$\boxed{U_{\tilde{\psi}^2}(z) - [U_{\tilde{\psi}}(z)]^2 = U_{\psi^2}(z) - [U_\psi(z)]^2} .$$

From this we see that $\eta(\tilde{\psi}) = \eta(\psi)$! So $\eta(\tilde{\psi}) = \eta(\psi) \le \sqrt{2}\,\|\psi\|_\infty$ by the statement at the beginning of this proof. Combining the last inequality with that statement, we are done.

The inequality $\eta(\varphi) \leq$ const. $\|\varphi\|_*$ will be proved in Section F - it is a consequence of some rather delicate work of Nirenberg and John, the discoverers of BMO-functions. The reverse inequality is, however, completely elementary.

Theorem. For φ real and 2π-periodic,

$$\|\varphi\|_* \leq K\eta(\varphi),$$

where K is a constant independent of φ.

Proof. If I is any interval, by Schwarz' inequality,

$$\frac{1}{|I|}\int_I |\varphi(t) - \varphi_I|\,dt \leq \left\{\frac{1}{|I|}\int_I [\varphi(t) - \varphi_I]^2 dt\right\}^{1/2} =$$

$$= \left\{\frac{1}{2|I|^2}\int_I\int_I [\varphi(t) - \varphi(s)]^2 dt\,ds\right\}^{1/2} ,$$

as is seen on multiplying out the integrands in the two last integrals.

Suppose $|I| \leq 2\pi$, wlog $I = [-\alpha,\alpha]$, $0 < \alpha \leq \pi$. Then use the formula

$$U_{\varphi^2}(r) - [U_\varphi(r)]^2 = \frac{1}{8\pi^2}\int_{-\pi}^{\pi}\int_{-\pi}^{\pi} [\varphi(s) - \varphi(t)]^2 \cdot \frac{1 - r^2}{1+r^2-2r\cos s} \cdot \frac{1 - r^2}{1+r^2-2r\cos t}\,ds\,dt,$$

taking for r the special value

$$r = 1 - \sin\frac{\alpha}{2}.$$

For this value of r, and $-\alpha \leq t \leq \alpha$,

$$\frac{1 - r^2}{1 + r^2 - 2r\cos t} = \frac{(1 + r)(1 - r)}{(1 - r)^2 + 4r\sin^2\frac{t}{2}} \geq \frac{1}{5\sin\frac{\alpha}{2}} \geq \frac{2}{5\alpha},$$

so

$$U_{\varphi^2}(r) - [U_\varphi(r)]^2 \geq \frac{1}{50\pi^2\alpha^2}\int_{-\alpha}^{\alpha}\int_{-\alpha}^{\alpha}[\varphi(s)-\varphi(t)]^2 ds\,dt =$$

$$= \left(\frac{2}{5\pi}\right)^2 \cdot \frac{1}{2|I|^2}\int_I\int_I [\varphi(s)-\varphi(t)]^2 ds\,dt\,.$$

Combining this with the inequality given at the beginning of the proof, we see that

$$\frac{1}{|I|}\int_I |\varphi(t) - \varphi_I|dt \leq \frac{5\pi}{2}\,\mathfrak{n}(\varphi)$$

if $|I| \leq 2\pi$.

Suppose finally that n is a positive integer and $2\pi n < |I| \leq 2\pi(n+1)$. Let $J \supseteq I$ be an interval of length $2\pi(n+1)$; then

$$\frac{1}{|I|}\int_I |\varphi(t) - \varphi_J|dt \leq \frac{n+1}{n}\cdot\frac{1}{|J|}\int_J|\varphi(t)-\varphi_J|dt \leq \frac{2}{|J|}\int_J|\varphi(t)-\varphi_J|dt.$$

But by the 2π-periodicity of φ, $\varphi_J = c = \frac{1}{2\pi}\int_0^{2\pi}\varphi(s)ds$, and

$$\frac{1}{|J|}\int_J |\varphi(t) - \varphi_J|dt = \frac{1}{2\pi}\int_{-\pi}^{\pi}|\varphi(t) - c|dt,$$

which is $\leq \frac{5\pi}{2}\,\mathfrak{n}(\varphi)$, as we have just proved.

Thus, by a Lemma of Section B.2,

$$\frac{1}{|I|}\int_I|\varphi(t)-\varphi_I|dt \leq \frac{2}{|I|}\int_I|\varphi(t)-\varphi_J|dt \leq \frac{4}{|J|}\int_J|\varphi(t)-\varphi_J|dt \leq 10\pi\mathfrak{n}(\varphi).$$

We have proved $\|\varphi\|_* \leq 10\pi\mathfrak{n}(\varphi)$ (and, incidently, almost completely worked out the exercise that was "left to the reader" in Section B.1 (Remark 2), if he or she has not already done it by now!). We are done.

Remark. Taken together, the above two theorems provide us with a new proof of the fact that $\|\varphi + \tilde{\psi}\|_* \leq C(\|\varphi\|_\infty + \|\psi\|_\infty)$, already shown in Section B. The splitting used in the proof of Section B, is, however, an important technique which is applied frequently in studying BMO. That's why we gave that proof.

Work in the next two sections is directly based on the use of the Garsia norm $\mathfrak{n}(\varphi)$. The original BMO norm $\|\varphi\|_*$ will be brought back into the picture in Section F.

D. Computations based on Green's theorem.

1°. Lemma. Let $W(z)$ be C_∞ in and on $\{|z| < 1\}$ and let $W(0) = 0$. Then

$$\int_{-\pi}^{\pi} W(e^{i\theta})d\theta = \iint_{|z|<1} [\log \frac{1}{|z|}] \nabla^2 W(z)\,dxdy.$$

Notation. As usual, $\nabla^2 W = W_{xx} + W_{yy}$.

Proof of Lemma. Writing $z = re^{i\theta}$ and taking ρ, $0 < \rho < 1$, we have by Green's theorem

$$\iint_{\rho<r<1} (\log \frac{1}{r}) \nabla^2 W\,dx\,dy = \iint_{\rho<r<1} [(\log \frac{1}{r})\nabla^2 W - W\nabla^2 \log \frac{1}{r}]dxdy =$$

$$= \int_0^{2\pi} [\log \frac{1}{r}\,\frac{\partial W}{\partial r} - W\,\frac{\partial \log(1/r)}{\partial r}]_{r=1}\,d\theta - \int_0^{2\pi} [\log \frac{1}{r}\frac{\partial W}{\partial r} - W\,\frac{\partial \log(1/r)}{\partial r}]_{r=\rho} \cdot \rho\,d\theta .$$

Since $W(0) = 0$, $W(\rho e^{i\theta}) = \mathcal{O}(\rho)$, and the second integral on the right goes to zero as $\rho \to 0$. The first integral on the right is just $\int_0^{2\pi} W(e^{i\theta})d\theta$. The lemma follows on making $\rho \to 0$.

Lemma. Let $W(z) = |z|W_1(z)$ with a function W_1 twice continuously differentiable in and on $\{|z| < 1\}$. Then

$$\int_{-\pi}^{\pi} W(e^{i\theta})d\theta = \iint_{|z|<1} (\log \frac{1}{|z|})\nabla^2 W(z)\,dxdy.$$

Proof. Just like that of the preceding lemma.

Notation. At this point, we introduce the (physicists') vector operator

$$\vec{\nabla} = \vec{i}\,\frac{\partial}{\partial x} + \vec{j}\,\frac{\partial}{\partial y}, \qquad \vec{i} \text{ and } \vec{j}$$

being unit vectors in the directions of the x and y axes respectively. We use "." to denote the dot product of two dimensional vectors.

Theorem. Let $u(z)$ and $V(z)$ be harmonic in $\{|z| < R\}$ where $R > 1$, and let $u(0) = 0$. Then

$$\int_{-\pi}^{\pi} u(e^{i\theta})V(e^{i\theta})\,d\theta = 2 \iint_{|z|<1} \left(\log \frac{1}{|z|}\right)(\vec{\nabla} u \cdot \vec{\nabla} V)\,dx\,dy.$$

Proof. $W(z) = u(z)V(z)$ satisfies the hypothesis of the first lemma above. Therefore the integral on the right equals $\iint_{|z|<1} (\log\frac{1}{|z|})\nabla^2(uV)\,dx\,dy$. However,

$$\nabla^2(uV) = (\nabla^2 u)V + u\nabla^2 V + 2\vec{\nabla} u \cdot \vec{\nabla} V = 2\vec{\nabla} u \cdot \vec{\nabla} V,$$

since $\nabla^2 u = \nabla^2 V = 0$ (harmonicity).

That does it.

2°. Lemma. Let $f(z)$ be analytic in $\{|z| < R\}$ where $R > 1$, and vanish only at the origin in that circle. Then $|f(z)|$ is C_∞ in $\{0 < |z| < R\}$ and

$$\nabla^2 |f| = \frac{\vec{\nabla} \Re f \cdot \vec{\nabla} \Re f}{|f|}, \qquad 0 < |z| < R.$$

(Due to P. Stein (not E. M. Stein!), ca. 1933.)

Proof. Writing $f = u + iv$ with u and v real and harmonic in $\{ |z| < R\}$, we have $|f| = (u^2+v^2)^{1/2}$ and this is C_∞ away from the origin because $u^2 + v^2 \neq 0$ if we're not at the origin.

We see that

$$\frac{\partial|f|}{\partial x} = \frac{uu_x + vv_x}{(u^2+v^2)^{1/2}},$$

$$\frac{\partial^2|f|}{\partial x^2} = \frac{u_x^2 + v_x^2}{(u^2 + v^2)^{1/2}} + \frac{uu_{xx} + vv_{xx}}{(u^2 + v^2)^{1/2}} - \frac{(uu_x + vv_x)^2}{(u^2 + v^2)^{3/2}};$$

similarly,

$$\frac{\partial^2|f|}{\partial y^2} = \frac{u_y^2 + v_y^2}{(u^2 + v^2)^{1/2}} + \frac{uu_{yy} + vv_{yy}}{(u^2 + v^2)^{1/2}} - \frac{(uu_y + vv_y)^2}{(u^2 + v^2)^{3/2}}.$$

Since $u_x = v_y$, $u_y = -v_x$, we have

$$(uu_x + vv_x)^2 + (uu_y + vv_y)^2 = (u^2 + v^2)(u_x^2 + u_y^2).$$

This relation, together with $u_{xx} + u_{yy} = 0$, $v_{xx} + v_{yy} = 0$, gives us

$$\frac{\partial^2|f|}{\partial x^2} + \frac{\partial^2|f|}{\partial y^2} = \frac{u_x^2 + u_y^2}{(u^2 + v^2)^{1/2}}$$

away from the origin, proving the lemma.

Theorem. Let $f(z)$ be analytic for $|z| < R$ where $R > 1$, and suppose that f has a simple zero at the origin and no other zeros in $|z| < R$.

Then

$$\boxed{\int_{-\pi}^{\pi} |f(e^{i\theta})|\,d\theta = \iint_{|z|<1} \left(\log\frac{1}{|z|}\right) \frac{\vec{\nabla}\Re f \cdot \vec{\nabla}\Re f}{|f(z)|}\,dxdy.}$$

Proof. Apply the second lemma of 1° with $W(z) = |f(z)| = |z||f(z)/z|$. Here $f(z)/z$ never vanishes in $|z| < R$ and is analytic there, so, by the work used in proving the above lemma, $|f(z)/z|$ is C_∞ for $\{|z| < R\}$. The identity established by the above lemma then gives us what we want.

3°. We return to the problem of finding an intrinsic characterization of the functions of the form $\varphi + \tilde{\psi}$, with φ and ψ real, 2π-periodic, and bounded. We already know from Section C.2 that if $F = \varphi + \tilde{\psi}$, then $\mathfrak{h}(F) < \infty$, and now set out along the road of proving that if $\mathfrak{h}(F) < \infty$, then F CAN be written as a sum $\varphi + \tilde{\psi}$ with φ, ψ real, bounded and of period 2π. According to the theorem in Section A, this will FOLLOW if we can show that

$$\lim_{R\to 1} \int_{-\pi}^{\pi} \Re f(Re^{i\theta})F(\theta)d\theta$$

exists for every $f \in H_1(0)$, and represents a LINEAR FUNCTIONAL on $\Re H_1(0)$, whenever $\mathfrak{h}(F) < \infty$.

We perform a series of successive reductions.

Lemma

$$\lim_{R\to 1} \int_{-\pi}^{\pi} \Re f(Re^{i\theta})F(\theta)d\theta$$

exists and represents a linear functional on $\Re H_1(0)$ if there exists a constant C such that, for any $f \in H_1(0)$ and any R, $0 < R < 1$,

$$\left| \int_{-\pi}^{\pi} \Re f(Re^{i\theta})F(\theta)d\theta \right| \le C\|f\|_1.$$

Proof. Let $f \in H_1(0)$ and let $\varepsilon > 0$. By what is essentially the theorem of the Brothers Riesz (Chapter II, Section B), there is an $R < 1$ such that $\int_{-\pi}^{\pi} |f(e^{i\theta}) - f(R'e^{i\theta})| d\theta < \varepsilon$ whenever $R < R' < 1$. Let now $R < R_1 < R_2 < 1$; then, if $R' = R_1/R_2$, $R < R' < 1$, so if $g(z) = f(z) - f(R'z)$, $g \in H_1(0)$ and $\|g\|_1 < \varepsilon$. Therefore, if the inequality in the hypothesis holds,

$$\left| \int_{-\pi}^{\pi} \Re\, g(R_2 e^{i\theta}) F(\theta) d\theta \right| < C\varepsilon,$$

i.e.,

$$\left| \int_{-\pi}^{\pi} \Re\, f(R_2 e^{i\theta}) F(\theta) d\theta - \int_{-\pi}^{\pi} \Re\, f(R_1 e^{i\theta}) F(\theta) d\theta \right| < C\varepsilon$$

whenever $R < R_1 < R_2 < 1$.

It follows that the desired limit exists; clearly it is in absolute value $\leq C\|f\|_1$.

Q.E.D.

Lemma. Let $F \in L_1(-\pi,\pi)$.

$$\lim_{R \to 1} \int_{-\pi}^{\pi} \Re\, f(Re^{i\theta}) F(\theta) d\theta$$

exists for all $f \in H_1(0)$, and represents a linear functional on $\Re H_1(0)$ if there is a constant C such that for all R and $\rho < 1$ and all $f \in H_1(0)$,

$$\left| \int_{-\pi}^{\pi} \Re\, f(Re^{i\theta}) U_F(\rho e^{i\theta}) d\theta \right| \leq C\|f\|_1.$$

Proof. If $R < 1$, $f(Re^{i\theta})$ is continuous, so, since (Chapter I!)

$$\int_{-\pi}^{\pi} |U_F(\rho e^{i\theta}) - F(\theta)| d\theta \longrightarrow 0$$

as $\rho \to 1$, we have, for any $R < 1$,

$$\int_{-\pi}^{\pi} \Re f(Re^{i\theta})F(\theta)d\theta = \lim_{\rho\to 1}\int_{-\pi}^{\pi} \Re f(Re^{i\theta})U_F(\rho e^{i\theta})d\theta.$$

The result now follows by the preceding lemma.

<u>Lemma</u>. Let $F \in L_1(-\pi,\pi)$. In order to prove that

$$\lim_{R\to 1}\int_{-\pi}^{\pi} \Re f(Re^{i\theta})F(\theta)d\theta$$

exists for all $f \in H_1(0)$, and represents a linear functional on $\Re H_1(0)$, it is enough to show that there exists a constant C such that, for all $R < 1$ and $\rho < 1$,

$$\left|\int_{-\pi}^{\pi} \Re f(Re^{i\theta})U_F(\rho e^{i\theta})d\theta\right| < C\|f\|_1$$

<u>whenever</u> $f \in H_1(0)$ <u>has precisely one simple zero at the origin, and no other zeros in</u> $\{|z| < 1\}$.

<u>Proof</u>. By a trick already mentioned in Chapter IV. Let $f \in H_1(0)$, then we can write $f(z) = zB(z)g(z)$, where $B(z)$ is a Blaschke product, $g(z)$ <u>has no zeros in</u> $|z| < 1$, and $\|g\|_1 = \|f\|_1$. We can now put

$$f_1(z) = \frac{1}{2}\, z(B(z) - 1)g(z)$$

$$f_2(z) = \frac{1}{2}\, z(B(z) + 1)g(z),$$

then f_1 and f_2 each have <u>one simple zero at the origin, and no others in the unit circle</u>. We have $f = f_1 + f_2$, so, if the inequality holds under the conditions stated in the hypothesis,

$$\left| \int_{-\pi}^{\pi} \Re f(Re^{i\theta}) U_F(\rho e^{i\theta})\, d\theta \right| \leq \sum_{k=1}^{2} \left| \int_{-\pi}^{\pi} \Re f_k(Re^{i\theta}) U_F(\rho e^{i\theta})\, d\theta \right| \leq$$

$$\leq C(\|f_1\|_1 + \|f_2\|_1) \leq 2C\|f\|_1,$$

because $\|f_1\|_1 \leq \|f\|_1$, $\|f_2\|_1 \leq \|f\|_1$.

Our result now follows by the preceding lemma.

We now combine the last of the above lemmas with the theorem of 1° to obtain:

Theorem. Let $F \in L_1(-\pi,\pi)$. In order to prove that

$$\lim_{R\to 1} \int_{-\pi}^{\pi} \Re f(Re^{i\theta}) F(\theta)\, d\theta$$

exists for all $f \in H_1(0)$, and represents a linear functional on $\Re H_1(0)$, it is enough to show that there exists a constant C with

$$\left| \iint_{|z|<1} \log \frac{1}{|z|} (\vec{\nabla} u \cdot \vec{\nabla} V)\, dx\,dy \right| \leq C\|f\|_1$$

for $u(z) = \Re f(Rz)$, $V(z) = U_F(\rho z)$ and all $R, \rho < 1$, whenever $f \in H_1(0)$ has a simple zero at 0 and no other zeros in $|z| < 1$.

E. Fefferman's theorem with the Garsia norm.

We continue, basing our work on the somewhat ungainly theorem in Section D.3. According to that result and the theorem of Section A, given a real 2π-periodic F with $\eta(F) < \infty$, we will have $F \in \Re L_\infty + \widetilde{\Re L_\infty}$ as soon as we prove, with a numerical constant K,

$$\boxed{\begin{array}{l} \left| \iint_{|z|<1} \log \frac{1}{|z|} (\vec{\nabla} u \cdot \vec{\nabla} V)\, dx\,dy \right| \leq K\|f\|_1 \eta(F) \quad \text{for} \\ u(z) = \Re f(Rz),\ V(z) = U_F(\rho z), \quad R \text{ and } \rho \text{ arbitrary } < 1, \end{array}}$$

whenever $f \in H_1(0)$ has a simple zero at 0 and no other zeros in $\{|z| < 1\}$.

The rest of this section, really the heart of the whole chapter, is devoted to the proof of the above boxed inequality.

<u>We continue to let</u> $u(z)$ <u>stand for</u> $\Re f(Rz)$ <u>and</u> $V(z)$ <u>for</u> $U_F(\rho z)$, <u>where</u> $f \in H_1(0)$ <u>and</u> R, ρ <u>are arbitrary positive numbers</u> < 1.

1°. Use of Schwarz' inequality.

We are interested in the case where $f \in H_1(0)$ has a simple zero at 0, and no others in the unit circle. Fefferman had the idea of applying Schwarz' inequality to the left hand member of the above boxed inequality, in such a way as to take advantage of the theorem in Section D.2.

We have, namely,

$$\left| \iint_{|z|<1} \log\frac{1}{|z|}(\vec{\nabla} u \cdot \vec{\nabla} V)\,dxdy \right| \leq$$

$$\leq \left[\iint_{|z|<1} \left(\log\frac{1}{|z|}\right)\frac{\vec{\nabla} u \cdot \vec{\nabla} u}{|f(Rz)|}\,dxdy \right]^{1/2} \left[\iint_{|z|<1} \log\frac{1}{|z|}(\vec{\nabla} V \cdot \vec{\nabla} V)\,|f(Rz)|\,dxdy \right]^{1/2} .$$

<u>The factor</u> $|f(Rz)|$, <u>which was not even present has thus been brought in!</u>

Since $u(z) = \Re f(Rz)$ with $f \in H_1(0)$ having a <u>simple zero</u> at 0, <u>and no others in the unit circle</u>, the theorem of Section D.2 says that

$$\iint_{|z|<1} \left(\log\frac{1}{|z|}\right)\frac{\vec{\nabla} u \cdot \vec{\nabla} u}{|f(Rz)|}\,dxdy = \int_{-\pi}^{\pi} |f(Re^{i\theta})|\,d\theta < \|f\|_1,$$

<u>so our boxed inequality will be proved as soon as we show that</u>

$$(*) \qquad \iint_{|z|<1} \log\frac{1}{|z|}(\vec{\nabla} V \cdot \vec{\nabla} V)\,|f(Rz)|\,dxdy \leq K\|f\|_1 (h(F))^2$$

for all $f \in H_1(0)$ of the aforementioned special form, with a purely numerical constant K.

2°. A measure to be proved Carleson.

In 1°, restriction to $f \in H_1(0)$ with a simple zero at 0, and no others in the unit circle, is on account of the requirement of the theorem in Section D.2. That is also why the third lemma of Section D.3 was needed.

Now, however, there is no obstacle to our proving (*) for all $f \in H_1(0)$. Observe that for $f \in H_1(0)$, $g(z) = f(Rz)/z$ is in H_1, and clearly $\|g\|_1 \leq \|f\|_1$. It is therefore enough to show that

$$\iint_{|z|<1} |z|\log\frac{1}{|z|}(\vec{\nabla V}\cdot\vec{\nabla V})|g(z)|\,dxdy \leq K(\eta(F))^2\|g\|_1$$

for all $g \in H_1$, with a numerical constant K.

Recall now the definition of Carleson measures (for the unit circle) in Chapter VIII, Section E. We see that

we have to prove that

$$|z|\log\frac{1}{|z|}(\vec{\nabla V}\cdot\vec{\nabla V})\,dxdy$$

is a Carleson measure with "Carleson constant" equal to a numerical multiple of $(\eta(F))^2$.

Part of this program is carried through rather easily.

Lemma. $\displaystyle\iint_{|z|<1/2} |z|\log\frac{1}{|z|}(\vec{\nabla V}\cdot\vec{\nabla V})|g(z)|\,dxdy \leq K'(\eta(F))^2\|g\|_1$ for all $g \in H_1$ with a numerical constant K'.

Proof. For $|z| \leq \frac{1}{2}$, $|z|\log\frac{1}{|z|} \leq e^{-1}$ and $|g(z)| \leq \frac{2}{\pi}\,\|g\|_1$.

Again, $V(z) = U_F(\rho z)$ where $0 \leq \rho < 1$, so $(\nabla V)(z) = \nabla U_F(\rho z) = \nabla U_{F_1}(\rho z)$, where $F_1(\theta) = F(\theta) - \frac{1}{2\pi}\int_{-\pi}^{\pi} F(t)dt$. Now by direct differentiation of Poisson's formula, we have, for $|z| \leq \frac{1}{2}$, $|\nabla U_{F_1}(\rho z)| \leq \frac{4}{\pi}\int_{-\pi}^{\pi} |F_1(t)|dt$, i.e.,

$$|\nabla U_{F_1}(\rho z)| \leq \frac{8}{2\pi}\int_{-\pi}^{\pi}\left|F(\theta) - \frac{1}{2\pi}\int_{-\pi}^{\pi} F(t)dt\right|d\theta$$

for $|z| \leq \frac{1}{2}$.

From the proof of the second theorem in Section C.2, we see that the quantity on the right in the above expression is $\leq 20\pi\, h(F)$. Using this, together with the other estimates, we find

$$\iint_{|z|<1/2} |z|\log\frac{1}{|z|}(\nabla V \cdot \nabla V)|g(z)|dxdy \leq \frac{800\pi}{e}(h(F))^2\|g\|_1$$

Q.E.D.

3°. We now come to the main work of this whole chapter, which is to prove that

$$\iint_{1/2<|z|<1} |z|\log\frac{1}{|z|}(\nabla V \cdot \nabla V)|g(z)|dxdy \leq K''(h(F))^2\|g\|_1$$

with a numerical constant K'' for all $g \in H_1$. This proof depends on the fundamental theorem on Carleson measures given in Chapter VIII, Section E.

Lemma. Let $0 < h \le \frac{1}{2}$ and let S_h be any curvelinear box of the form $1 - h \le r < 1$, $\theta_0 < \theta < \theta_0 + h$.

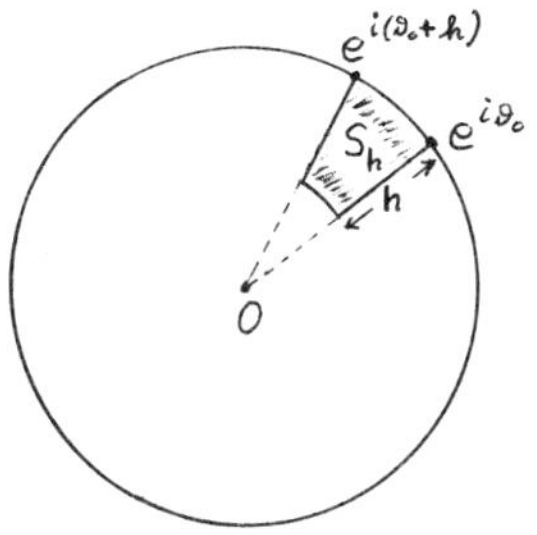

Then, with a numerical constant C,

$$\iint_{S_h} |z|\log\frac{1}{|z|}(\underset{\to}{\nabla}V \cdot \underset{\to}{\nabla}V)\,dxdy \le C(\eta(F))^2 \cdot h.$$

Proof. Wlog $\theta_0 = -\frac{h}{2}$, so that

$$S_h = \{re^{i\theta};1 - h \le r < 1, -\frac{h}{2} \le \theta \le \frac{h}{2}\}.$$

In that case we estimate the integral in question by first making a change of variable $z \longrightarrow \zeta$, where $\frac{i - \zeta}{i + \zeta} = z$. This is a conformal mapping of $\{|z| < 1\}$ onto $\Im\zeta > 0$. As usual, we write $\zeta = \xi + i\eta$.

Put $V(z) = w(\zeta)$ for z corresponding to ζ in the above fashion.

We have, for $\frac{1}{2} \le |z| < 1$,

$$|z|\log\frac{1}{|z|} \le \frac{1}{2}\log\frac{1}{|z|^2} = \frac{(1 - |z|^2)}{2}[1 + \frac{1}{2}(1 - |z|^2)^2 + \frac{1}{3}(1 - |z|^2)^3 + \dots] \le$$

$$\le \frac{(1 - |z|^2)}{2|z|^2} \le 2(1 - |z|^2).$$

In terms of $\zeta = \xi + i\eta$, $|z|^2 = [\xi^2 + (\eta - 1)^2]/[\xi^2 + (\eta + 1)^2]$, so $1 - |z|^2 = \dfrac{4\eta}{\xi^2 + (1+\eta)^2} \leq 4\eta$, and finally

$$|z|\log\frac{1}{|z|} \leq 8\eta \quad \underline{\text{for}} \quad \frac{1}{2} \leq |z| < 1.$$

Let now S_h correspond to the set Σ_h in the ζ-plane under the mapping $z \longrightarrow \zeta$:

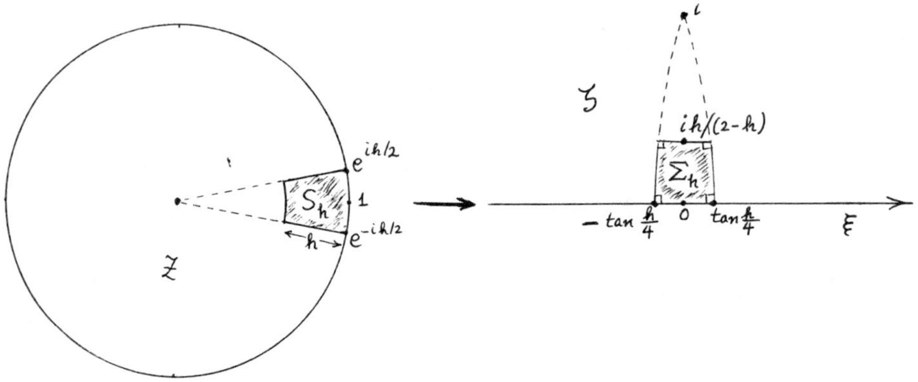

Then, since the mapping is conformal, the quantity we are trying to estimate,

$$\iint_{S_h} |z|\log\frac{1}{|z|}\left\{\left(\frac{\partial V}{\partial x}\right)^2 + \left(\frac{\partial V}{\partial y}\right)^2\right\} dxdy,$$

equals

$$\iint_{\Sigma_h} |z| \log \frac{1}{|z|} \left\{ \left(\frac{\partial w}{\partial \xi}\right)^2 + \left(\frac{\partial w}{\partial \eta}\right)^2 \right\} d\xi d\eta$$

which is

$$\leq \iint_{\Sigma_h} 8\eta \left\{ \left(\frac{\partial w}{\partial \xi}\right)^2 + \left(\frac{\partial w}{\partial \eta}\right)^2 \right\} d\xi d\eta$$

by the above boxed inequality.

Since $0 \leq h \leq \frac{1}{2}$, Σ_h is included in the square

$$B_h = \{(\xi, \eta); -\frac{h}{2} < \xi < \frac{h}{2}, \ 0 < \eta < h\},$$

as some simple calculations show. (Actually, it is manifest without any calculation that for $0 \leq h \leq \frac{1}{2}$, Σ_h is always included in B_{Ch} for some numerical constant C, and the reader may content himself with this evident fact. We take here the permissible value $C = 1$ in order to combat the proliferation of numerical constants in our formulas!) Here is the situation:

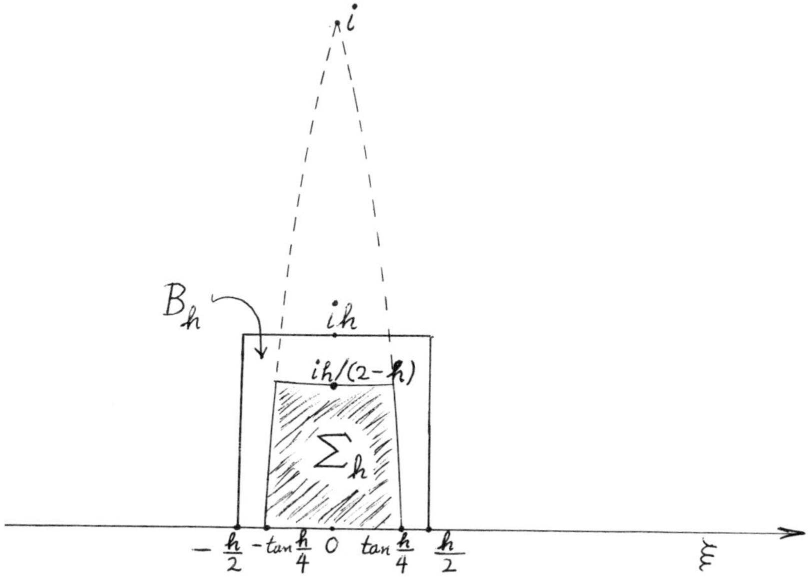

We therefore have

$$\iint_{S_h} |z|\log\frac{1}{|z|}\,|\vec{\nabla} V|^2 dxdy \le 8\iint_{B_h} \eta[w_\xi^2 + w_\eta^2]d\xi d\eta,$$

and our lemma will be proved when we show that the quantity on the right is $\le K''(\mathfrak{n}(F))^2\cdot h$ with a numerical constant K''.

Now we use a trick. B_h is entirely contained in the semi-circle $|\zeta| < 2h$, $\eta > 0$, and on B_h, $1 - (|\zeta|/2h) > 1/4$. Therefore

$$\iint_{B_h} \eta(w_\xi^2 + w_\eta^2)d\xi d\eta \le 4 \iint_{\substack{|\zeta|<2h,\\ \eta>0}} \left(1 - \frac{|\zeta|}{2h}\right)\eta(w_\xi^2 + w_\eta^2)d\xi d\eta.$$

The factor $[1-(|\zeta|/2h)]\eta$ in the right-hand integrand vanishes on the boundary of the region of integration, and that will help us, as we shall see in a moment.

We have $w(\zeta) = V(z)$ with $z = \frac{i-\zeta}{i+\zeta}$, so $w(\zeta)$ is harmonic for $\Im\zeta > 0$. (Recall that we are using $V(z)$ to denote $U_F(\rho z)$ throughout this whole section, with ρ a fixed but arbitrary positive number < 1. Recall also that

$$U_F(z) = \frac{1}{2\pi}\int_{-\pi}^{\pi} \frac{1-|z|^2}{|z-e^{it}|^2}F(t)dt,$$

the Poisson integral of F.) On account of the harmonicity of $w(\zeta)$, we have the identity (TRICK!)

$$\boxed{\left(\frac{\partial^2}{\partial\xi^2} + \frac{\partial^2}{\partial\eta^2}\right)w^2 = 2(w_\xi^2 + w_\eta^2)}$$

(The second edition of Zygmund's book already contains an application of this identity to the study of a related question.) Our problem thus reduces to the estimate of

$$J = 2 \iint\limits_{\substack{|\zeta|<2h \\ \eta>0}} \left(1 - \frac{|\zeta|}{2h}\right) \eta \left(\frac{\partial^2 w^2}{\partial \xi^2} + \frac{\partial^2 w^2}{\partial \eta^2} \right) d\xi d\eta$$

in terms of h and $\eta(F)$.

In Section C.1, $\eta(F)$ was defined as

$$\sup_{|z|<1} \{U_{F^2}(z) - [U_F(z)]^2\}^{1/2} ,$$

so if we put $P(z) = U_{F^2}(z) - [U_F(z)]^2$ for $|z| < 1$, we have

$$\boxed{0 \leq P(z) \leq (\eta(F))^2.}$$

Now for $z = \frac{i - \zeta}{i + \zeta}$,

$$[w(\zeta)]^2 = [U_F(\rho z)]^2 = U_{F^2}(\rho z) - P(\rho z),$$

<u>where the term</u> $U_{F^2}(\rho z)$ <u>on the right is harmonic</u>. <u>Therefore, if we put</u>

$$b(\zeta) = P(\rho z),$$

<u>we have</u>

$$\left(\frac{\partial^2}{\partial \xi^2} + \frac{\partial^2}{\partial \eta^2} \right) w^2 = - \left(\frac{\partial^2 b}{\partial \xi^2} + \frac{\partial^2 b}{\partial \eta^2} \right) ,$$

and the integral we have to estimate boils down to

$$J = -2 \iint\limits_{\substack{|\zeta|<2h \\ \eta>0}} \left(1 - \frac{|\zeta|}{2h}\right) \eta \left(\frac{\partial^2 b}{\partial \xi^2} + \frac{\partial^2 b}{\partial \eta^2} \right) d\xi d\eta ,$$

where

$$\boxed{0 \le b(\zeta) \le (\eta(F))^2.}$$

Write $\nabla_\zeta^2 = \frac{\partial^2}{\partial\xi^2} + \frac{\partial^2}{\partial\eta^2}$. Applying Green's theorem to the integral just written, we find

$$J = -2 \iint_{\substack{|\zeta|<2h \\ \eta>0}} b(\zeta)\nabla_\zeta^2\left\{\eta\left(1 - \frac{|\zeta|}{2h}\right)\right\} d\xi\, d\eta + 2\int_\Gamma \left\{b(\zeta)\frac{\partial\{(1-|\zeta|/2h)\}}{\partial n_\zeta} - \eta\left(1 - \frac{|\zeta|}{2h}\right)\frac{\partial b(\zeta)}{\partial n_\zeta}\right\}|d\zeta|,$$

where Γ is this contour,

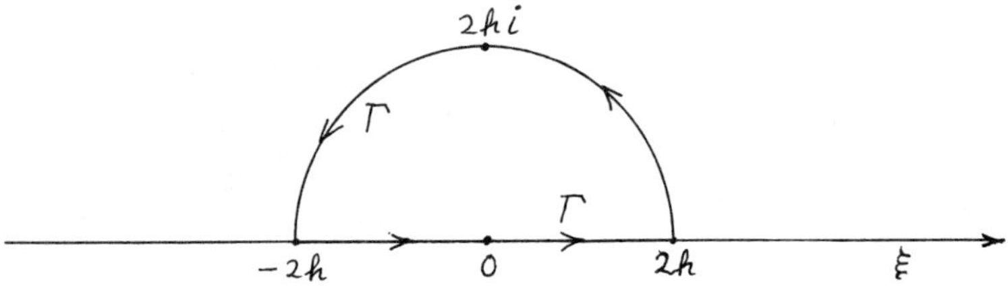

and $\partial/\partial n_\zeta$ denotes differentiation with respect to distance in the direction of the outward normal to Γ at ζ.

The line integral around Γ is negative. Indeed, $\eta(1 - |\zeta|/2h) \equiv 0$ on Γ (now we see why the extra factor $1 - |\zeta|/2h$ was brought in!); also, $\eta(1 - |\zeta|/2h) > 0$ inside Γ, so clearly $\partial\{\eta(1 - |\zeta|/2h)\}/\partial n_\zeta \le 0$ on Γ. Therefore $b(\zeta)\frac{\partial\{\eta(1 - |\zeta|/2h)\}}{\partial n_\zeta} \le 0$ on Γ in view of the above boxed inequality, and the line integral around Γ does come out to be ≤ 0. (If the reader has trouble keeping track of signs here, he or she may simply compute

$$\int_\Gamma |b(\zeta)| \left| \frac{\partial\{\eta(1-|\zeta|/2h)\}}{\partial n_\zeta} \right| |d\zeta|$$

using polar coordinates. A value $\leq$ const. $h \sup_\zeta |b(\zeta)| \leq \text{const.}(\eta(F))^2 h$ will be found.)

We thus have

$$J \leq -2 \iint_{\substack{|\zeta|\leq 2h \\ \eta>0}} b(\zeta)\nabla_\zeta^2 \left\{\eta\left(1 - \frac{|\zeta|}{2h}\right)\right\} d\xi d\eta .$$

To evaluate this, use the polar coordinates $\zeta = \sigma e^{i\varphi}$; then

$$\nabla_\zeta^2 = \frac{1}{\sigma}\frac{\partial}{\partial\sigma}\left(\sigma\frac{\partial}{\partial\sigma}(\quad)\right) + \frac{1}{\sigma^2}\frac{\partial^2}{\partial\varphi^2}$$

and $\eta(1-|\zeta|/2h) = (1-\sigma/2h)\sigma\sin\varphi$, from which we easily find

$$\nabla_\zeta^2\{\eta(1-|\zeta|/2h)\} = -\frac{3}{2h}\sin\varphi,$$

which, substituted into the above integral, yields

$$J \leq \frac{3}{h}\int_0^\pi\int_0^{2h} b(\sigma e^{i\varphi})\sin\varphi\,\sigma\,d\sigma\,d\varphi \leq$$

$$\leq \frac{3}{h}(\eta(F))^2 \int_0^\pi\int_0^{2h} \sin\varphi\,\sigma\,d\sigma\,d\varphi = 12(\eta(F))^2 h.$$

We are at the end of the calculation. Going back to what we started with, we find

$$\iint_{S_h} |z|\log\frac{1}{|z|}|\vec{\nabla}V|^2 dxdy \leq 32J \leq 384(\eta(F))^2 h,$$

and the lemma is proved.

Combining the lemma just proven with the theorem on Carleson measures for the unit circle, given at the end of Chapter VIII, Section E, we find:

Theorem. There is a numerical constant K'' such that

$$\iint_{1/2<|z|<1} |z|\log\frac{1}{|z|}(\nabla V\cdot\nabla V)|g(z)|dxdy \le K''(\eta(F))^2\|g\|_1$$

for all $g \in H_1$.

4°. Fefferman's theorem with $\eta(F)$.

Combining the theorem in 3° with the lemma of 2°, we see that there is a numerical constant K such that

$$\iint_{|z|<1} |z|\log\frac{1}{|z|}|\nabla V\cdot\nabla V||g(z)|dxdy \le K(\eta(F))^2\|g\|_1$$

for all $g \in H_1$. In particular, inequality (*) at the end of 1° holds. The work in 1° thus shows that the boxed inequality at the very beginning of this Section is valid, whence, by the theorem at the end of Section D.3, we obtain:

Theorem. Let F be real-valued and periodic, of period 2π. If $\eta(F) < \infty$,

$$\lim_{R\to 1}\int_{-\pi}^{\pi} \Re f(Re^{i\theta})F(\theta)d\theta$$

exists for every $f \in H_1(0)$, and represents a linear functional on $\Re H_1(0)$.

Scholium. The norm of the linear functional on $\Re H_1(0)$ furnished by the above theorem is $\le \text{const}\cdot\eta(F)$.

Indeed, the proofs of the lemmas in Section D.3 show that the norm of the functional in question is $\le 2C$, where C is the constant figuring in the statement of the theorem in Section D.3. The boxed inequality at the very beginning of this section is the same as the one in the theorem, Section D.3, with $C = K\eta(F)$, K being a numerical constant. And we proved the boxed inequality in $1^\circ - 3^\circ$.

We can, however, say more than this, namely, $\eta(F)$ is also $\leq$ a constant times the norm of the above linear functional!

Indeed, call

$$Lf = \lim_{R\to 1} \int_{-\pi}^{\pi} \Re f(Re^{i\theta})F(\theta)d\theta$$

for $f \in H_1(0)$, and denote the norm of L by $|||L|||$. The Hahn-Banach argument in Section A shows that we can find $\varphi + i\psi$ periodic of period 2π in L_∞ with φ and ψ real, and

$$\|\varphi + i\psi\|_\infty \leq |||L|||$$

such that

$$Lf = \Re \int_{-\pi}^{\pi} [\varphi(t) + i\psi(t)]f(e^{it})dt,$$

for $f \in H_1(0)$. By the lemma in Section A,

$$Lf = \lim_{R\to 1} \int_{-\pi}^{\pi} \Re f(Re^{it})[\varphi(t) + \tilde{\psi}(t)]dt$$

so that

$$F(t) = \varphi(t) + \tilde{\psi}(t) + c$$

with a constant c, according to the Scholium at the very end of Section A.

With the first lemma of Section C.1, we see that

$$\eta(F) = \eta(\varphi + \tilde{\psi}),$$

but by Section C.2, the quantity on the right is $\leq \sqrt{2}\,(\|\varphi\|_\infty + \|\psi\|_\infty) \leq$ $\leq 2\sqrt{2}\,\|\varphi + i\psi\|_\infty \leq 2\sqrt{2}\,|||L|||$.

Let us combine the above theorem with its scholium, the one in Section A, and the first one of Section C.2. We get

Garsia's Version of Fefferman's Theorem: If $F(\theta)$ is real and periodic of period 2π, the following three conditions on F are equivalent:

i) $L\Re f = \lim_{R\to 1} \int_{-\pi}^{\pi} \Re f(Re^{i\theta})F(\theta)d\theta$ exists for every $f \in H_1(0)$, and L is a linear functional on $\Re H_1(0)$.

ii) $F(\theta) = \varphi(\theta) + \tilde{\psi}(\theta)$, with φ, ψ real, 2π-periodic, and bounded.

iii) $\mathfrak{h}(F) < \infty$.

Moreover, the norms $|||L|||$ (of L as a linear functional on $\Re H_1(0)$) and $\mathfrak{h}(F)$ are equivalent, $\mathfrak{h}(F) \leq \sqrt{2}\,(\|\varphi\|_\infty + \|\psi\|_\infty)$ with φ and ψ the bounded functions in ii), and we can find φ_1, ψ_1 in L_∞, periodic of period 2π, and real valued, such that

$$F(\theta) = \varphi_1(\theta) + \tilde{\psi}_1(\theta) + c,$$

c a constant, with

$$\|\varphi_1\|_\infty + \|\psi_1\|_\infty \leq M\cdot\mathfrak{h}(F),$$

M being a numerical constant.

F. Fefferman's theorem with the BMO norm.

We now show that in the theorems of Section E.4, the condition $\mathfrak{h}(F) < \infty$ can be replaced by $\|F\|_* < \infty$, i.e., by the fact that $F \in$ BMO. Since we already saw in Section C.2 that $\|F\|_* \leq$ const. $\mathfrak{h}(F)$, we have here to establish the reverse inequality, $\mathfrak{h}(F) \leq$ const. $\|F\|_*$, proving *equivalence* of the norms $\mathfrak{n}(\ )$ and $\|\ \|_*$. Proof of the reverse inequality depends on some deep work of Nirenberg and John.

1°. Theorem of Nirenberg and John.

Recall first of all the definition from Section B:

$$\|F\|_* = \sup_I \int_I |F(t) - F_I|\,dt$$

with I ranging over the set of *intervals*.

Lemma. If $F \in \mathrm{BMO}$ and if I and J are intervals with *the same midpoint* and $I \subset J$, then

$$|F_I - F_J| \le 2\{\log_2 \frac{|J|}{|I|} + 1\}\|F\|_*.$$

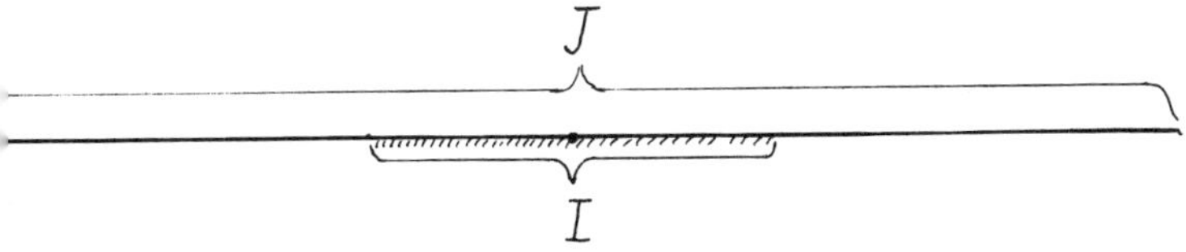

(Here, $\log_2$ means logarithm to the base 2.)

Proof. First of all, if $|J| \le 2|I|$,

$$|F_I - F_J| = \frac{1}{|I|}\left|\int_J (F(t) - F_J)\,dt\right| \le \frac{1}{|I|}\int_I |F(t) - F_J|\,dt \le$$

$$\le \frac{2}{|J|}\int_J |F(t) - F_J|\,dt \le 2\|F\|_*,$$

and the result is true in this case. (Here, we only used the relations $I \subset J$, $|J| \leq 2|I|$.)

In the general case, argue by induction. If $2^n|I| < |J| \leq 2^{n+1}|I|$ and the result is proved for intervals J' with $|J'| \leq 2^n|I|$, take an interval J' having the same midpoint as I, but half the length of J, so that $I \subset J' \subset J$. By the induction hypothesis,

$$|F_I - F_{J'}| \leq 2\|F\|_* \left\{\log_2\frac{|J'|}{|I|} + 1\right\} .$$

Since $|J| = 2|J'|$, by the argument given above, $|F_J - F_{J'}| \leq 2\|F\|_*$, so

$$|F_J - F_I| \leq |F_J - F_{J'}| + |F_{J'} - F_I| \leq 2\|F\|_* + 2\|F\|_* \left\{\log_2\frac{|J'|}{|I|} + 1\right\} =$$

$$= 2\|F\|_* \left\{\log_2\frac{|J|}{|I|} + 1\right\} ,$$

and the result holds for intervals J with $2^n|I| < |J| \leq 2^{n+1}|I|$.

We are done.

BMO actually first appeared when the following result was published. In it, the function F is assumed to be real valued.

Theorem (Nirenberg and John). For any interval I and positive integer n,

$$|\{x \in I;\ F(x) - F_I > 4\|F\|_* n\}| \leq 2^{-n}|I|.$$

Proof (Garnett). By making a change of scale, we may reduce our situation to one where $\|F\|_* \leq \frac{1}{2}$, and, under this assumption, it is enough to prove that

$$|\{x \in I;\ F(x) - F_I > 2n\}| \leq 2^{-n}|I|.$$

A change of variable now allows us to take $I = [0,1]$; there is also no loss of generality in taking F_I to be zero.

These reductions made, we shall work with the dyadic subintervals of $[0,1] = I$. These are those of the very special form $[k/2^n, (k+1)/2^n]$, with k an integer and n a non-negative integer.

> Dyadic intervals have the very nice property that if two of them overlap, one must be contained in the other.

Let $E = \{x \in I; F(x) > 2n\}$. By Lebesgue's theorem (differentiation of indefinite integrals), almost every $x \in E$ is contained in some small dyadic interval J with $F_J > 2n$. Since $F_I = 0$, there must, for almost every $x \in E$, be a dyadic interval J containing x of greatest possible length such that $F_J > 1$. Let such a J of greatest possible length containing x be called $J(x)$. After throwing away from E a set of measure zero, we will have

$$E \subseteq \bigcup_{x \in E} J(x).$$

Let $\mathcal{E} = \bigcup_{x \in E} J(x)$. $\mathcal{E}$ can actually be written as a non-overlapping union of $J(x)$. Indeed, if $x, y \in E$ and $J(x)$ and $J(y)$ overlap, then $J(x) = J(y)$. For, by the above boxed remark, either $J(x) \subseteq J(y)$ or $J(y) \subseteq J(x)$. If, say, $J(y) \supseteq J(x)$ and $|J(y)| > |J(x)|$, then $x \in J(y)$ and $F_{J(y)} > 1$ together contradict the fact that $J = J(x)$ is a dyadic interval of maximum length containing x such that $F_J > 1$. So if $J(y) \supseteq J(x)$ then $|J(y)| = |J(x)|$, so $J(y)$ must in fact equal $J(x)$.

We see that we can extract a sequence of points $x_i \in E$ such that the $J(x_i)$ do not overlap and

$$\bigcup_{i=1}^{\infty} J(x_i) = \mathcal{E}.$$

(This sequence, and the union in question, may in fact be finite. A countable set of the $J(x)$ already serves to cover $\mathcal{E}$ because each $J(x)$ contains an open set.) The above dissection of $\mathcal{E}$ into a countable number of non-overlapping dyadic intervals is incidentally called a Calderón-Zygmund decomposition.

Now $|\mathcal{E}| < \frac{1}{2}|I|$. Indeed, since $F_I = 0$,

$$\frac{1}{|I|}\int_{\mathcal{E}} F(x)dx \le \frac{1}{|I|}\int_{\mathcal{E}} |F(x) - F_I|dx \le \frac{1}{|I|}\int_I |F(x) - F_I|dx \le \|F\|_* \le \frac{1}{2},$$

whilst

$$\frac{1}{|I|}\int_{\mathcal{E}} F(x)dx = \frac{1}{|I|}\sum_i \int_{J(x_i)} F(t)dt$$

because the $J(x_i)$ don't overlap, and this equals

$$\frac{1}{|I|}\sum_i |J(x_i)|F_{J(x_i)} > \frac{1}{|I|}\sum_i |J(x_i)| = \frac{|\mathcal{E}|}{|I|},$$

so $|\mathcal{E}|/|I| < \frac{1}{2}$.

Since $E \subseteq \mathcal{E}$, our theorem is now proved for the case $n = 1$.

To prove the theorem for $n > 1$, we proceed by induction. I say that for each i, $F_{J(x_i)} \le 2$. Indeed, given $J(x_i)$, let J' be the dyadic interval $\subseteq I$ containing $J(x_i)$ and having twice its length. (There is only one such:

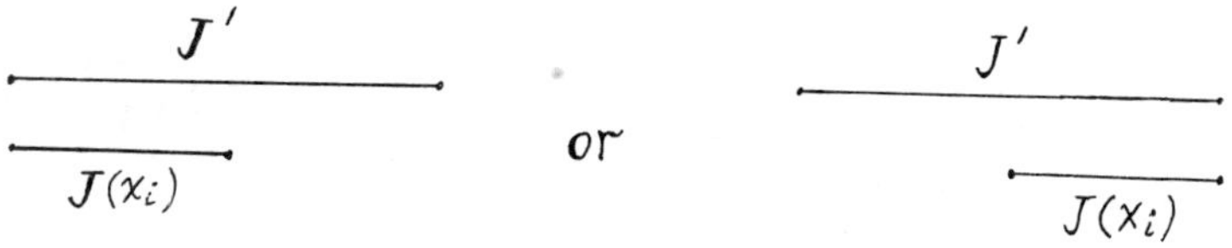

Then, by the maximal property according to which $J(x_i)$ was chosen,

$$F_{J'} \leq 1.$$

By the first part of the proof of the above lemma, with $J = J'$ and $I = J(x_i)$, we now have $|F_{J'} - F_{J(x_i)}| \leq 2\|F\|_* \leq 1$, so $F_{J(x_i)} \leq 2$, as claimed.

Let now $E \cap J(x_i) = E_i$. Since the $J(x_i)$ don't overlap, and cover $\mathcal{E} \supseteq E$,

$$E = \bigcup_i E_i, \quad \text{and} \quad |E| = \sum_i |E_i|.$$

If $x \in E_i$, $F(x) > 2n$, so, since $F_{J(x_i)} \leq 2$,

$$F(x) - F_{J(x_i)} > 2(n-1).$$

We see that we can make an induction step at this point. For, as just seen,

$$E_i \subseteq \{x \in J(x_i);\ F(x) - F_{J(x_i)} > 2(n-1)\},$$

so, if the theorem is true with $n-1$ in place of n,

$$|E_i| \leq \left(\tfrac{1}{2}\right)^{n-1} |J(x_i)|.$$

Therefore, since the $J(x_i)$ are non-overlapping and add up to $\mathcal{E}$,

$$|E| = \sum_i |E_i| \leq \left(\tfrac{1}{2}\right)^{n-1} \sum_i |J(x_i)| = \left(\tfrac{1}{2}\right)^{n-1} |\mathcal{E}| <$$
$$< \left(\tfrac{1}{2}\right)^{n-1} \cdot \tfrac{1}{2}\, |I| = \left(\tfrac{1}{2}\right)^{n} |I|,$$

and the theorem holds with n.

We are done.

Corollary. If $F \in BMO$ and I is an interval,

$$\frac{1}{|I|}\int_I [F(x) - F_I]^2 dx \le C\|F\|_*^2,$$

with C a numerical constant.

Remark. This is like a reversed Schwarz inequality!

Proof. Let $m(\lambda) = |\{x \in I;\ |F(x) - F_I| > \lambda\}|$; by the theorem of Nirenberg and John just proven, for every non-negative integer n,

$$m(\lambda) \le 2 \times (\tfrac{1}{2})^n |I| \quad \text{if} \quad \lambda > 4\|F\|_* n$$

(the extra factor of 2 coming from the fact that we are looking at the measure of the set of x where

$$F(x) - F_I < -\lambda \ \underline{\text{or}}\ F(x) - F_I > \lambda).$$

Therefore

$$m(\lambda) \le 2|I| \exp\left\{- \left[\frac{\lambda}{4\|F\|_*}\right] \log 2\right\},$$

where $[t]$ denotes the greatest integer $\le t$.

By Chapter VIII, Section A, we now have

$$\int_I |F(x) - F_I|^2 dx = 2\int_0^\infty \lambda m(\lambda) d\lambda \le 4|I| \int_0^\infty \lambda \exp\left\{-\log 2\left[\frac{\lambda}{4\|F\|_*}\right]\right\} d\lambda =$$

$$= 64|I|\,\|F\|_*^2 \int_0^\infty s\, e^{-[s]\log 2}\, ds = C\|F\|_*^2 |I|$$

with a numerical constant C. This does it.

Remark. Clearly much stronger inequalities for integrals taken over I and involving $|F(x) - F_I|$ are valid.

2°. Equivalence of BMO norm and Garsia norm.

Lemma. Let $f(t) \geq 0$. Then there is a numerical constant C such that

$$\int_{-\pi}^{\pi} \frac{1 - r^2}{1 + r^2 - 2r\cos t}\, f(t)dt \leq C \int_0^{\pi} \frac{(1-r)s^2}{[(1-r)^2 + s^2]^2} \left(\frac{1}{2s} \int_{-s}^{s} f(t)dt \right) ds.$$

Proof. $\dfrac{1 - r^2}{1 + r^2 - 2r\cos t} = \dfrac{(1+r)(1-r)}{(1-r)^2 + 4r\sin^2\frac{t}{2}}$, and the right hand side is clearly $\leq k\dfrac{1-r}{(1-r)^2 + t^2}$ for $0 \leq r < 1$ and $-\pi \leq t \leq \pi$, where k is a suitable numerical constant.

Write $1 - r = u$. Then, if $f(t) \geq 0$,

$$\int_0^{\pi} \frac{1-r^2}{1+r^2-2r\cos t}\, [f(t) + f(-t)]dt \leq k \int_0^{\pi} \frac{u}{u^2+t^2}\, [f(t) + f(-t)]dt$$

$$\leq k \int_0^{\infty} \frac{u}{u^2+t^2}\, [f(t) + f(-t)]dt.$$

The last expression equals

$$k\int_0^{\infty} [f(t) + f(-t)] \int_t^{\infty} \frac{2us}{(u^2+s^2)^2}\, dsdt = k\int_0^{\infty} \frac{2us}{(u^2+s^2)^2} \int_0^{s} [f(t) + f(-t)]\, dtds =$$

$$= 4k \int_0^{\infty} \frac{us^2}{(u^2+s^2)^2} \left\{ \frac{1}{2s} \int_{-s}^{s} f(t)dt \right\} ds,$$

an expression of the required form.

Theorem. If F is real and periodic, of period 2π, we have

$$\eta(F) \leq K\|F\|_*$$

with a numerical constant K.

Proof. Using once more the notation introduced at the beginning of Section C, we have

$$(h(F))^2 = \sup_{|z|<1} \{U_{F^2}(z) - (U_F(z))^2\};$$

it suffices to estimate $U_{F^2}(z) - (U_F(z))^2$ in terms of $\|F\|_*$. Since $F(x)$ and its translates $F_h(x) = F(x-h)$ clearly have the same BMO norm $\|\ \|_*$, there is no loss of generality in taking $z = r$ with $0 < r < 1$. By the first lemma of Section C.1,

$$U_{F^2}(r) - (U_F(r))^2 = \frac{1}{8\pi^2}\int_{-\pi}^{\pi}\int_{-\pi}^{\pi} \frac{(1-r^2)^2[F(s)-F(t)]^2 ds dt}{(1+r^2-2r\cos s)(1+r^2-2r\cos t)} .$$

Since $[F(s) - F(t)]^2 \geq 0$, we can apply the above lemma twice to estimate the above right-hand double integral, and find it to be

$$\leq C^2 \int_0^{\infty}\int_0^{\infty} \frac{us^2}{(u^2+s^2)^2}\cdot\frac{ut^2}{(u^2+t^2)^2}\left[\frac{1}{4st}\int_{-s}^{s}\int_{-t}^{t}[F(\sigma)-F(\tau)]^2 d\sigma d\tau\right] ds dt,$$

where $u = 1-r$ and C is a numerical constant.

Now, if I and J are intervals with the same midpoint, we have:

$$\left[\frac{1}{|J|}\cdot\frac{1}{|I|}\int_J\int_I [F(\sigma)-F(\tau)]^2 d\sigma d\tau\right] \leq \left[\frac{1}{|J|}\cdot\frac{1}{|I|}\int_J\int_I [F(\sigma)-F_I]^2 d\sigma d\tau\right]^{1/2} +$$

$$+ \left[\frac{1}{|I|}\cdot\frac{1}{|J|}\int_I\int_J [F(\tau)-F_J]^2 d\tau d\sigma\right]^{1/2} + \left[\frac{1}{|I|}\cdot\frac{1}{|J|}\int_I\int_J [F_I-F_J]^2 d\tau d\sigma\right]^{1/2} =$$

$$= \left[\frac{1}{|I|}\int_I [F(\sigma)-F_I]^2 d\sigma\right]^{1/2} + \left[\frac{1}{|J|}\int_J [F(\tau)-F_J]^2 d\tau\right]^{1/2} + |F_I - F_J|.$$

By the Corollary to the Nirenberg-John theorem (1°), this is

$$\leq 2\sqrt{C}\,\|F\|_* + |F_I - F_J|,$$

and by the lemma at the beginning of 1°, the expression just found is in turn

$$\leq 2\sqrt{C}\,\|F\|_* + 2\|F\|_*\left[1 + \left|\log_2 \frac{|J|}{|I|}\right|\right] .$$

We see that for s and $t > 0$,

$$\left[\frac{1}{4st}\int_{-s}^{s}\int_{-t}^{t}[F(\sigma) - F(\tau)]^2 d\sigma d\tau\right]^{1/2} \leq 2c\|F\|_* + 2\|F\|_*(1 + |\log_2(s/t)|).$$

Plugging this expression into the quadruple integral obtained at the end of the preceding paragraph, we see that

$$U_{F^2}(r) - (U_F(r))^2 \leq K\int_0^\infty\int_0^\infty \frac{u^2s^2t^2\|F\|_*^2[1 + \log(s/t))^2]}{(u^2+s^2)^2(u^2+t^2)^2}\, dsdt,$$

where K is a suitable numerical constant and $u = 1 - r$. Making the change of variable $\frac{s}{u} = x$, $\frac{t}{u} = y$, the integral on the right becomes

$$K\|F\|_*^2 \int_0^\infty\int_0^\infty \frac{x^2y^2[1 + (\log x - \log y)^2]dxdy}{(1+x^2)^2(1+y^2)^2}$$

$$\leq K\|F\|_*^2 \int_0^\infty\int_0^\infty \frac{x^2y^2[1 + 2(\log x)^2 + 2(\log y)^2]\, dxdy}{(1+x^2)^2(1+y^2)^2} .$$

But the double integral in this last expression is clearly finite, and has a numerical value independent of u, hence of r. Therefore, with a numerical constant L,

$$[U_{F^2}(r) - [U_F(r)]^2]^{1/2} \leq L\|F\|_*,$$

and finally, $\eta(F) \leq L\|F\|_*$.

Q.E.D.

Corollary. The norms $\|\ \|_*$ and $\eta(\ )$ are equivalent.

Proof. By the above theorem and the second one of Section C.2.

3. Fefferman's theorem.

Combining the equivalence of $\mathfrak{h}$ and $\| \ \|_*$ established in 2° with Garsia's version of the Fefferman theorem (Section E.4), we obtain the fundamental theorem about BMO, namely:

Fefferman's Theorem. The dual of $\Re H_1(0)$ is BMO/(constant functions). If $F(\theta)$ is real, of period 2π, and locally integrable, the following three conditions on F are equivalent:

i) $L\Re f = \lim_{R\to 1} \int_{-\pi}^{\pi} \Re f(Re^{i\theta})F(\theta)d\theta$ exists for every $f \in H_1(0)$ and L is a linear functional on $\Re H_1(0)$.

ii) $F(\theta) = \varphi(\theta) + \tilde{\psi}(\theta)$ with φ and ψ real 2π-periodic, and bounded.

iii) $F \in$ BMO.

In case $F \in$ BMO, the norm of the linear functional L given by i) is equivalent to $\|F\|_*$, the BMO-norm of F, and we can find bounded functions $\varphi_1(\theta)$, $\psi_1(\theta)$, periodic of period 2π, such that

$$F(\theta) = \varphi_1(\theta) + \tilde{\psi}_1(\theta) + \text{const.},$$

with

$$\|\varphi_1\|_\infty + \|\psi_1\|_\infty \le A\|F\|_* \le B(\|\varphi_1\|_\infty + \|\psi_1\|_\infty),$$

A and B being numerical constants. Thus, BMO $= \Re L_\infty + \widetilde{\Re L_\infty}$.

Remark. At the time these lectures were being given, one could still say that, for an arbitrary $F \in$ BMO, no method was known for actually constructing φ and $\psi \in L_\infty$ such that $\varphi + \tilde{\psi} = F$, and that the only known proof for this decomposition was indirect, being based on combination of the Hahn-Banach duality argument of Section A with the inequalities established in Sections D and E and the present section. Of course, the Hahn-Banach theorem does have a constructive side to it, at least insofar as linear functionals over separable Banach spaces are considered. Provided the dual we are looking at is identified with a space of functions, careful examination of what the steps used in proving the Hahn-Banach theorem signify concretely in a given situation can sometimes lead to an explicit construction procedure for the function corresponding to the linear functional in question. Be that as it may, the existence proof given here is certainly not constructive as it stands.

A constructive procedure for obtaining the decomposition has, however, been found; it is due to Peter Jones. He was in fact working on it at the same time that this chapter was being covered in the lectures (late Spring of 1978), and had obtained his result by the time the lectures were finished!

G. Representation in terms of radially bounded measures.

Fefferman's theorem, given in the preceding section, can be combined with the radial maximal function characterization of $\Re H_1$ (Chapter VIII, Section D.2) to yield, by duality, a curious representation of functions in BMO. This representation has some resemblance to the decomposition $\text{BMO} = \Re L_\infty + \widetilde{\Re L_\infty}$, and was important in the development of the subject because Carleson published a constructive procedure for getting it in 1976, when a constructive method for decomposing arbitrary BMO-functions into a sum of something in $\Re L_\infty$

and something else in $\Re\widetilde{L}_\infty$ <u>was not yet known</u> (see remark at end of Section F.3). It is still interesting in its own right, and shows a connection between the material of Chapter VIII, Section D and the present chapter.

1° We will be interested in signed measures ν on $\{|z| \leq 1\}$ which, roughly speaking, can be put in the form

$$d\nu(re^{i\theta}) = d\mu_\theta(r)d\theta,$$

with $\int_0^1 |d\mu_\theta(r)|$ <u>bounded for</u> $0 \leq \theta < 2\pi$. In order to avoid measure-theoretic niceties and digressions, we simply take the

<u>Definition</u>. A measure ν on $\{|z| \leq 1\}$ is called <u>radially bounded</u> if there is a constant C such that, for any <u>sector</u> (Nota bene!) S_h of the form

$$S_h = \{re^{i\theta};\ 0 \leq r \leq 1,\ \theta_0 \leq \theta \leq \theta_0 + h\}$$

We have $\iint_{S_h} |d\nu(z)| \leq Ch.$

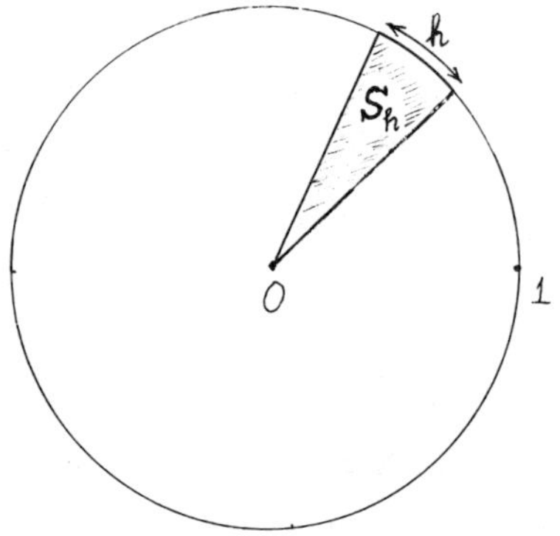

The smallest permissible value of C for which the above relation holds (for all S_h) is denoted by $\|\nu\|^*$.

$\|\ \|^*$ is a <u>norm</u> for radially bounded measures, and the <u>ordinary measure norm</u>, $\|\nu\| = \iint_{|z|\leq 1} |d\nu(z)|$

is clearly $\leq 2\pi\|\nu\|^*$. By the last theorem in Chapter VIII, Section E, a radially bounded measure is surely Carleson.

Lemma. A radially bounded measure ν can be written as a sum $\mu+\sigma$, where:

i) μ is radially bounded, and carried on the open circle $\{|z| < 1\}$.

ii) σ is carried on $\{|z| = 1\}$, and $d\sigma(e^{i\theta}) = s(\theta)d\theta$ with $s \in L_\infty$.

Proof. Let μ simply be the restriction of ν to $\{|z| < 1\}$, and σ the restriction of ν to $\{|z| = 1\}$. Then, if I is any arc of $\{|z| = 1\}$,

$$\int_I |d\sigma| \leq \|\nu\|^* \cdot \text{length } I,$$

so σ is absolutely continuous with respect to linear Lebesgue measure on $\{|z| = 1\}$ and has a bounded density with respect to the latter, i.e., $d\sigma(e^{i\theta}) = s(\theta)d\theta$ with $s \in L_\infty$.

Definition. If ν is radially bounded and $\nu = \mu+\sigma$ with μ supported on $\{|z| < 1\}$ and σ on $\{|z| = 1\}$, and if $d\sigma(e^{i\theta}) = s(\theta)d\theta$, then we write

$$P_\nu(\theta) = s(\theta) + \frac{1}{2\pi}\iint_{|z|<1} \frac{1-|z|^2}{|z-e^{i\theta}|^2}\,d\mu(z).$$

Remark. The integral on the right converges absolutely for almost all θ, and yields a function in $L_1[-\pi,\pi]$. This is indeed true for any finite measure μ on $\{|z| < 1\}$.

Indeed,

$$\frac{1}{2\pi}\int_{-\pi}^{\pi}\iint_{|z|<1}\frac{1-|z|^2}{|z-e^{i\theta}|^2}|d\mu(z)|d\theta = \frac{1}{2\pi}\iint_{|z|<1}\int_{-\pi}^{\pi}\frac{1-|z|^2}{|z-e^{i\theta}|^2}d\theta|d\mu(z)| = \iint_{|z|<1}|d\mu(z)|.$$

We will eventually prove in this section that BMO coincides with the set of P_ν for ν ranging over the family of radially bounded measures. In the present subsection, let us just show that $P_\nu \in$ BMO whenever ν is radially bounded.

Lemma. If $U(z)$ is continuous on $\{|z| \le 1\}$ and harmonic in $\{|z| < 1\}$, then

$$\int_{-\pi}^{\pi} U(e^{i\theta})P_\nu(\theta)d\theta = \iint_{|z|\le 1} U(z)d\nu(z) ,$$

whenever ν is radially bounded.

Proof. Putting $\nu = \mu + \sigma$ where μ is the restriction of ν to $\{|z| < 1\}$ and $d\sigma(e^{i\theta}) = s(\theta)d\theta$, $s \in L_\infty$, we have by definition $P_\nu(\theta) = P_\mu(\theta) + s(\theta)$, where $\int_{-\pi}^{\pi} U(e^{i\theta})s(\theta)d\theta = \int_{-\pi}^{\pi} U(e^{i\theta})d\sigma(e^{i\theta})$. Again, by Poisson's formula (Chapter I!), for $|z| < 1$,

$$U(z) = \frac{1}{2\pi}\int_{-\pi}^{\pi}\frac{1-|z|^2}{|z-e^{i\theta}|^2}U(e^{i\theta})d\theta,$$

from which, by Fubini's theorem,

$$\iint_{|z|<1} U(z)d\mu(z) = \int_{-\pi}^{\pi} U(e^{i\theta})P_\mu(\theta)d\theta.$$

We're done.

Theorem. If ν is radially bounded, $P_\nu(\theta)$ is in BMO, and $\|P_\nu\|_* \le C\|\nu\|^*$ with a numerical constant C.

Proof. By the first lemma of Section D.3 and the Scholium in Section E.4, it is enough to show that there is a numerical constant C such that

$$\int_{-\pi}^{\pi} \Re f(Re^{i\theta})P_\nu(\theta)d\theta \le C\|f\|_1\|\nu\|^*$$

for all $f \in H_1(0)$ and all R, $0 \le R < 1$.

By the above lemma,

$$\int_{-\pi}^{\pi} \Re f(Re^{i\theta})P_\nu(\theta)d\theta = \iint_{|z|<1} \Re f(Rz)d\nu(z),$$

since $\Re f(Rz)$ is harmonic for $|z| < 1$ and continuous for $|z| \le 1$. By this continuity, the integral on the right can be approximated by a Riemann sum over <u>sectorial elements of the form</u>

$$S_h = \{re^{i\theta};\ 0 \le r \le 1,\ \theta_0 \le \theta \le \theta_0 + h\}.$$

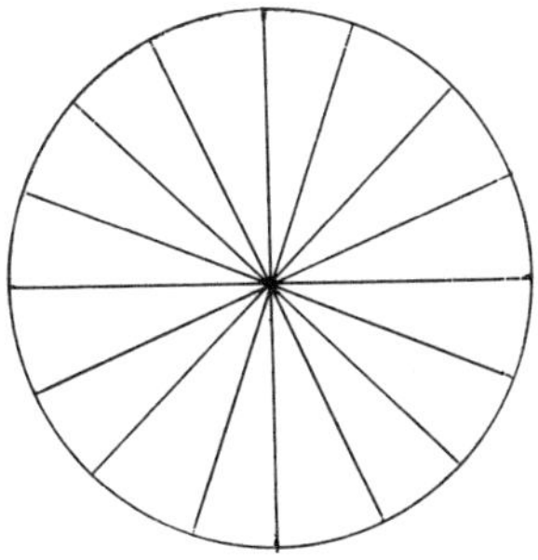

For large N, let $h = \frac{2\pi}{N}$ and write $S_h(n) = \{re^{i\theta};\ 0 \le r \le 1, nh \le \theta \le (n+1)h\}$ for $n = 0,1,\ldots,N-1$. Take $z_n \in S_h(n)$, then, if N is <u>sufficiently large</u>, $\sum_{n=0}^{N-1} \Re f(Rz_n)\nu(S_h(n))$ differs from $\iint_{|z|<1} \Re f(Rz)d\nu(z)$ by less than ε. If $z_n = r_n e^{i\theta_n}$ (with $nh \le \theta_n \le (n+1)h$), the finite sum just written is in absolute value

$$\le \sum_{n=0}^{N-1} \sup_{0\le r\le 1} |\Re f(Re^{i\theta_n})| \iint_{S_n(n)} |d\nu(z)| \le$$

$$\le \|\nu\|^* \sum_{n=0}^{N-1} \sup_{0\le r\le 1} |\Re f(Rre^{i\theta_n})| \cdot \frac{2\pi}{N}$$

by definition of $\|\ \|^*$.

Since $\sup_{0\leq r\leq 1}|\Re f(Rre^{i\theta})|$ is, for $R < 1$, a continuous function of θ, the last expression written is within ε of

$$\|\nu\|^* \int_0^{2\pi} \sup_{0\leq r\leq 1} |\Re f(Rre^{i\theta})|d\theta,$$

if N is large. Squeezing ε, we see that

$$\left| \iint_{|z|\leq 1} \Re f(Rz)d\nu(z) \right| \leq \|\nu\|^* \int_0^{2\pi} \sup_{0\leq r\leq 1} |\Re f(Rre^{i\theta})|d\theta.$$

We therefore have

$$\left| \int_{-\pi}^{\pi} \Re f(Re^{i\theta})P_\nu(\theta)d\theta \right| \leq \|\nu\|^* \int_{-\pi}^{\pi} \sup_{0\leq r\leq 1} |\Re f(re^{i\theta})|d\theta$$

for every $f \in H_1(0)$ and all $R < 1$, using the formula at the beginning of this proof. By a corollary at the end of Section C.3 in Chapter VIII,

$$\int_{-\pi}^{\pi} \sup_{0\leq r\leq 1} |\Re f(re^{i\theta})|d\theta \leq C \|f\|_1$$

for $f \in H_1$. Substituting this into the previous inequality, we obtain

$$\left| \int_{-\pi}^{\pi} \Re f(Re^{i\theta})P_\nu(\theta)d\theta \right| \leq C\|\nu\|^* \|f\|_1,$$

and the proof is complete.

<u>Remark</u>. This result only depends on the "easy part" of Fefferman's theorem.

2°. We now set out to establish the converse of the theorem in 1°.

Lemma. Let $\|\nu_k\|^* \leq M$, $k = 1,2,3,\ldots$. Then there is a subsequence $\{\nu_{k_j}\}$ of $\{\nu_k\}$ and a radially bounded ν, $\|\nu\|^* \leq M$, with

$$d\nu_{k_j} \xrightarrow[j]{} d\nu \quad (w^*)$$

on $\{|z| \leq 1\}$.

Proof. By a remark at the beginning of 1°, $\|\nu_k\| \leq 2\pi\|\nu_k\|^* \leq 2\pi M$, so some subsequence of the ν_k converges w^* to a measure ν on $\{|z| \leq 1\}$. It is easy to verify that $\|\nu\|^* \leq M$.

Theorem. Let $F(\theta)$, real and 2π-periodic, belong to BMO. There exists a radially bounded real measure ν with

$$F(\theta) = P_\nu(\theta) + \text{constant}.$$

We can take ν so as to satisfy

$$\|\nu\|^* \leq K\|F\|_*,$$

where K is a numerical constant.

Remark. Putting μ = restriction of ν to $\{|z| < 1\}$ we have, by a lemma of 1°,

$$F(\theta) = \text{const} + s(\theta) + \frac{1}{2\pi} \iint_{|z|<1} \frac{1-|z|^2}{|z-e^{i\theta}|^2}\, d\mu(z)$$

with $s(\theta) \in L_\infty$. This looks a little like the basic decomposition

$$F(\theta) = \varphi(\theta) + \tilde{\psi}(\theta)$$

with φ and $\psi \in L_\infty$.

Proof By the hard part of Fefferman's theorem (Section F.3), the function $F \in BMO$ defines a linear functional L on $\Re H_1(0)$ according to the formula

$$L(\Re f) = \lim_{R \to 1} \int_{-\pi}^{\pi} \Re f(Re^{i\theta}) F(\theta)\, d\theta, \qquad f \in H_1(0).$$

We have $|||L||| \leq A\|F\|_*$ with a numerical constant A.

Let, for each $M > 0$,

$$K_M = \{P_\nu;\ \nu \text{ real},\ \|\nu\|^* \leq M|||L|||\}.$$

K_M is a convex set.

I claim that, for sufficiently large M, L is in the w^* closure, over $\Re H_1(0)$, of the linear functionals corresponding to the P_ν in K_M.

If ν is radially bounded, it is convenient to denote by L_ν the linear functional on $\Re H_1(0)$ corresponding to P_ν; it is given by the formula

$$L_\nu(\Re f) = \lim_{R \to 1} \int_{-\pi}^{\pi} \Re f(Re^{i\theta}) P_\nu(\theta)\, d\theta, \qquad f \in H_1(0).$$

We are to prove that if M is large enough, L is in the w^* closure of the L_ν, $\|\nu\|^* \leq M|||L|||$.

For given M, suppose that L is not in the w^* closure of the L_ν, $\|\nu\|^* \leq M|||L|||$. Then there is a w^*-closed hyperplane separating L from these L_ν. That is, there is a $g \in \Re H_1(0)$ with $L_\nu(\Re g) < L(\Re g)$ whenever $\|\nu\|^* \leq M|||L|||$. Now the set of L_ν under consideration contains 0 and is taken onto itself on multiplication by -1. Therefore $|L_\nu(\Re g)| < L(\Re g)$ for $\|\nu\|^* \leq M|||L|||$, and without loss of generality we can take $L(\Re g) = 1$; otherwise, use a suitable positive multiple of g in place of g.

Thus, for $\|\nu\|^* \le M|||L|||$, $|L_\nu(\Re g)| < 1$ whilst $L(\Re g) = 1$.

For the moment, write $M|||L||| = B$. We can find ν with $\|\nu\|^* = B$ such that $L_\nu(\Re g)$ is as close as we please to

$$B \cdot \int_{-\pi}^{\pi} \sup_{0 \le r < 1} |\Re g(re^{i\theta})| d\theta .$$

Indeed, we can, by Lebesgue's monotone convergence theorem first find an $R < 1$ such that

$$B \cdot \int_{-\pi}^{\pi} \sup_{0 \le r < 1} |\Re g(Rre^{i\theta})| d\theta$$

is already within, say, ε of the preceding expression. The function $g_R^+(\theta) = \sup_{0 \le r < 1} |\Re g(Rre^{i\theta})|$ now being continuous, we see that the sum

$$B \sum_{n=1}^{N} g_R^+ \left(\frac{2\pi}{N} n\right) \cdot \frac{2\pi}{N}$$

is in turn within ε of the previous integral if N is sufficiently large. Let, for $\theta_n = \frac{2\pi}{N} n$, the actual maximum

$$\sup_{0 \le r \le 1} |\Re g(Rre^{i\theta_n})| = g_R^+(\theta_n)$$

be attained for $r = r_n$. Then the previous sum equals

$$B \sum_{n=1}^{N} |\Re g(Rr_n e^{i\theta_n})| \cdot \frac{2\pi}{N} .$$

$\Re g(Rz)$ is uniformly continuous for $|z| \le 1$. Therefore, if N is large, $\Re g(Rr_n e^{i\theta})$ oscillates by less than $\frac{\varepsilon}{2\pi B}$ when θ runs from

$\frac{2\pi}{N}(n-1)$ to $\frac{2\pi}{N}n$, no matter what the value of $r_n \in [0,1]$ may be. That is,

$$\left|\frac{2\pi}{N}\,\Re g(Rr_n e^{i\theta_n}) - \int_{\theta_{n-1}}^{\theta_n} \Re g(Rr_n e^{i\theta})\,d\theta\right| < \frac{\varepsilon}{2\pi B}$$

for $n = 1,2,\ldots,N$, no matter what the r_n are, as long as N is large. Let $\varepsilon_n = \operatorname{sgn} \Re g(Rr_n e^{i\theta_n})$. We conclude that

$$B \sum_{n=1}^{N} \int_{\theta_{n-1}}^{\theta_n} \Re g(Rr_n e^{i\theta}) \cdot \varepsilon_n d\theta$$

differs at most by ε from

$$B \sum_{n=1}^{N} |\Re g(Rr_n e^{i\theta_n})| \cdot \frac{2\pi}{N} ,$$

hence at most by 3ε from

$$B \int_{-\pi}^{\pi} \sup_{0 \le r < 1} |\Re g(re^{i\theta})|\,d\theta,$$

if N is large.

For such a large N, which we now fix, let us define a measure ν on $\{|z| \le R\}$ as follows: In the sector $\theta_{n-1} < \theta < \theta_n$, ν is carried on the circular arc $|z| = Rr_n$, and on that arc,

$$d\nu(Rr_n e^{i\theta}) = B\ \varepsilon_n d\theta,\ \ \theta_{n-1} < \theta < \theta_n.$$

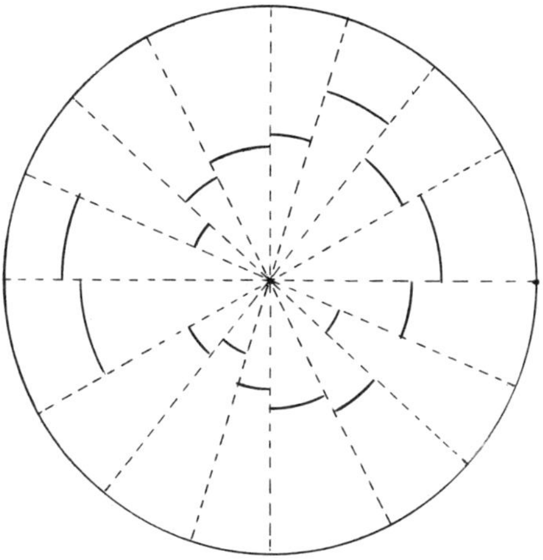

Clearly,

$$B\sum_{n=1}^{N}\int_{\theta_{n-1}}^{\theta_n}\Re g(Rr_n e^{i\theta})\varepsilon_n d\theta = \iint_{|z|\le R}\Re g(z)d\nu(z),$$

and it is evident that $\|\nu\|^* = B$.

Since ν is supported on $\{|z| \le R\}$,

$$\iint_{|z|\le 1}\Re g(z)d\nu(z) = \lim_{r\to 1}\iint \Re g(rz)d\nu(z) = \lim_{r\to 1}\int_{-\pi}^{\pi}\Re g(re^{i\theta})P_\nu(\theta)d\theta,$$

the last relation holding by a lemma in 1°, since $\Re g(rz)$ is continuous for $|z| \le 1$ and harmonic for $|z| < 1$. That is,

$$L_\nu(\Re g) = \iint_{|z|\le 1}\Re g(z)d\nu(z).$$

We have thus found a ν, $\|\nu\|^* = B$, such that $L_\nu(\Re g)$ is within 3ε of

$$B\int_{-\pi}^{\pi} \sup_{0\le r<1} |\Re g(re^{i\theta})|\,d\theta,$$

proving our assertion.

Because of this, $|L_\nu(\Re g)| < 1$ for $\|\nu\|^* \le M|||L|||$ implies that

$$M|||L||| \int_{-\pi}^{\pi} \sup_{0\le r<1} |\Re g(re^{i\theta})|\,d\theta \le 1.$$

> Now, by Chapter VIII, Section D.2, there is a numerical constant K such that
>
> $$\|g\|_1 \le K\int_{-\pi}^{\pi} \sup_{0\le r<1} |\Re g(re^{i\theta})|\,d\theta$$
>
> for $g \in H_1(0)$.

On account of this, we see that $|||L|||\ \|g\|_1 \le K/M$.

However, we also had $L(\Re g) = 1$. Therefore $|||L|||\ \|g\|_1 \ge 1$, $1 \le K/M$, that is

$$M \le K.$$

We see that L certainly is in the w^* closure of the L_ν for $\|\nu\|^* \le M|||L|||$ provided that $M > K$, the numerical constant furnished by the result of Fefferman and Stein in Section D.2 of Chapter VIII. That means that, for such M, we can find a sequence of measures ν_k, $\|\nu_k\|^* \le M|||L|||$, with

$$L_{\nu_k} \xrightarrow[k]{} L \quad (w^*).$$

The lemma at the beginning of this subsection now shows that a subsequence of the $d\nu_k$ converge w^*, AS MEASURES, to some $d\nu$ supported on $\{|z| \leq 1\}$, and that also $\|\nu\|^* \leq M|||L|||$. Without loss of generality,

$$d\nu_k \xrightarrow[k]{} d\nu \quad (w^*)$$

as measures.

Recall that L was the linear functional on $\Re H_1(0)$ corresponding to the BMO function F:

$$L(\Re f) = \lim_{R \to 1} \int_{-\pi}^{\pi} \Re f(re^{i\theta}) F(\theta) d\theta.$$

I claim that

$$F(\theta) = P_\nu(\theta) + \text{const.}$$

For each $n = 1,2,3,\ldots$ the function z^n is in $H_1(0)$, so each of the functions $r^n \left\{ {\sin \atop \cos} \right\} n\theta$ is in $\Re H_1(0)$. Take any such function and call it $U(z)$. Because $U(z)$ is continuous in $\{|z| \leq 1\}$, we here have simply

$$LU = \lim_{r \to 1} \int_{-\pi}^{\pi} U(re^{i\theta}) F(\theta) d\theta = \int_{-\pi}^{\pi} U(e^{i\theta}) F(\theta) d\theta,$$

Similarly, for each ν_k,

$$L_{\nu_k} U = \int_{-\pi}^{\pi} U(e^{i\theta}) P_{\nu_k}(\theta) d\theta.$$

By a lemma in 1°, due to continuity of $U(z)$ in $\{|z| \leq 1\}$ and its harmonicity on the interior thereof,

$$\int_{-\pi}^{\pi} U(e^{i\theta})P_{\nu_k}(\theta)d\theta = \iint_{|z|\leq 1} U(z)d\nu_k(z).$$

Since $U(z)$ is <u>continuous</u> on $|z| \leq 1$ and $d\nu_k \xrightarrow[k]{} d\nu$ (w^*) <u>as measures</u>,

$$\iint_{|z|\leq 1} U(z)d\nu_k(z) \xrightarrow[k]{} \iint_{|z|\leq 1} U(z)d\nu(z) \;;$$

the integral on the right <u>is</u>, however, $\int_{-\pi}^{\pi} U(e^{i\theta})P_\nu(\theta)d\theta$ by the lemma already used. That is,

$$L_{\nu_k} U \xrightarrow[k]{} \int_{-\pi}^{\pi} U(e^{i\theta})P_\nu(\theta)d\theta.$$

However, by w^* convergence of the <u>functionals</u> L_{ν_k} to L,

$$L_{\nu_k} U \xrightarrow[k]{} LU = \int_{-\pi}^{\pi} U(e^{i\theta})F(\theta)d\theta.$$

In fine,

$$\int_{-\pi}^{\pi} U(e^{i\theta})P_\nu(\theta)d\theta = \int_{-\pi}^{\pi} U(e^{i\theta})F(\theta)d\theta$$

whenever $U(e^{i\theta})$ is of the form $\cos n\theta$ or $\sin n\theta$ for $n = 1,2,3,\ldots$. It follows that

$$F(\theta) - P_\nu(\theta) = \text{const.}$$

Here, $\|\nu\|^* \leq M|||L||| \leq MA\|F\|_*$ where A is a numerical constant and M is any number larger than K, the numerical constant furnished by the theorem of Fefferman and Stein given in Section D.2 of Chapter VIII.

The theorem is completely proved.

Remark. Together with the theorem of 1°, we see that BMO is identical with the set of transforms $P_\nu(\theta)$ for radially bounded ν, and that the norms $\|P_\nu\|_*$ and $\|\nu\|^*$ are equivalent, if two functions in BMO which differ by a constant are considered to be the same.

Problem N° 10.

Let $\Phi(z)$ be C_∞ in $\{|z| < 1\}$, NOT NECESSARILY HARMONIC THERE, and suppose that the radial boundary value $\Phi(e^{i\theta})$ exists a.e., and that
$\int_{-\pi}^{\pi} |\Phi(e^{i\theta}) - \Phi(\rho e^{i\theta})| d\theta \to 0$ as $\rho \to 1$.

If $|\vec{\nabla}\Phi| dx\,dy = \sqrt{\Phi_x^2 + \Phi_y^2}\, dx\,dy$ is a Carleson measure on $\{|z| < 1\}$, show that $\Phi(e^{i\theta})$ is BMO.

Hint: With $F \in H_1(0)$, $R < 1$, $F(Rz) = u(z) + iv(z)$ $(u,v$ real) and $\Phi_\rho(z) = \Phi(\rho z)$, $\rho < 1$, first show that

$$\int_{-\pi}^{\pi} u(e^{i\theta})\Phi_\rho(e^{i\theta})d\theta = \iint_{|z|<1} \frac{1}{r}\left(u\frac{\partial\Phi_\rho}{\partial r} - \frac{v}{r}\frac{\partial\Phi_\rho}{\partial\theta}\right) dx\,dy.$$

BIBLIOGRAPHY

A. Books

Bari, N. K. *Trigonometricheskii͡e ri͡ady*. State Pub. Hse. for Physico-Math. Lit., Moscow, 1961.

_________. *A Treatise on Trigonometric Series* (English translation of previous item). Pergamon, New York, 1964.

Browder, A. *Introduction to Function Algebras*. Benjamin, New York, 1969.

Duren, P. *Theory of H_p Spaces*. Academic Press, New York, 1970.

Gamelin, T. *Uniform Algebras*. Prentice-Hall, Englewood Cliffs, 1969.

Garnett, J. *Bounded Analytic Functions*. Academic Press, New York, to appear shortly.

Golusin, G. M. *Geometrische Funktionentheorie* (German translation of first edition of following item). Deutscher V. der Wiss., Berlin, 1957.

Goluzin, G. M. *Geometricheskai͡a teoriia funkstiĭ kompleksnovo peremennovo*. Second edition, Nauka, Moscow, 1966.

____________. *Geometric Theory of Functions of a Complex Variable* (English translation of above item). *Amer. Math. Soc.*, Providence, 1969.

Guzmán, M. de. *Differentiation of Integrals in* $\mathbb{R}^n$. *Lecture Notes in Math.*, N° 481. Springer, Berlin, 1975.

Helson, H. *Lectures on Invariant Subspaces*. Academic Press, New York, 1964.

Hoffman, K. *Banach Spaces of Analytic Functions*. Prentice Hall, Englewood Cliffs, 1962.

Katznelson, Y. *An Introduction to Harmonic Analysis*. Wiley, New York, 1968.

Privalov, I. I. *Granichnyi͡e svoĭstva analiticheskikh funktsiĭ*. State Pub. Hse. for Tech. and Theor. Lit., Moscow, 1950.

Priwalow, I. I. *Randeigenschaften analytischer Funktionen* (German translation of above item). Deutscher V. der Wiss., Berlin, 1956.

Sarason, D. *Function Theory on the Unit Circle*. Department of Math., Virginia Poly. Inst. and State Univ., Blacksburg, 1978.

Stein. E. M. *Singular Integrals and Differentiability Properties of Functions*. Princeton Univ. Press, 1970.

Titchmarsh, E. C. *Introduction to the Theory of Fourier Integrals*. Oxford, Second edition, 1948.

Tsuji, M. *Potential Theory in Modern Function Theory*. Maruzen, Tokyo, 1959.

Zygmund, A. *Trigonometrical Series*. First edition, Monografje Matematyczne, Warsaw, 1935. Reprinted by Chelsea, New York, 1952.

__________. *Trigonometric Series*. (Greatly augmented and revised re-edition of above item.) Two volumes, Cambridge, 1959.

B. Research Papers and Articles

Adami͡an, V. Ni͡evyrozhdennyi͡e unitarnyi͡e stseplleni͡ia poluunitarnykh operatorov. *Funkts. Analiz i Prilozh*. 7 : 4(1973), pp. 1-16.

Adamian, V., Arov, D., and Kreĭn, M. G. O beskonechnykh Gankelevykh matritsakh i obobshchennykh zadachakh Karateodori-Feĭera i F. Rissa. Funkst. Analiz i Prilozh. 2 : 1(1968), pp. 1-19.

____________________. Beskonechnyie Gankelevy matritsy i obobshchennyîe zadachi Karateodori-Feĭera i Shura. Funkst. Analiz i Prilozh. 2 : 4(1968), pp. 1-17.

____________________. Analytic properties of Schmidt pairs for a Hankel operator and the generalized Schur-Takagi problem. Math. USSR Sbornik 15 (1971), pp. 31-73. (English translation of article appearing in Mat. Sbornik 86 (128), (1971), pp. 31-73.)

____________________. Infinite Hankel block matrices and related extension problems. AMS Translations (2) 111 (1978), pp. 133-156. (English translation of article appearing in Izv. Akad. Nauk Armîan. SSR., Ser. Mat. 6 (1971), pp. 87-112.)

Amar, E. Sur un théorème de Mooney relatif aux fonctions analytiques bornées. Pacific J. Math. 49 (1973), pp. 311-314.

Behrens, M. The maximal ideal space of algebras of bounded analytic functions on infinitely connected domains. Trans. A.M.S. 161 (1971), pp. 359-380.

__________. Interpolation and Gleason parts in L-domains. Trans. A.M.S., to appear.

Bernard, A., Garnett, J. and Marshall, D. Algebras generated by inner functions. J. Funct. Analysis 25 (1977), pp. 275-285.

Beurling, A. On two problems concerning linear transformations in Hilbert space. Acta Math., 81 (1949), pp. 239-255.

Burkholder, D., Gundy, R. and Silverstein, M. A maximal function characterization of the class H_p. Trans. AMS 157 (1971), pp. 137-153.

Carleson, L. On bounded analytic functions and closure problems. Arkiv för Mat. 2 (1952), pp. 283-291.

__________. An interpolation problem for bounded analytic functions. Amer. J. Math. 80 (1958), pp. 921-930.

__________. A representation formula for the Dirichlet integral. Math. Zeitsch. 73 (1960), pp. 190-196.

__________. Interpolation by bounded analytic functions and the corona problem. Annals of Math. 76 (1962), pp. 547-559.

__________. Maximal functions and capacities. Annales Inst. Fourier, Grenoble 15 (1965), pp. 59-64.

__________. On convergence and growth of partial sums of Fourier series. Acta Math. 116 (1966), pp. 135-157.

__________. Two remarks on H_1 and BMO. Advances in Math. 22 (1976), pp. 269-277.

Carleson, L. and Garnett, J. Interpolating sequences and separation properties. J. d'Analyse Math. 28 (1975), pp. 273-299.

Carleson, L. and Jacobs, S. Best uniform approximation by analytic functions. Arkiv för Math. 10 (1972), pp. 219-229.

Chang, S-Y. A characterization of Douglas subalgebras. Acta Math. 137 (1976), pp. 81-89.

__________. Structure of subalgebras between L_∞ and H_∞. Trans. AMS 227 (1977), pp. 319-332.

Chang, S-Y. and Garnett, J. Analyticity of functions and subalgebras of L_∞ containing H_∞. To appear.

Chang, S.-Y. and Marshall, D. Some algebras of bounded analytic functions containing the disk algebra, in Banach Spaces of Analytic Functions, Lecture Notes in Math. No. 604. Springer, Berlin, 1977, pp. 12-20.

Coifman, R. and Fefferman, C. Weighted norm inequalities for maximal functions and singular integrals. Studia Math. 51 (1974), pp. 241-250.

Davie, A., Gamelin, T. and Garnett, J. Distance estimates and pointwise bounded density, Trans. A.M.S. 175 (1973), pp. 37-68.

Douglas, R. G. Toeplitz and Wiener-Hopf operations in H_∞ + C. Bull A.M.S. 74 (1968), pp. 895-899.

____________. On the spectrum of a class of Toeplitz operators. J. Math. and Mechanics 18(1968), pp. 433-435.

____________. On the spectrum of Toeplitz and Wiener-Hopf operators, in Proceedings of the Conference on Abstract Spaces and Approximation, Oberwohlfach, 1968. I.S.N.M. 10, Birkhäuser V., Basel, 1969.

Douglas, R. G. and Rudin, W. Approximation by inner functions. Pacific J. Math. 31 (1969), pp. 313-320.

Douglas, R. G. and Sarason, D. A class of Toeplitz operators. Indiana Univ. Math. J. 20 (1971), pp. 891-895.

Earl, J. On the interpolation of bounded sequences by bounded functions. J. London Math. Soc. (2)2 (1970), pp. 544-548.

Fatou, P. Séries trigonométriques et séries de Taylor. Acta Math. 30 (1906), pp. 335-400.

Fefferman, C. and Stein, E. M. H_p spaces of several variables. Acta Math. 129 (1972), pp. 137-193.

Gamelin, T. Localization of the corona problem. Pacific J. Math. 34 (1970), pp. 73-81.

Gamelin, T. The Shilov boundary of $H_\infty(U)$. Amer. J. Math. 96 (1974), pp. 79-103.

Garnett, J. Interpolating sequences for bounded harmonic functions. Indiana Univ. Math. J. 21 (1971), pp. 187-192.

________. Two remarks on interpolation by bounded analytic functions, in Banach Spaces of Analytic Functions. Lecture Notes in Math. N° 604. Springer, Berlin, 1977, pp. 32-40.

________. Harmonic interpolating sequences, L_p, and BMO. Annales Inst. Fourier Grenoble 28 (1978), pp. 215-228.

Garnett, J. and Jones, P. The distance in BMO to L_∞. Annals of Math. 108 (1978), pp. 373-393.

Garnett, J. and Latter, R. The atomic decomposition for Hardy spaces in several complex variables. Duke Math. J. 45 (1978), pp. 815-845.

Gleason, A. and Whitney, H. The extension of linear functionals on H_∞. Pacific J. Math. 12 (1962), pp. 163-182.

Helson, H. and Lowdenslager, D. Prediction theory and Fourier series in several variables. Acta Math. 99 (1958), pp. 165-202.

________________. Prediction theory and Fourier series in several variables, II. Acta Math. 106 (1961), pp. 175-213.

Helson, H. and Sarason, D. Past and future. Math. Scand. 21 (1967), pp. 5-16.

Helson, H. and Szegö, G. A problem in prediction theory. Annali di Mat. Pura ed Applicata 4, 51 (1960), pp. 107-138.

Hoffman, K. Bounded analytic functions and Gleason parts. Annals of Math. 86 (1967), pp. 74-111.

Hunt, R., Muckenhoupt, B., and Wheeden, R. Weighted norm inequalities for the conjugate function ahd Hilbert transform. Trans. A.M.S. 176 (1973), pp. 227-251.

John, F. and Nirenberg, L. On functions of bounded mean oscillation. Comm. Pure Appl. Math. 14 (1961), pp. 415-426.

Jones, P. Carleson measures and the Fefferman-Stein decomposition of BMO($\mathbb{R}$). Annals of Math., to appear.

________. Ratios of interpolating Blaschke products. To appear.

________. Extension theorems for BMO. Indiana Univ. Math. J., to appear.

________. Bounded holomorphic functions with all level sets of infinite length. Michigan Math. J. To appear.

________. A complete bounded complex submanifold of C^3. Proc. A.M.S., to appear.

________. Factorization of A_p weights. To appear.

Kahane, J.-P. Another theorem on bounded analytic functions. Proc. A.M.S. 18 (1967), pp. 827-831.

Khavin, V. P. Obobshchenie teoremy Privalova-Zigmunda o module nîepreryvnosti soprîazhennoĭ funktsii. Izv. Akad. Nauk Armîan. S.S.R. 6 (1971), pp. 252-258 and pp. 265-287.

Khavin, V. P. Slabaîà polnota prostranstva $L_1/H_1(0)$. Vestnik Lenningrad. Univ. 13 (1973), pp. 77-81.

Khavinson, S. On some extremal problems of the theory of analytic functions. A.M.S. Translations (2) 32 (1963), pp. 139-154. (English translation of article appearing in Moskov. Univ. Uchen. Zapiski, 148 Matem. 4 (1951), pp. 133-143, with historical appendix by the translator, A. Shields.)

Khavinson, S. Extremal problems for certain classes of analytic functions in finitely connected regions. A.M.S. Translations (2) 5 (1957), pp. 1-33. (English translation of article appearing in Mat. Sbornik 36, (78) (1955), pp. 445-478.

Koosis, P. On functions which are mean periodic on a half-line. Comm. Pure Appl. Math. 10 (1957), pp. 133-149.

________. Interior compact spaces of functions on a half-line. Comm. Pure Appl. Math. 10 (1957), pp. 583-615.

________. Weighted quadratic means of Hilbert transforms. Duke Math. J. 38 (1971), pp. 609-634.

________. Moyennes quadratiques de transformées de Hilbert et fonctions de type exponentiel. C.R. Acad. Sci. Paris 276 (1973), pp. 1201-1204.

________. Sommabilité de la fonction maximale et appartenance à H_1. C.R. Acad. Sci. Paris 286 Ser. A (1978), pp. 1041-1043.

________. Sommabilité de la fonction maximale et appartenance à H_1. Cas de plusieurs variables. C.R. Acad. Sci. Paris, Sér. A, 288 (1979), pp. 489-492.

Latter, R. H. A decomposition of $H_p(\mathbb{R}^n)$ in terms of atoms. Studia Math. 62 (1979), pp. 92-101.

Lax, P. Remarks on the preceding paper. Comm. Pure Appl. Math. 10 (1957), pp. 617-622.

______. Translation invariant subspaces. Acta Math. 101 (1959), pp. 163-178.

Lee, M. and Sarason, D. The spectra of some Toeplitz operators. J. Math. Anal. Appl. 33 (1971), pp. 529-543.

de Leeuw, K. and Rudin, W. Extreme points and extremum problems in H_1. Pacific J. Math. 8 (1958), pp. 467-485.

Lindelöf, E. Sur la représentation conforme d'une aire simplement connexe sur l'aire d'un cercle, in Quatrième Congrès des Mathématiciens Scandinaves, Stockholm (1916), pp. 59-90.

Marshall, D. Blaschke products generate H_∞. Bull. A.M.S. 82 (1976), pp. 494-496.

__________. Subalgebras of L_∞ containing H_∞. Acta Math. 137 (1976), pp. 91-98.

Mooney, M. A theorem on bounded analytic functions. Pacific J. Math. 43 (1972), pp. 457-463.

Muckenhoupt, B. Weighted norm inequalities for the Hardy maximal function. Trans. A.M.S. 165 (1972), pp. 207-226.

__________. Hardy's inequality with weights. Studia Math. 44 (1972), pp. 31-38.

Nevanlinna, R. Über beschränkte Funktionen die in gegebenen Punkten vorgeschrieben Werte annehmen. Ann. Acad. Sci. Fenn. 13 (1919), N° 1.

__________. Über beschränkte analytische Funktionen. Ann. Acad. Sci. Fenn. 32 (1929), N° 7.

Newman, D. J. Pseudo-uniform convexity in H_1. Proc. A.M.S. 14 (1963), pp. 676-679.

Riesz, F. and M. Über die Randwerte einer analytischen Funktion, in Quatrième Congrès des Mathematiciens Scandinaves, Stockholm (1916), pp. 27-44.

Rogosinski, W. and Shapiro, H. S. On certain extremum problems for analytic functions. Acta Math. 90 (1953), pp. 287-318.

Sarason, D. Generalized interpolation in H_∞. Trans. A.M.S. 127 (1967), pp. 179-203.

_________. Approximation of piecewise continuous functions by quotients of bounded analytic functions. Canad. J. Math. 24 (1972), pp. 642-657.

_________. An addendum to "Past and future". Math. Scand. 30 (1972), pp. 62-64.

_________. Algebras of functions on the unit circle. Bull. A.M.S. 79 (1973), pp. 286-299.

_________. Functions of vanishing mean oscillation. Trans. A.M.S. 207 (1975), pp. 391-405.

_________. Algebras between L_∞ and H_∞, in Lecture Notes in Math. N° 512. Springer, Berlin, 1976, pp. 117-129.

_________. Toeplitz operators with semi almost-periodic symbols. Duke Math. J. 44 (1977), pp. 357-364.

_________. Toeplitz operators with piecewise quasicontinuous symbols. Indiana Univ. Math. J. 26 (1977), pp. 817-838.

Shapiro, H. S. and Shields, A. On some interpolation problems for analytic functions. Amer. J. Math. 83 (1961), pp. 513-532.

Srinivasan, T. and Wang, J-K. On closed ideals of analytic functions. Proc. A.M.S. 16 (1965), pp. 49-52.

Stein, E. M. A note on the class L log L. Studia Math. 32 (1969), pp. 305-310.

Uchiyama, A. The construction of certain BMO functions. To appear.

Varopoulos, N. Sur un problème d'interpolation. C.R. Acad. Sci. Paris 274 Sér. A (1972), pp. 1539-1542.

____________. BMO functions and the $\bar{\partial}$-equation. Pacific J. Math. 71 (1977), pp. 221-273.

____________. A remark on BMO and bounded harmonic functions. Pacific J. Math. 73 (1977), pp. 257-259.

Widom, H. Toeplitz matrices, in Topics in Real and Complex Analysis, edited by I. I. Hirschman. Mathematical Association of America, 1965, pp. 197-209.

________. Toeplitz operators on H_p. Pacific J. Math. 19 (1966), pp. 573-582.

Ziskind, S. Interpolating sequences and the Shilov boundary of $H_\infty(\Delta)$. J. Funct. Analysis 21 (1976), pp. 380-388.

APPENDIX

The Corona Theorem

A. Homomorphisms of H_∞ and maximal ideals

H_∞ is actually a Banach algebra over $\mathbb{C}$, because if f and $g \in H_\infty$, $fg \in H_\infty$ and $\|fg\|_\infty \leq \|f\|_\infty \|g\|_\infty$. Because H_∞ has this multiplicative structure, it is natural to consider the algebraic multiplicative homomorphisms of H_∞ onto $\mathbb{C}$.

Let $L : H_\infty \to \mathbb{C}$ be such a homomorphism. Since H_∞ contains the multiplicative identity 1, we must have $L(1) = 1$. If $f \in H_\infty$ and λ is any complex number of modulus $> \|f\|_\infty$, the function $(\lambda - f(z))^{-1}$ belongs to H_∞, so since $(\lambda - f(z))^{-1}(\lambda - f(z)) = 1$, taking L of both sides shows that $L(\lambda - f)$ can't be zero. Letting λ range over all complex numbers of modulus $> \|f\|_\infty$, we see that $|L(f)| \leq \|f\|_\infty$; an algebraic multiplicative homomorphism L of H_∞ onto $\mathbb{C}$ is necessarily continuous, and of norm ≤ 1 as a linear functional on the Banach space H_∞. It is in fact of norm equal to 1 because $L(1) = 1$. The set of such L is obviously a w^*-closed subset of the unit sphere in H_∞'s dual; as such it is w^* compact.

For a multiplicative homomorphism L of H_∞ onto $\mathbb{C}$, the set $\mathfrak{m}$ of elements of H_∞ taken onto 0 by L is a maximal (proper) ideal in H_∞ because $\mathbb{C}$ is a field. So to every homomorphism corresponds a maximal ideal. Conversely, to every maximal ideal corresponds a homomorphism of H_∞ onto $\mathbb{C}$.

Indeed, if m is a proper ideal in H_∞, so is $\bar{m}$, its norm closure. For if $f \in H_\infty$ and $\|1 - f\|_\infty < 1$, then $f^{-1} \in H_\infty$, so m cannot contain f without $1 = f^{-1}f$ also being in m. So, m being proper makes $\|1 - m\|_\infty \geq 1$, and $1 \notin \bar{m}$. From this it is manifest that if m is a (proper) maximal ideal, we must already have $m = \bar{m}$.

Take any maximal ideal m. Since it is norm-closed, the quotient ring H_∞/m is a (complete !) Banach algebra over $\mathbb{C}$. It is a field because m is maximal. But now a celebrated theorem of Gelfand tells us that the only complete normed field over $\mathbb{C}$ is $\mathbb{C}$ itself! So in fact, H_∞/m is isomorphic to $\mathbb{C}$, and the canonical homomorphism L of H_∞ onto H_∞/m is in fact one of H_∞ onto $\mathbb{C}$; we can define $L(f)$ as the unique complex number λ for which $\lambda - f \in m$ - there must be one because H_∞/m is isomorphic to $\mathbb{C}$.

In this way the set of multiplicative homomorphisms L of H_∞ onto $\mathbb{C}$ is in natural one-to-one correspondence with the set of maximal ideals m in H_∞.

If m is such a maximal ideal and L is the multiplicative homomorphism of H_∞ corresponding to it, it is customary to write

$$h(m) \text{ for } L(h).$$

This we do henceforth.

The set of maximal ideals m is denoted by $\mathfrak{M}$. We take for $\mathfrak{M}$ the topology of pointwise convergence of maximal ideals m (as multiplicative homomorphisms) over H_∞ - $\mathfrak{M}$ is then compact for the reasons stated at the beginning of this section.

It now becomes natural to look at H_∞ as a Banach algebra of functions on its set $\mathfrak{M}$ of maximal ideals, associating to each $f \in H_\infty$ the function $f(m)$, $m \in \mathfrak{M}$. $\mathfrak{M}$ is called the maximal ideal space of H_∞. This approach, a rather abstract one, has proven quite fruitful; it is in fact the main point of view adopted in books like Gamelin's. One complication is, however, that the space $\mathfrak{M}$ is very large - so vast is it, in fact, that it has many bizarre properties.

There is, however, a simple subset of $\mathfrak{M}$ ready at hand. If $|z| < 1$, the point evaluation

$$f \to f(z)$$

is a homomorphism of H_∞ onto $\mathbb{C}$! So each point z in the open unit circle corresponds in obvious fashion to a certain maximal ideal, namely the ideal of functions $f \in H_\infty$ vanishing at z. We denote that maximal ideal by z, also. If we do this, then we can consider the open unit circle $\{|z| < 1\}$ as a subset of $\mathfrak{M}$.

The natural question now arises:

Is $\{|z| < 1\}$ w^* dense in $\mathfrak{M}$? If the answer is yes, there is some hope of being able to arrive at a more concrete description of the very complicated space $\mathfrak{M}$. The conjecture that the response is positive was known as the corona conjecture. To prove it or disprove it was the celebrated corona problem.

Carleson solved the corona problem in 1962, in the positive sense. His proof of what has come to be known as the corona theorem was based on an intricate geometrical construction, of combinatorial character. The construction itself has proved useful for the study of other problems, especially in the hands of Garnett and his students. It is, however, so difficult to master as to discourage many from attempting to go through the details in the proof of the corona theorem.

This state of affairs has been changed by the work of T. Wolff, done in the spring of 1979. That is why we are able to present a complete proof of the corona theorem in this appendix.

The corona theorem has two equivalent formulations:

1° If $m \in \mathfrak{M}$ there is a net $\{z_\alpha\}$, $|z_\alpha| < 1$, with $z_\alpha \xrightarrow[\alpha]{} m$ in $\mathfrak{M}$.

2° If $f_1, \ldots, f_n \in H_\infty$ and

$$\sup_k |f_k(z)| \geq \text{some } \delta > 0$$

for all z, $|z| < 1$, there exist functions $g_1, \ldots, g_n \in H_\infty$ such that $f_1 g_1 + f_2 g_2 + \ldots + f_n g_n \equiv 1$ on $\{|z| < 1\}$.

Let us prove the equivalence.

If 2° is true, let $m \in \mathfrak{M}$ and suppose there is no net of z, $|z| < 1$, which tends w^* to m. By the definition of the w^* topology, there exist $h_1, \ldots, h_n \in H_\infty$ and a $\delta > 0$ such that for each z, $|z| < 1$, at least one of the inequalities

$$|h_k(z) - h_k(m)| \geq \delta,$$

$k = 1, \ldots, n$, must hold. Call $f_k(z) = h_k(z) - h_k(m)$; $f_k \in H_\infty$. Then all the $f_k(m)$ are zero, but for each z, $|z| < 1$, some $|f_k(z)|$ is

$\geq \delta > 0$. Thence, by 2°, we get $g_1,\ldots,g_n \in H_\infty$ with $g_1f_1 + \ldots + g_nf_n \equiv 1$. So $g_1(m)f_1(m) + \ldots + g_n(m)f_n(m)$ must $= 1$! But each $f_k(m) = 0$, a contradiction.

Suppose now that 1° is <u>true</u>, and let $f_1,\ldots,f_n$ satisfy the hypothesis of 2°. If there are <u>no</u> $g_1,\ldots,g_n \in H_\infty$ with

$$g_1f_1 + \ldots g_nf_n \equiv 1,$$

the set of all sums $g_1f_1 + \ldots + g_nf_n$ constitutes a <u>proper ideal</u> in H_∞. Since $1 \in H_\infty$, this proper ideal <u>must be contained in some maximal ideal</u>, say m, by the usual application of Zorn's lemma. <u>Then</u> surely $f_k(m) = 0$ for each k. By the truth of 1°, we now get a net $\{z_\alpha\}$ of points, $|z_\alpha| < 1$, with $z_\alpha \underset{\alpha}{\longrightarrow} f_k(m) = 0$ for $k = 1,\ldots,n$. The hypothesis of 2° is <u>now contradicted</u>, so there must in fact <u>be</u> $g_1,\ldots,g_n \in H_\infty$ with $g_1f_1 + \ldots + g_nf_n \equiv 1$.

Knowing that the corona theorem is true, mathematicians, especially Hoffman, <u>have indeed</u> been able to obtain a <u>fairly complete description</u> of $\mathfrak{M}$, by getting at the $m \in \mathfrak{M}$ with nets of points from $\{|z| < 1\}$. In this investigation the <u>interpolating sequences</u> studied in Chapter IX turn out to be of special importance.

B. <u>The $\bar{\partial}$-equation.</u>

Wolff's proof of the corona theorem makes systematic use of the two differential operators

$$\partial = \frac{1}{2}\left(\frac{\partial}{\partial x} - i\frac{\partial}{\partial y}\right), \quad \bar{\partial} = \frac{1}{2}\left(\frac{\partial}{\partial x} + i\frac{\partial}{\partial y}\right).$$

A C_∞ function $f(z)$ is analytic if and only if $\bar{\partial} f \equiv 0$, and then $f'(z) = \partial f(z)$. The Laplacian ∇^2 is equal to $4\partial\bar{\partial}$.

Let $g(x)$ be C_∞ and of compact support. Can we solve the equation

$$\bar{\partial} f(z) = g(z)$$

for $|z| < 2$, say, with a C_∞ function f (not necessarily of compact support)? We can - one solution is

$$f(z) = \frac{1}{\pi} \iint_{C} \frac{g(\zeta)\, d\xi\, d\eta}{z - \zeta} ,$$

where $\zeta = \xi + i\eta$.

To see this, let us make the change of variable $z - \zeta = w = u + iv$, so that the above double integral goes over into

$$\frac{1}{\pi} \iint_{C} \frac{g(z - w)}{w} \, du\, dv .$$

Since g is of compact support, we can, if we restrict ourselves to $|z| < 2$, rewrite the above integral as

$$\frac{1}{\pi} \iint_{|w| < R} g(z - w) \, \frac{du\, dv}{w}$$

with some very large R. Here g is C_∞ and

$$\iint_{|w| < R} \frac{du\, dv}{|w|} < \infty$$

so the previous expression - call it $f(z)$ - can be differentiated under the integral sign, yielding

$$\bar{\partial} f(z) = \frac{1}{\pi} \iint_{|w|<R} \bar{\partial} g(z-w) \, \frac{du\,dv}{w} \quad .$$

I claim that the integral on the right equals $g(z)$. Wlog, take $z = 0$, then we have to evaluate

$$\lim_{\rho \to 0} \frac{1}{\pi} \iint_{\rho<|w|<R} \bar{\partial} g(-w) \, \frac{du\,dv}{w} \quad .$$

Here, we can replace $\bar{\partial}$ by $-\frac{1}{2}\left(\frac{\partial}{\partial u} + i\frac{\partial}{\partial v}\right)$. Then, since

$$\left(\frac{\partial}{\partial u} + i\frac{\partial}{\partial v}\right)\left(\frac{1}{w}\right) \equiv 0,$$

the previous expression becomes

$$\frac{i}{2\pi} \lim_{\rho \to 0} \iint_{\rho<|w|<R} \left[\frac{\partial}{\partial u}\left(i\,\frac{g(-w)}{w}\right) - \frac{\partial}{\partial v}\left(\frac{g(-w)}{w}\right)\right] du\,dv.$$

By Green's theorem, this expression is just

$$-\frac{i}{2\pi} \lim_{\rho \to 0} \int_{|w|=p} g(-w) \, \frac{du + idv}{w} + \frac{i}{2\pi} \int_{|w|=R} g(-w) \, \frac{du + idv}{w} \quad .$$

Now, R being very large, and g of compact support, the line integral around $|w| = R$ is zero. On putting $w = \rho e^{i\varphi}$, that around $|w| = \rho$ is seen to be just

$$\frac{1}{2\pi} \int_0^{2\pi} g(-\rho e^{i\varphi}) \, d\varphi,$$

and as $\rho \to 0$, this just tends to $g(0)$. So our f does satisfy $\bar{\partial} f(0) = g(0)$, and in the same way we get $\bar{\partial} f(z) = g(z)$. The f given by our double integral is evidently C_∞, since g is.

In what follows we will be interested in solutions f of $\bar{\partial} f(z) = g(z)$, valid in some slightly larger circle than the unit one, for which $\sup_\theta |f(e^{i\theta})|$ is not too large. For this we have the

Lemma (Wolff). Let $h(z)$ be $\mathcal{C}_\infty$ in the circle $\{|z| < R\}$, where $R > 1$. Suppose that, in $\{|z| < 1\}$,

$$|z| \log \frac{1}{|z|} \, |h(z)|^2 \, dx\, dy$$

and

$$|z| \log \frac{1}{|z|} \, |\partial h(z)| \, dx\, dy$$

are Carleson measures, with "Carleson constants" A and B respectively. Then we can find a function $v(z)$, $\mathcal{C}_\infty$ in some circle $\{|z| < R'\}$ with $R' > 1$, such that

$$\bar{\partial} v(z) = h(z), \qquad |z| < 1 ,$$

whilst $|v(e^{i\theta})| < 9\,(\sqrt{A} + B)$.

Note. For the definition of Carleson measures and their properties, see Chapter VIII, Section E.

Proof of Lemma (As simplified by Varopoulous and Garnett). We may first suppose $h(z)$ to be redefined for $|z| > \frac{1+R}{2}$, say, so as to make it $\mathcal{C}_\infty$ in all of $\mathbb{C}$ and of compact support. Then the formula described above certainly gives us a $\mathcal{C}_\infty$ function v_0 satisfying

$$\bar{\partial} v_0(z) = h(z).$$

The trouble is, of course, that $|v_0(e^{i\theta})|$ may get quite large.

If $f(z)$ is analytic in some circle $\{|z| < R'\}$ with $R' > 1$, we of course also have

$$\bar{\partial} v = h$$

with $v = v_0 - f$ in that circle, because $\bar{\partial} f \equiv 0$. The idea is to choose f so that $|v(e^{i\varphi})|$ does not get too big.

In the notation of Chapter VII, Section A, $v_0(e^{i\theta})$ certainly belongs to $\mathcal{C}$. If there is an $f \in \mathcal{G}$ $(= H_\infty \cap \mathcal{C})$ such that $\|v_0 - f\|_\infty = \sup_\theta |v_0(e^{i\theta}) - f(e^{i\theta})| < d$, say, then, if $r < 1$ is sufficiently close to 1, $\|v_0 - f_r\|_\infty$ is still $< d$ with $f_r(z) = f(rz)$, and f_r is $\mathcal{C}_\infty$ in $\{|z| < 1/r\}$. So what we want to find is $\|v_0 - \mathcal{G}\|_\infty$; for any d larger than $\|v_0 - \mathcal{G}\|_\infty$, we will have a solution v, fulfilling the desired conditions and such that $|v(e^{i\theta})| \le d$.

By the duality theory of Chapter VII, Section A,

$$\|v_0 - \mathcal{G}\|_\infty = \sup\left\{\left|\int_0^{2\pi} v_0(e^{i\theta})F(e^{i\theta})d\theta\right|;\ F \in H_1(0) \text{ and } \|F\|_1 \le 1\right\}.$$

We may clearly restrict the set of $F \in H_1(0)$ over which the above sup is taken to those having an analytic continuation into some larger circle $\{|z| < R_F\}$, with $R_F > 1$ depending on F.

For such F, we can apply Green's theorem as in Chapter X, Section D.1, getting

$$\int_0^{2\pi} v_0(e^{i\theta})F(e^{i\theta})d\theta = \iint_{|z|<1} \log\frac{1}{|z|}\,\nabla^2(v_0(z)F(z))\,dx\,dy,$$

because $v_0(z)$ is, at any rate, $\mathcal{C}_\infty$ in the unit circle and up to its boundary. We have $\nabla^2(v_0F) = 4\partial\bar{\partial}(v_0F) = 4\bar{\partial}v_0\partial F + 4F\partial\bar{\partial}v_0$, because

$\bar{\partial}F$ and $\partial\bar{\partial}F$ vanish identically. Also, we are supposed to have $\bar{\partial}v_0 = h$, so that $\partial\bar{\partial}v_0 = \partial h$. Therefore $\nabla^2(v_0F) = 4hF' + 4\partial hF$, and the above double integral breaks down to

$$4\iint_{|z|<1} F(z)\partial h(z)\log\frac{1}{|z|}\,dx\,dy + 4\iint_{|z|<1} F'(z)h(z)\log\frac{1}{|z|}\,dx\,dy .$$

Of these two terms, the first is in absolute value

$$\leq 4\iint_{|z|<1} \left|\frac{F(z)}{z}\right| |\partial h(z)| \; |z| \;\log\frac{1}{|z|}\,dx\,dy .$$

Here, since $F \in H_1(0)$, $F(z)/z$ belongs to H_1 and has the same H_1-norm as F. So, by the hypothesis, and the definition of Carleson measures, our first term is in absolute value $\leq 4B\|F\|_1$.

Regarding the second term, we apply Schwarz' inequality just as in Chapter X, Section E.1, and see that the second term is in modulus

$$\leq 4\sqrt{\iint_{|z|<1} \log\frac{1}{|z|}\frac{|F'(z)|^2}{|F(z)|}\,dx\,dy \;\iint_{|z|<1} \left|\frac{F(z)}{z}\right| |h(z)|^2 |z| \log\frac{1}{|z|}\,dx\,dy} .$$

As in Chapter X, Section D.3, we see that it is sufficient to estimate our second term for $F \in H_1(0)$ having only a simple zero at the origin, and no others, since any $F \in H_1(0)$ can be written as the sum of two such, each with norm no bigger than its own. But for such F,

$$\iint_{|z|<1} \log\frac{1}{|z|}\frac{|F'(z)|^2}{|F(z)|}\,dxdy \;=\; \|F\|_1$$

by a theorem of Chapter X, Section D.2. Again,

$$\iint_{|z|<1} \left|\frac{F(z)}{z}\right| |h(z)|^2 |z| \log\frac{1}{|z|}\,dx\,dy \;\leq A\|F\|_1$$

for $F \in H_1(0)$, by the hypothesis.

We see that, for general $F \in H_1(0)$, analytic over the closed ircle $\{|z| \leq 1\}$,

$$\left| 4 \iint_{|z|<1} F'(z)h(z)\log\frac{1}{|z|}\,dx\,dy \right| \leq 8\sqrt{A}\ \|F\|_1$$

with 8 on the right and not 4 because of the restriction to the ind of F for which Schwarz' inequality was applied).

Putting the above two estimates together we see that

$$\left| \int_0^{2\pi} v_0(e^{i\theta})F(e^{i\theta})d\theta \right| \leq 8(\sqrt{A} + B)\|F\|_1$$

or $F \in H_1(0)$ which are analytic over the closed unit circle. Therefore $\|v_0 - \alpha\|_\infty \leq 8(\sqrt{A} + B)$, and the lemma is proved.

Wolff's proof of the corona theorem

heorem (Carleson, 1962). Let $f_1,\ldots,f_n \in H_\infty$; suppose $\|f_k\|_\infty \leq 1$ for $= 1,\ldots,n$ and that for some $\delta > 0$ we have

$$\sup_k |f_k(z)| > \delta$$

or all z, $|z| < 1$. There is a number $M(\delta,n)$ depending only on δ and n such that

$$g_1f_1 + g_2f_2 + \ldots + g_nf_n \equiv 1 \quad \text{on} \quad \{|z| < 1\}$$

with some functions $g_k \in H_k$ satisfying

$$\|g_k\|_\infty \leq M(\delta,n).$$

Proof (Wolff, 1979). The main analytical idea is contained in the case $n = 2$, which we proceed to treat first.

It is sufficient to prove the theorem with functions analytic in some slightly larger circle than the unit one standing in place of the f_k. Indeed, once that is done, it will apply to the functions $f_k^{(r)}(z) = f_k(rz)$ where $r < 1$, giving us some $g_k^{(r)} \in H_\infty$ with $g_1^{(r)}f_1^{(r)} + \ldots + g_n^{(r)}f_n^{(r)} \equiv 1$ in $|z| < 1$. The bounds on the $\|g_k^{(r)}\|_\infty$ furnished by the theorem do not depend on r. Therefore, on making $r \to 1$, a normal family argument will give us $g_k \in H_\infty$ (with $\|g_k\|_\infty$ satisfying the same bounds) such that

$$g_1f_1 + \ldots + g_nf_n = 1.$$

Here, $n = 2$, so that we have f_1 and f_2 with $\|f_1\|_\infty \leq 1$, $\|f_2\|_\infty \leq 1$, and, for every z, $|z| < 1$,

$$|f_1(z)| > \delta \quad \text{or} \quad |f_2(z)| > \delta.$$

Let $U(w)$ be a C_∞ function, depending only on $|w|$, with

$$U(w) \equiv 0 \quad \text{for} \quad |w| \leq \frac{\delta}{2},$$

$$U(w) \equiv 1 \quad \text{for} \quad |w| \geq \delta,$$

and $0 \leq U(w) \leq 1$ elsewhere. Put

$$\varphi_k(z) = \frac{U(f_k(z))}{U(f_1(z)) + U(f_2(z))}$$

for $k = 1,2$. The functions $\varphi_k(z)$ are clearly C_∞ on some circle $\{|z| > R\}$ with $R > 1$, and $\varphi_1(z) + \varphi_2(z) \equiv 1$.

We have

$$\varphi_k(z) \equiv \begin{cases} 0 & \text{where } |f_k(z)| < \frac{\delta}{2} \\ 1 & \text{where } |f_k(z)| > \delta. \end{cases}$$

Observe that the hypothesis makes $\varphi_2(z) = 1$ wherever $\varphi_1(z) = 0$ and $\varphi_1(z) = 1$ wherever $\varphi_2(z) = 0$.

Now

$$\frac{\varphi_1}{f_1} \cdot f_1 + \frac{\varphi_2}{f_2} \cdot f_2 \equiv 1.$$

The trouble is, of course, that φ_1/f_1 and φ_2/f_2 are <u>not analytic</u>! Here, using an idea that goes back to Hörmander, we look for some new function $v(z)$ which will <u>make</u>

$$g_1 = \frac{\varphi_1}{f_1} + vf_2 \quad \text{and} \quad g_2 = \frac{\varphi_2}{f_2} - vf_1$$

<u>analytic</u> in $|z| < 1$. <u>Any</u> such v will <u>automatically</u> give us

$$g_1f_1 + g_2f_2 \equiv 1$$

in $|z| < 1$.

For analyticity of g_1 and g_2 we need $\bar{\partial}g_1 = \bar{\partial}g_2 = 0$ in $|z| < 1$. Since $\bar{\partial}f_1 = \bar{\partial}f_2 = 0$, we get the conditions

$$\frac{\bar{\partial}\varphi_1}{f_1} + f_2\bar{\partial}v = 0, \qquad \frac{\bar{\partial}\varphi_2}{f_2} - f_1\bar{\partial}v = 0.$$

$\varphi_1 + \varphi_2 \equiv 1$, so $\bar{\partial}\varphi_1 + \bar{\partial}\varphi_2 \equiv 0$, therefore the two conditions are <u>compatible</u>, and equivalent to the single one

$$\bar{\partial}v = \frac{\bar{\partial}\varphi_2}{f_1f_2} .$$

Observe that on the open set where $|f_1(z)| < \delta/2$, $\varphi_2(z) \equiv 1$ so $\bar{\partial}\varphi_2(z) \equiv 0$; on the open set where $|f_2(z)| < \delta/2$, $\varphi_2(z) \equiv 0$ so $\bar{\partial}\varphi_2(z) \equiv 0$. Therefore

$$\left|\frac{\bar{\partial}\varphi_2}{f_1 f_2}\right| \leq \frac{4}{\delta^2}\,|\bar{\partial}\varphi_2|$$

on $\{|z| < 1\}$, and

$$h(z) = \frac{\bar{\partial}\varphi_2(z)}{f_1(z)f_2(z)}$$

is a nice C_∞ function defined on some circle $\{|z| < R\}$ with $R > 1$.

Now apply the lemma of Section B! We are looking for solutions v to $\bar{\partial}v = h$ on some circle slightly larger than the unit one; as we have just seen, for $|z| < 1$,

$$|h(z)| \leq \frac{4}{\delta^2}\,|\bar{\partial}\varphi_2(z)| = \frac{4}{\delta^2}\,\frac{|U(f_1(z))\bar{\partial}U(f_2(z)) - U(f_2(z))\bar{\partial}U(f_1(z))|}{|U(f_1(z)) + U(f_2(z))|^2} .$$

Since $U(f_1(z)) + U(f_2(z)) \geq 1$, this last expression is $\leq K_\delta(|f_1'(z)| + |f_2'(z)|)$ by the chain rule, where $K_\delta = 4\delta^{-2} \sup_w |\text{grad } U(w)|$ depends only on δ. Therefore

$$|h(z)|^2 |z| \log\frac{1}{|z|} \leq 2\,K_\delta^2\,(|f_1'(z)|^2 + |f_2'(z)|^2)\,|z| \log\frac{1}{|z|} .$$

But $f_1, f_2 \in H_\infty$ are, in particular, harmonic in $\{|z| < 1\}$, and $|f_1(z)| < 1$, $|f_2(z)| < 1$ there.

The lemmas in Sections E.2, E.3 of Chapter X now show that

$$(|f_1'(z)|^2 + |f_2'(z)|^2)\,|z| \log\frac{1}{|z|}\,dx\,dy$$

is a Carleson measure whose Carleson constant can be taken $\leq$ some pure number.

Therefore

$$|h(z)|^2 |z| \log \frac{1}{|z|}\, dx\, dy$$

is a Carleson measure whose Carleson constant, A_δ, can be taken to depend only on δ. (It is remarkable that the above boxed statement can be proved directly, without appealing to Chapter X, Section E or the theorem on Carleson measures! Gamelin and Davie first noticed this - see exercise at the end of this appendix.)

Now let us look at

$$\partial h = \frac{\partial\bar{\partial}\varphi_2}{f_1 f_2} - \frac{\bar{\partial}\varphi_2}{f_1 f_2}\left[\frac{f_1'}{f_1} + \frac{f_2'}{f_2}\right] .$$

Of the two terms on the right, the second vanishes identically on the open set where $|f_1|$ or $|f_2|$ is $< \delta/2$, and on the complement of that set it is in absolute value

$$\leq 8\delta^{-3} \sup_w |\text{grad } U(w)| (|f_1'(z)| + |f_2'(z)|)^2 .$$

The first term also vanishes identically on the open set just mentioned, and on its complement equals

$$\frac{1}{4 f_1(z) f_2(z)} \nabla^2 \left\{\frac{U(f_2(z))}{U(f_1(z)) + U(f_2(z))}\right\}.$$

But $\nabla^2 f_1(z) = \nabla^2 f_2(z) \equiv 0$, so the expression just written only involves $f_1'(z)$ and $f_2'(z)$ and is clearly in modulus $\leq C_\delta(|f_1'(z)|^2 + |f_2'(z)|^2)$, with C_δ depending only on δ. We see in this way that

$$|\partial h(z)|\,|z| \log \frac{1}{|z|} \leq L_\delta(|f_1'(z)|^2 + |f_2'(z)|^2)|z| \log \frac{1}{|z|} .$$

By Sections E.2, E.3 of Chapter X, this makes $$\lvert \partial h(z)\rvert\, \lvert z\rvert \log \frac{1}{\lvert z\rvert}\, dx\,dy$$ a Carleson measure with Carleson constant, B_δ, depending only on δ since $\lvert f_1(z)\rvert \le 1$, $\lvert f_2(z)\rvert \le 1$ in $\{\lvert z\rvert < 1\}$.

The lemma thus gives us a $v(z)$, $\mathscr{C}_\infty$ in some circle slightly larger than the unit one, with $\bar{\partial} v = h$ in $\{\lvert z\rvert < 1\}$ and $\lvert v(e^{i\theta})\rvert \le 9(\sqrt{A_\delta} + B_\delta)$. The functions

$$g_1 = \frac{\varphi_1}{f_1} + vf_2 \quad \text{and} \quad g_2 = \frac{\varphi_2}{f_2} - vf_1$$

will be in H_∞ - even in $\mathscr{Q}$ - and satisfy $g_1 f_1 + g_2 f_2 \equiv 1$ in $\{\lvert z\rvert < 1\}$. Clearly, for $\lvert z\rvert \le 1$,

$$\lvert \varphi_1(z)/f_1(z)\rvert \le \frac{2}{\delta}, \qquad \lvert \varphi_2(z)/f_2(z)\rvert \le \frac{2}{\delta},$$

so for $k = 1,2$,

$$\lvert g_k(e^{i\theta})\rvert \le \frac{2}{\delta} + 9(\sqrt{A_\delta} + B_\delta),$$

i.e.,

$$\lVert g_k\rVert_\infty \le \frac{2}{\delta} + 9(\sqrt{A_\delta} + B_\delta),$$

proving the theorem for $n = 2$.

For $n > 2$, the situation is more complicated algebraically. Given $f_1, \ldots, f_n$ with $\lVert f_k\rVert_\infty \le 1$ and $\sup_k \lvert f_k(z)\rvert > \delta$ for all z in the unit circle, we take the function $U(w)$ used above and put, for $k = 1,2,\ldots,n$,

$$\varphi_k(z) = \frac{U(f_k(z))}{U(f_1(z)) + U(f_2(z)) + \ldots + U(f_n(z))} .$$

ach φ_k vanishes identically on the set where $|f_k| < \frac{\delta}{2}$, and

$$\sum_1^n \varphi_k(z) \equiv 1$$

'or $|z| < 1$.

Assuming, as we may, each $f_k(z)$ to be analytic over the closed init circle, we search for analytic functions g_k of the form

$$g_k = \frac{\varphi_k}{f_k} + \sum_j v_{kj} f_j$$

vith certain, as yet unknown, functions v_{kj} which we require to satisfy $v_{kj} = -v_{jk}$, $v_{kk} \equiv 0$, so as to automatically have

$$g_1 f_1 + \ldots + g_n f_n \equiv 1.$$

For analyticity of the g_k, we need $\bar{\partial} g_k \equiv 0$, which means that the v_{kj} must satisfy

$$\frac{\bar{\partial}\varphi_k}{f_k} + \sum_j f_j \, \bar{\partial} v_{kj} = 0.$$

This holds if, for instance,

$$\bar{\partial} v_{kj} = \frac{\varphi_k}{f_k f_j} \bar{\partial}\varphi_j - \frac{\varphi_j}{f_j f_k} \bar{\partial}\varphi_k,$$

as may be verified directly using the relations $\varphi_1 + \ldots + \varphi_n \equiv 1$, $\bar{\partial}\varphi_1 + \ldots + \bar{\partial}\varphi_n \equiv 0$. Here, we first solve each of the equations

$$\bar{\partial} w_{kj} = \frac{\varphi_k}{f_k f_j} \bar{\partial} \varphi_j ,$$

and then put $v_{kj} = w_{kj} - w_{jk}$, so that v_{kj} will equal $-v_{jk}$ automatically. The lemma, applied in the same way as for the case $n = 2$, shows us that we can get solutions $w_{kj}(z)$, $\mathcal{C}_\infty$ on a circle slightly larger than the unit one, and satisfying $|w_{kj}(e^{i\theta})| \leq M_\delta$, a number depending only on δ. The (analytic) g_k obtained from these w_{kj} via the v_{kj} will satisfy

$$\|g_k\|_\infty \leq \frac{2}{\delta} + 2(n - 1)M_\delta ,$$

and $g_1 f_1 + g_2 f_2 + \ldots + g_n f_n \equiv 1$ on $\{|z| < 1\}$.

The corona theorem is completely proved!

Exercise. Let $f \in H_\infty$. Show directly, without appealing to the lemmas of Chapter X, Sections E.2, 3 or the theorem at the end of Chapter VIII, Section E, that

$$|z| \log \frac{1}{|z|} \, |f'(z)|^2 dx\, dy$$

is a Carleson measure.

Hint: One may suppose that $f(z)$ is analytic in a slightly larger circle than $\{|z| < 1\}$, and has no zeros there. If $F \in H_1(0)$ has only a simple zero at the origin and is analytic in a circle slightly larger than $\{|z| < 1\}$, apply the inequality

$$|(f')^2 F| \leq 2|f| \frac{|(fF)'|^2}{|fF|} + 2|f|^2 \frac{|F'|^2}{|F|}$$

(idea of A. M. Davie).